KNOWING

"Still and all," added Pencroff, "there's so much knowledge in the world. What a fat book could be made, Mr. Cyrus, out of everything we know!"

"And an even fatter one out of everything we don't," answered Cyrus Smith.

—JULES VERNE, *The Mysterious Island*

KNOWING

The Nature of Physical Law

Michael Munowitz

OXFORD
UNIVERSITY PRESS

OXFORD
UNIVERSITY PRESS

Oxford University Press, Inc., publishes works that
further Oxford University's objective of excellence
in research, scholarship, and education.

Oxford   New York
Auckland   Cape Town   Dar es Salaam   Hong Kong   Karachi
Kuala Lumpur   Madrid   Melbourne   Mexico City   Nairobi
New Delhi   Shanghai   Taipei   Toronto

With offices in
Argentina   Austria   Brazil   Chile   Czech Republic   France   Greece
Guatemala   Hungary   Italy   Japan   Poland   Portugal   Singapore
South Korea   Switzerland   Thailand   Turkey   Ukraine   Vietnam

Copyright © 2005 by Michael Munowitz

Published by Oxford University Press, Inc.
198 Madison Avenue, New York, NY 10016
www.oup.com

Oxford is a registered trademark of Oxford University Press

Library of Congress Cataloging-in-Publication Data
Munowitz, M. (Michael)
Knowing : the nature of physical law / Michael Munowitz.
p. cm.
Includes index.
ISBN-13: 978-0-19-516737-5 (cloth : acid-free paper)
ISBN-10: 0-19-516737-6 (cloth : acid-free paper)
1.   Physical laws—Popular works. 2. Physics—Popular works. I. Title.
QC24.5.M864 2005
530—dc 22
2005008494

1 3 5 7 9 8 6 4 2
Printed in the United States of America
on acid-free paper

Illustrations, book design, layout, and typography by the author

Contents

A Closer Look...

vii

Preface

Philosophy is written in this grand book, the universe, which stands continually open to our gaze.... It is written in the language of mathematics.

—GALILEO[1]

Traduttore, Traditore.
Translation is betrayal.

—ITALIAN PROVERB

If the book of nature, as Galileo suggests, is written in the language of mathematics, then the present volume—which uses words and pictures in place of equations—cannot even claim to be a translation. No book can. Mathematics, of all the languages, stands alone. All languages are equally effective and equally expressive, except one. All languages can be translated, except one.

Mathematics can be simplified, analogized, paraphrased, and put into words, but nothing comes close to the original. Mathematics cannot be translated with any real degree of fidelity.

Isaac Newton couldn't do it. Newton, who revealed the secret of the Moon in orbit, made only incomplete progress with his mechanics until he invented a new language in which to express it: calculus.

Michael Faraday couldn't do it. One of the greatest experimental scientists ever to step into a laboratory, the nineteenth-century Faraday discovered many of the electric and magnetic phenomena that power the

---

[1] Galileo Galilei, *The Assayer* (1623), Stillman Drake, translator, in *Discoveries and Opinions of Galileo*. Anchor Books (1957), pp. 237–238.

modern world today. He went further than anyone before him (and, incidentally, he was a master of thinking and writing in clear, plain English), but he lacked the mathematics to go beyond. It fell to another genius, James Clerk Maxwell, to write down four majestically terse equations and put all the pieces together at last. The equations told Maxwell what no amount of words could: that there must be such a thing as electromagnetic waves, and they must all travel in a vacuum at an unchanging 186,000 miles per second. And so it was. Radio waves were discovered in the laboratory twenty years later, and X rays not long after that.

Albert Einstein couldn't do it. Struggling for many years to formulate a new theory of gravity, he was stuck. He needed finally to discover a dusty mathematical tome in which a pure mathematician, concerned only with the intellectual pleasure of abstract reasoning, had already worked out the rules for doing geometry in a curved space. After that it was easy, at least in hindsight. Mastering the language of non-Euclidean geometry, Einstein gave us the general theory of relativity, out of which grew the idea of the Big Bang and the birth of the universe from a speck of nothingness.

Nobody can do it. Neither Newton, nor Maxwell, nor Einstein, nor the inventors of quantum mechanics in the early part of the twentieth century, nor the inventors of string theory in the early part of the twenty-first— nobody who truly seeks to understand the physical laws of nature can dispense with mathematics. It is not a mere convenience. It is not jargon. It is not window dressing. Mathematics is the uniquely powerful language with which we interpret the book of nature, and if there is some other royal road to understanding, then the greatest minds among us have yet to find one.

For all that, though, a work as wondrous as the universe must not remain a closed book, accessible only to those with the inclination and wherewithal to master a difficult language. One should not have to become a professional mathematician or physicist simply to appreciate what is arguably the greatest joy there is: the joy of knowing (just a little bit) how everything is put together. If not, what's a heaven for?

Here, then, just for the fun of it, is my selective paraphrase of the book of nature, intended for anyone who wants a glimpse into the way the world works. A world of pushes and pulls, of certainty and chance, of constancy and change. A world where mass becomes energy and space-time never rests. A world where apples fall to the ground yet the Moon does not, a world of quarks and photons and electrons, a world where the rules are strict but surprise is king. Relativity. Conservation. Gravity. Newton's laws. Electricity and magnetism. Quantum mechanics. Heat and work. Energy. Entropy. Equilibrium. The arrow of time. Chaos. The Big Bang. Dark matter. Dark energy. Superstrings. In twelve chapters of words and pictures (no equations, promise) and an optional commentary and glossary, here is my ruthlessly abridged version of *what* there is to know and how a scientist takes up the intellectual challenge of knowing it.

This is not a course in school. There are no formulas to memorize, no problems to solve, no exams to dread, and no grades to grub. There are no practical rewards to be had, but maybe there is something more.

Have fun.

1

GREAT EXPECTATIONS

The prize, beyond measure, is to know all there is to know: to grasp the workings of everything in the universe, large and small, at every place and at every time. That tacit hope, still to be realized, animates the whole of our science.

A vain hope? A hope never to be realized? So it would seem in a universe both unimaginably large and unimaginably small, a universe so immense as to paralyze the mind. Look and see. Look outward and see a wasteland pockmarked by stars too many to count, vast beyond reason, unending. Look inward and see, just as vast, a miniature universe in which a drop of water already holds more molecules than a galaxy contains stars. And look, inward still, into consciousness itself, and see how lifeless particles combine unconsciously to produce the most complicated structures of all: sentient beings (us) informed with the capacity to wonder. Where even to begin?

We begin with an act of faith, with this bit of wishful thinking vindicated later by experience: that, amidst the apparent chaos and complexity, there is really something to know. We assume first that such a universe even lends itself to finite description; that it can be reduced to parts and blueprints; that a relationship exists between structure and condition; that, moment to moment, the condition changes in some apprehensible way.

Look closely at the weave of the world and see, as if in a tapestry, a frugal simplicity masquerading as complexity. Go beyond the finished work, which can only dazzle, to find the pattern hidden within. Take apart the tapestry strand by strand, color by color, stitch by stitch. Find the regularity. Find the rules. There *must* be rules. Nothing can be as complex as the universe first appears, and nothing deceives the mind more than complexity.

Step by step, one small piece of the world at a time, we look and we learn. We look to the heavens and ask the questions *where* and *when*. Where is the Moon today? Where was it yesterday? Where might it be tomorrow? We learn precisely what questions to ask, what simplifications to make, what course to plot.

Earthbound, too, we look and we learn. A ball rolls down a slope, and we ask our questions again. How much ground does the ball cover in one second? In two seconds? In three? What happens when the slope is made steeper? What happens when the ball is made larger? Is what we see, in any way, like Earth and Moon?

With practice and with just the right gadgets, we learn to look into realms where the eye cannot. We discover the fine grain of energy and matter, the tiny bits and pieces called quarks and gluons and photons and protons and neutrons and electrons that join together to form Earth and Moon, ball and slope, you and me. We find, if only we look deep enough, a structure common to all. Little by little, the visible universe takes shape in all its complexity; little by little, the pieces fall into place.

A wonder, too, for they fall into place without purpose or plan, yet seemingly without fail. The world comes together and moves ahead with uncanny certainty in the face of a nature manifestly blind, where random chance decides every turn. No hand guides the particles of matter, but fall into place they do, again and again; and they do so, not least, simply because they can. Particles attract and repel. They exert influence. They accept influence. They come together and they move apart.

Trapped in a web of opposing forces, pushed and pulled in different directions, particles find a balance. Quarks clump into protons and neutrons. Neutrons and protons and electrons clump into atoms, a few dozen kinds in all. Groups of atoms—sometimes two, sometimes three, sometimes thousands—clump into molecules, each one united in an electrical embrace. Atoms and molecules, in multitudes that conjure images of the sand on the shore, blend together imperceptibly and fill the land, sea, and sky.

And so, with ruler in one hand and clock in the other, we set out metaphorically to map the world. We take inventory of space and time. Where a particle appears, we enter it in a ledger. Where a force exerts its push or pull, we note the strength and direction. Where a condition of any kind alters the quality of space, we record the effect point to point, instant to instant. Doing so, we see a material universe impregnated with energy, a universe, from supernovae to quarks, that observation shows to evolve from state to state under the government of just a few laws. Such is our task: to describe the states; to discover the laws; to learn, within limits, how the world works.

An abstraction, maybe, but from abstractions like these we develop our missions to Mars, our computers, our pharmaceuticals, our genomic maps, our entire science and technology. We begin to expose a unity and coherence otherwise masked in the tapestry of nature. We begin to trace the intellectual threads that bind together the whole of human knowledge. We begin to realize that the great questions in physics, chemistry, and biology have more in common than meets the eye.

Now to suggest that there are common threads is not to ignore the obvious differences between the different fields, nor is it to wish away the necessary division of scientific labor. Details matter, of course, and details make all the difference in the exacting work of interrogating nature. The details of a dying star differ enough from those of a dying cell to ensure that astrophysicists and cell biologists work in different university departments, attend different conferences, and speak mutually unintelligible dialects of a universal language. A chemist studying an enormously large protein molecule, with its many thousands of atoms, uses tools and models different from those of a chemist studying the ultrafast transfer of energy in a molecule of only six atoms. They are working on different problems. They use different equipment. They read different journals.

At some level, though, they all think alike. They frame their most basic questions in similar ways, and they observe how nature enforces the same overarching laws throughout the universe. If they are natural lawyers of a sort, then they are lawyers working in different specialties and dealing with all manner of detail—but the myriad local statutes that regulate dying stars and dying cells, Earth and Moon, ball and slope (not to mention you and me) conform without exception to a far broader, far more concise global constitution.

As observers we dream globally, but we pay closest attention locally. Practical necessity demands no less. We content ourselves with a small part of the universe, hoping to discover exactly how it is put together and what it is doing. By trial and error, by looking and learning, we happen upon the specific numbers, words, or other attributes needed to say: "This is it, precisely. With this minimal information we can put together our little system and let it run, if only in our minds. Restricting ourselves to this one part of the whole, to this one specific purpose, we know all there is to know."

With Earth and Moon, say,

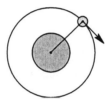

experience teaches us to measure the position, velocity, and mass of one body relative to the other. Knowing just those few numbers at any instant, we can predict what will happen in the next. And if that is all we want—to track the lunar orbit as if it were driven by a machine—then we do not need to know that Earth and Moon are themselves built from innumerable quarks and electrons, nor do we need to know the location of every star in the Milky Way. To realize this one objective (no small achievement in itself), a handful of numbers tells us all we need to know at any particular moment. We can write them down on a card and file the information away in a cubbyhole. Think of it as representing one possible mechanical state of Earth and Moon, one of infinitely many:

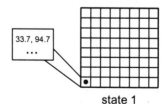

state 1

Then we look a second later and we find that everything has changed. The Moon has moved. Its position is different, its velocity is different, and our numbers are different. They go into a different cubbyhole:

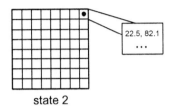

state 2

Something, we say, has caused the Moon to move. Call it an angel if you like, or call it the force of gravity or the curvature of space-time, but something—some kind of influence—has transformed the system of Earth and Moon from one mechanical state into another:

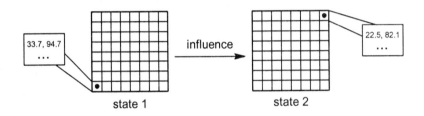

State 2 is born of state 1, and the world moves on.

State 1. Influence. State 2. Given this most general of templates, filling in the blanks as needed, we have a way to construe every conceivable action in the universe. Not just Earth and Moon, but ball and slope and electron and proton and all the rest can be made to fit the pattern. The tricky business is to pick a system, describe it fully, and follow its history under some particular form of influence.

Tricky, indeed. Different systems demand different descriptions; different particles respond to different influences. What works for Earth and Moon fails utterly for electron and proton, and what works for electron and proton fails just as miserably for gluon and quark. There are all kinds of surprises. An enormously complex system often behaves simply and predictably, whereas a small, apparently simple system may behave complexly and chaotically. There is no single answer. There is no magic formula, no single interpretation, no one specific recipe that describes the way *everything* works all at once.

Yet we are undeterred. We have to fill in the blanks for each system,

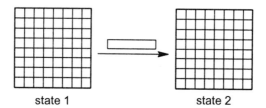

but our universal template of state–influence–state remains applicable in all its wonderful vagueness and generality. For Earth and Moon, for electron and proton, for almost any phenomenon we choose, the paradigm of state and influence stubbornly persists in one form or another.

Thus encouraged, in the chapters to follow we shall refine our approach further and ask more sharply:

1. How large is the system? Does it contain just one or two parts, like Earth and Moon, or do tens or thousands or trillions or zillions of actors all play a role? Do the actors act strictly as individuals, each an island unto itself, or do they forge a whole different from the sum of its parts?

2. What particular values (numbers, not words), will lend quantitative substance to the symbolic states? What specifically do we need to know at a given instant? The height of a tall building? The force of a powerful locomotive? The path of a speeding bullet? What information, given to us today, will allow us to describe the system tomorrow?

3. Whatever it is, how well can we know it? Is an exact specification within our reach, free from all doubt? Can we determine that an electron in some atom is moving at exactly 1,000,000.000000000 meters per second, no more and no less, or must we allow for the possibility of 1,000,000.000000001 meters per second, or 999,999.999999999 meters per second, or something else entirely? If so, what kind of a world might take

shape before us? Will a small change at the beginning produce only a small change in the end,

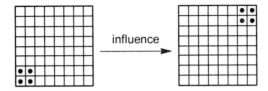

or will the outcome deviate wildly with even the slightest initial disturbance?

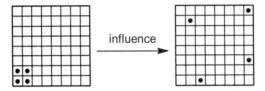

4. What meaning, too, shall we ascribe to the arrow of influence? By what course and under what compulsion will our corner of the world evolve from one time to another? Is it a one-way trip, never to be retraced,

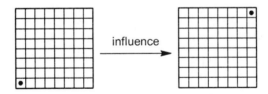

or is the road from future to past equally valid?

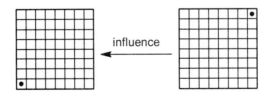

Ask...those questions and more. For to answer them, if ever we could, would be to know all there is to know.

2

TIES THAT BIND

Suppose that matter were suddenly to lose the ability to interact, its particles unable to exchange influence. There would be no attractions. There would be no repulsions. There would be no lumps, no clumps, no organized structures of any kind. There would be no Earth, no Moon, no stars. No atoms, no molecules, no mountains, no valleys, no us.

A world without interactions, so hard to imagine, is a world without preferences, a world without differences, a world without *cause*. When interactions disappear, one particle has no cause to change its position relative to another, since all positions are alike. If there are no interactions, there are no differences. There is no agency to effect a change.

With interactions gone, a particle at rest remains at rest. No agency forces it to move. A particle in motion moves uniformly in a straight line, always in the same direction, always with the same speed. No agency forces it to change; no agency alters its path.

In a world without interactions, each particle becomes a world in itself: a solitary system, forever isolated, unaffected by its neighbor. Indeed, if particle 1 cannot recognize and respond to the presence of particle 2—no matter how near or far—then particle 1 really has no neighbor, and the notions of "near" and "far" also have no meaning. If the bodies truly do not interact, then the same condition prevails when they stand at opposite ends of the universe as when they touch. This configuration,

●   ●

is no different from this one:

●●

For a world to contain structures more complicated than single particles, there must be a distinction between near and far. The particles must have a way to interact.

Let there be interactions. Let one particle be able to influence another, to push it away or bring it near. Let the effect be different at different distances, so as to give meaning to each position in space. Let there be different kinds of influence and thus different kinds of particles, each responding in its own way to some particular agency. And then— whatever these agencies may turn out to be, whatever the particles may turn out to be, whatever the groupings of particles may turn out to be— there will be this one essential element of design governing them all: the potential to be *different*. The possibility will arise that a certain arrangement taking shape under a given influence (symbolized by the broken lines below),

might be more or less likely to endure than some other arrangement under the same agency:

For when such possibility is allowed to exist, nature can fashion a connected universe from a store of individual parts.

The Potential To Be Different

Jack and Jill (for our present purposes, a pair of interacting particles) approach from a great distance, initially just out of eyeshot:

Not yet aware of each other, they walk undirected by any special influence. They are acquaintances, on generally good terms, but at this degree of separation they act as two disconnected individuals.

A few steps closer, Jack and Jill start to take notice. They perceive a change in their environment and respond proportionately. Wishing to discuss some matter of mutual interest, perhaps, they are encouraged to draw near. Ever so slightly, their movements acquire a greater urgency. Each member of the pair, as if walking down an incline, feels drawn to the other:

With the distance between them still fairly large, the downhill slope is gentle at first. Even so, a potential association starts to take hold. The gap begins to close.

Jack and Jill continue to walk. Closer and closer they come, and with each step the sense of attraction increases. The symbolic slope grows steeper:

Now at some point, we know, the contact inevitably becomes too close for comfort. Jack and Jill arrive at an invisible but almost palpable barrier, beyond which any closer approach begins to feel unnatural. Although with sufficient motivation they can cross the barrier, to do so under the current circumstances would be akin to walking steeply uphill:

They feel driven instead to take the easy way, to step back, to be repelled. Choosing downhill over uphill, they increase the distance between them:

A little too far? They draw slightly closer. A little too near? They retreat once more. Near and far, back and forth, attracting and repelling, Jack and

Jill eventually find a distance at which they can stand comfortably. It is a position of equilibrium, where the force of repulsion exactly balances the force of attraction:

With that, our two formerly separated particles have become one. Endowed with the potential to interact, Jack and Jill have transformed themselves into Jack–Jill, a union stabilized by a balance of opposing forces. The arrangement will last until some stronger influence comes along to change it.

Put all these changing slopes together now into a single drawing, one that will show at a glance the full range of Jack and Jill's potential to interact. From the actors standing far apart (at the right-hand side of the pattern below) to the actors nearly nose to nose (at the far left), the sloping push–pull of their relationship shapes up in the following way:

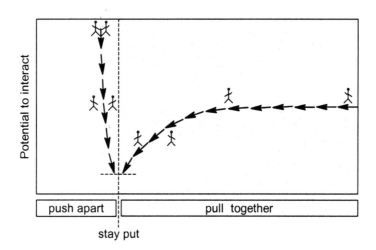

Jack and Jill move apart when the separation between them is too small, and they come together when the separation is too large. The steeper the slope, the stronger is the push or pull.

Smoothing out the curve, taking ever smaller steps, we thus draw out a summary profile of interaction—a profile to which we shall return again and again:

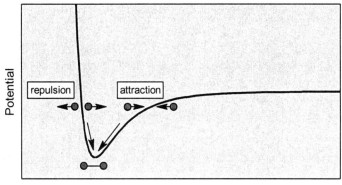

Separation

It tells us this: Like a stone rolling around a valley, our two-particle system follows a slope of influence different for different configurations. At the extreme right, where the hill flattens into a plateau, a step closer together brings little change. The tendency to interact is small here, since to be "far apart" is much the same as to be "*very* far apart." For particles nearer to the valley, however, the effect becomes either steeply attractive or steeply repulsive. Small changes in separation then make more of a difference in potential, and the system finds itself more likely to roll down to the valley floor. There, with the potential at its lowest value, the curve is flat.

So much for the story of Jack and Jill. It is a trivial meeting on the street somewhere, an encounter of no real importance, maybe not even to the two principals. The contours of that experience, though, are repeated all over the universe every instant of every day. Pieces of matter randomly come together and move apart, finding stability in a balance between attractions and repulsions. Replace Jack and Jill by particles of a different sort—a neutron and a proton, or two atoms of nitrogen, or the double helix of DNA, or the Moon in orbit around the Earth, to cite just a few of many, many possible systems—and we observe a similar pattern in each, an opposition of forces expressed in different ways. Without the tension between repulsions (when particles are close together) and attractions (when they are far apart), the constituents of matter would crunch together or fly asunder without limit. It is not the explanation for everything, by no means, but it is a Big Idea nonetheless.

Matter and Its Endowments

From what well, we wonder, might these space-differentiating, matter-organizing, universe-shaping attractions and repulsions spring up? What tools and tricks might enable nature to turn disconnected units into interconnected groups? What might it mean for particles to fall into different classes,

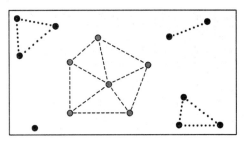

each with its own protocol for making connections? What might it mean for one pair of particles to repel or attract more strongly than another at a given distance, or for certain kinds of particles to interact not at all with certain others?

It would mean, first, that the potential actors surely must have some essential quality—some quality inherent in their very being—that empowers one body to recognize and respond to another as an eligible partner. People, acting with particle-like simplicity, do it all the time. They form families. They form bowling leagues. They form religious congregations. They form labor unions, political parties, corporations, predatory cartels, towns, cities, countries, transnational alliances. They form associations as disparate as polo teams and marauding armies, yet all of these networks, different as they are, have in common a similar if generalized organizational principle. Each is an association of fundamental units brought together by a recognizable source of communal interest. Each is an association of individual *persons* with a shared kinship, belief, affiliation, or any other attribute by which an actor can announce its presence to the world at large.

Appropriately endowed, the actors must then have a way to make that presence felt. They must communicate with other actors similarly disposed to interact. Each unit must have the capacity to broadcast a signal from its own source and receive a signal from somewhere else.

For people? For people, the possible mechanisms are as unlimited as the individual sources themselves. People talk at the dinner table. They sponsor tournaments. They attend worship services. They work in assembly plants. They meet in Vienna. They publish newsletters. They use the telephone. They send e-mail. They march in formation.

For electrons? For electrons and protons and the various other bits of matter that organize into bowling leagues and predatory cartels and everything else, both the fundamental endowments (like mass and electric charge, to name two) and the means of communication (like gravitational and electromagnetic fields) prove to be surprisingly but satisfyingly few: subtle, intangible, tantalizing...but few. Turning to them, next, we shall begin to glimpse how nature makes much out of little.

Gravity and Mass

Gravity we know firsthand: matter attracts matter; stuff is drawn to stuff. The apple falls from the tree. The Moon orbits the Earth. Our feet remain on the ground. For all of that, blame gravity.

The more stuff (the more *mass*), the greater is the attraction. The tug may be strong over here and weak over there, but nature grants equal gravitational opportunity to all. Everything endowed with mass, from a barely substantial particle to a collapsing star, is pulled toward everything else endowed with mass. Suns do it. Moons do it. Quarks and electrons do it. They all do it. All it takes is mass.

To have mass is simply *to be*. To have mass is to be a piece of matter, and to be a piece of matter is to have an affinity for every other piece of matter. A quark has properties different from an electron, and an electron has properties different from a neutron, and a neutron has properties different from a proton, but each of them—along with every other material particle—has a mass. Each has a certain essential substance. Each contributes a certain quantity of matter to the universe.

We come to know mass not by what it is, but by what it does. Ask a weightlifter. Alexei feels unmistakably how the barbell poised above his chest is attracted to all of the matter (the whole mass of the Earth) lying below his back:

Living dangerously, caught in the crossfire between two interacting masses, he must use his own energy to produce an opposing force and thereby thwart a potentially crushing union of iron and Earth. He has to support the weight.

Double the mass, and the gravitational attraction doubles as well. A 200-pound barbell contains twice the mass—twice the matter, twice the number of elementary particles—of a 100-pound barbell. It interacts twice as strongly with the fixed mass of the Earth, and Alexei must exert twice the force to keep his ribs intact:

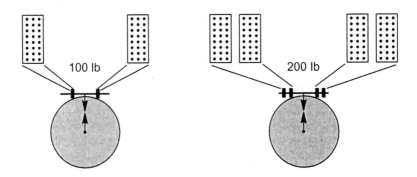

Keep going. Triple the mass. Quadruple the mass. Quintuple the mass. Multiply the mass of the original 100-pound barbell thousandfold, millionfold, billionfold, trillionfold, quadrillionfold, quintillionfold, sextillionfold. Increase the amount of matter to the equivalent of 1,620,000,000,000,000,000,000 such barbells, and an overburdened Alexei will eventually have a mass equal to that of the Moon to support. Were Atlas-Alexei then to balance this lunar mass just above the surface of the Earth, like a barbell, he would need to apply 162 thousand million million million pounds of force against gravity.

A titanic task, no doubt, to come between Earth and Moon, but the arithmetic is simple and unforgiving. The more matter, the stronger is the gravitational attraction. If one barbell sustains such-and-such a pull at such-and-such a distance, then two barbells sustain exactly twice as much. Three barbells, three times as much. Four barbells, four times as much; five barbells, five times as much; and so it goes, all the way up to a whole Moon's worth of barbells, and more. The attraction scales one-for-one with the amount of mass. We can count on it.

We can count on it, precisely, because Earth, Moon, barbells, and everything else all take shape from the same few building blocks: the same limited set of generic particles, rearranged in unlimited ways. Big things always come from little things, and each elementary particle adds a small but definite mass to the total. Anywhere in the universe, as long as Earth and Moon remain intact, each of the two bodies has its fixed amount of matter, its fixed amount of *stuff*, its fixed number of electrons and protons and neutrons. Every particle makes its prescribed contribution to the whole, like bricks in a wall or tiles in a mosaic. Bit by bit, Earth and Moon thus acquire their full endowment of mass and exert their cumulative gravitational influence. Mass attracts mass. The particles of the Earth attract the particles of the Moon, and the particles of the Moon attract the particles of the Earth in equal measure.

But the strength of that attraction does not depend on mass alone. Like the encounter of Jack and Jill, the gravitational pull weakens as the source (Earth or Moon) grows more distant from the receiver (Moon or Earth). Far apart, *very* far apart,

the potential for interaction is only slight, and the slope of influence is nearly flat. Closer together, the attraction becomes steadily stronger

until finally the two bodies touch:

And somewhere in the middle, between the extremes of gravitational indifference and gravitational excess, Earth and Moon do find a stable balance. Month after month, we see proof of it in the night sky. The Moon orbits the Earth, round and round, drawn close and yet somehow kept away.

Later we shall explore that balance of forces more closely, but for now let us take away at least the germ of a possibility, the *potential*, for a mass-based association between these two clumps of matter. Strong when the bodies are near and weak when the bodies are far, the potential for gravitational attraction differs at every separation. The force that develops (think of it as an "inclination" to come together) is reflected in the steepness of the change in potential:

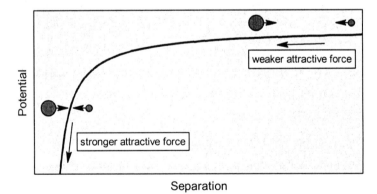

The attraction goes far beyond Earth and Moon, too. Gravity is a universal force of nature, an inclination not just for Earth and Moon but for every speck of matter—however far away—to draw a bit closer, and closer, and closer still. Nothing is exempt. Effective over even astronomical distances, gravity gives nature the potential to pull together the cosmos on a large scale. Gravity enables clouds of gas to form stars, and stars to form galaxies, and galaxies to form clusters of galaxies. In our own neighborhood, it regulates the motion of planets about the Sun. It governs the tides. It sheathes the Earth in a protective atmosphere, drawing from just enough terrestrial mass to keep a life-sustaining blanket of oxygen and other gases overhead.

The long arm of gravity knows no bounds and makes no exceptions. Gravitational potential weakens with distance but never vanishes completely, and its universal reach gives gravity a dominant role in shaping the macrocosm. Everything, everywhere, is affected by the mass of everything everywhere else.

Yet dominance on a grand scale is tempered by weakness on a small scale, because particle by particle, bit by bit, the influence of gravity proves so faint as to appear irrelevant. For although gravity shapes the stars and galaxies, gravity is powerless to put together even the most basic starting ingredients of a star itself: simple associations of just a few protons, neutrons, and electrons to form hydrogen and helium. The particles, taken one at a time, have too little mass to muster up an appreciable gravitational potential.

Small masses make for small attractions, and protons, neutrons, and electrons have small masses indeed. It would take, for example, more than a billion trillion trillion trillion trillion protons or neutrons (1 followed by 57 zeros) to amount to the mass of one Sun, and even a heavenly body as puny as the Moon has the mass of 44 trillion trillion trillion trillion such particles. That Sun, Moon, and Earth interact by gravity is a testament not to the intrinsic strength of gravity (it is not strong at all), but rather to the sheer numbers of particles contained within each body. What they lack as individuals, the component particles make up for in bulk. In unity there is strength.

Turning inward, then, looking to what goes *into* Earth and Sun and Moon, we start to consider the small stuff. One proton and one electron, interacting over an astonishingly small distance (five billionths of a centimeter, two billionths of an inch), come together to form an atom of hydrogen, simplest of the chemical elements:

hydrogen atom

Gravity has nothing to do with it. Were mass the only source of interaction, then protons, neutrons, and electrons alike would suffer neither attraction nor repulsion. They are lightweights. They are featherweights. They are *less* than featherweights. Their endowment of mass is woefully insufficient to make much use of gravity. The electron, after all, has a mass less than one billionth of a billionth of a billionth of a gram, and a proton has a mass only about 2000 times greater. Put together enough of them and we might have the makings of a star, maybe, but one pair at a time will never make a single hydrogen atom—not if gravity is the sole means of support.

No, gravity does not give us hydrogen. Gravity does not give us helium. Gravity does not give us carbon, or nitrogen, or oxygen, or silicon, or phosphorus, or any of the other several dozen kinds of atoms that go into everything the eye can see. Nor does gravity give us the molecules built from those atoms: not the small molecules, like H_2O, and not the large ones, like DNA. Gravity does not bind a molecule of oxygen to a molecule of hemoglobin in the blood, and gravity does not bring together a trillion trillion molecules of water to form an ice cube. For those things, the small things, there must be something else.

There is. We call it the electric interaction (more generally, the electromagnetic interaction), and it arises from an endowment of matter known as the electric charge. Standing still, an electrically charged particle throws up an electric potential to which other charged particles can respond. In motion, the same charge produces a magnetic potential to which other *moving* charges can respond.

Electric or magnetic, charge gives rise to both. Whether we say "electric potential" (because we perceive a charge to be at rest) or "magnetic potential" (because we perceive a charge to be in motion), the difference lies solely in our point of view. The source is one.

From the world of mass we descend next into the world of charge, ready to see our most familiar surroundings in a new light. Let there be electric charge.

Electric Charge and Potential

If not for the occasional accident, like a bolt from the blue, we might never suspect it. We might never suspect that the familiar world of everyday matter hangs in a precise yet precarious balance, surviving only as a neutralized mixture of explosive forces. But hang in the balance it does. Every drop of the ocean, every grain of sand, every blade of grass, everything within the reach of our senses—everything, all of it, does a delicate dance of opposites. Every macroscopic combination of atoms brings together vast numbers of two very different kinds of actor, equal in number yet opposite in one particular endowment. We could label the pairs yin and yang or heads and tails or whatever else might vividly express a diametric relationship, but instead we shall honor the tradition begun by Benjamin Franklin with his kite in the storm. We shall label them "positive" and "negative" electric charges.

In one camp we have the positive particles, the nuclei of atoms. Each elemental nucleus contains a prescribed number of protons, and each proton bears a charge of +1. In the other camp we have the negative particles: the electrons outside the nucleus, each electron bearing a charge of −1, one electron for every proton. Built from equal numbers of protons and electrons, the world and its atoms thus have no net charge as a whole:

The math $(1 - 1 = 0)$ is elementary. The implications are profound.

We discover, first of all, that nature needs both types, positive and negative, to fashion the dogs and cats and Atlantic Oceans of this world. Neither electrical camp could exist stably without the other. Particles endowed with the same kind of charge, whether positive against positive or negative against negative, have the potential only to repel. The closer together they stand, the steeper is the tendency to move away:

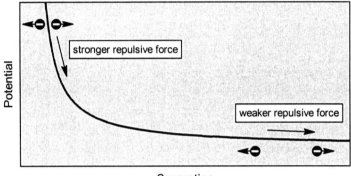

Separation

Left to themselves, acting out a primordial realization of "likes repel," the similarly charged particles would blow apart. No matter how distant the separation, there would always be a tendency to move away just a bit more.

The other side of the coin, of course, is "opposites attract," and so they do. Sliding down a mirror image of the like–like potential, a negative particle and a positive particle tend to move ever closer:

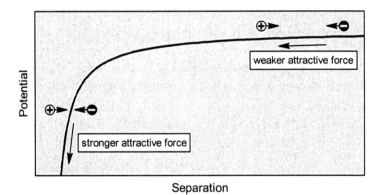

Separation

Even from far away there exists an inclination, a force, that brings the two together.

Compare now these last two diagrams (for electric charge) with the one on page 18 (for mass) and note some striking similarities. Just as the gravitational potential weakens with distance, the electric potential weakens in exactly the same way. Where the curve is nearly flat, as at the far right, the inducement to move is small. The potential at large distances barely differs between one separation and the next, and the widely spaced particles have scant cause to move in or out. Where the curve is steep, however, the actors realize a large change in potential for even the smallest change in position. They respond to an appropriately large force, be it a repulsive push or an attractive pull.

Like gravity, too, that push or pull depends on the strength of the source, although the electric potential takes its source from charge rather than mass. Opposite charges attract, like charges repel, and larger charges exert greater influence. Other than that, the electric potential scales with charge in the same way as the gravitational potential scales with mass.

We substitute "charge" for "mass" and start to climb the ladder. Large mass, small mass, the effect is the same. What matters is the charge, not the mass. Double the charge, double the electric potential. Triple the charge, triple the potential. Quadruple the charge, quadruple the potential.

And so on? As a practical matter, no: not without limit. The ladder of charge, as it exists in nature, offers far fewer rungs than the ladder of mass. Opposite charges attract, but like charges repel. Positives tend not to associate exclusively with other positives, and negatives tend not to associate exclusively with other negatives. We never see, for example, a million billion trillion zillion protons (bursting apart from all the repulsions) facing off against a million billion trillion zillion electrons (also ready to burst):

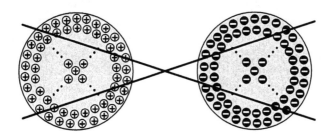

Instead, the opposite charges mix and mingle. Two by two, positive and negative particles interact to form composite structures with no net charge at all.

What develops is a microscopic world in which a positive charge is never far away from a negative charge,

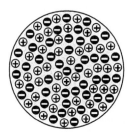

a world in which the bits of charge are fantastically small and fantastically many. Our eyes take in the finished product as if from a distance, blurring its intrinsically fine grain into an apparently smooth neutrality:

Our sight is too coarse to make out the individual charges. We become aware of them only when the balance is slightly disturbed, as in the flash of electric energy between a lightning bolt and the ground, or in the spark of "static electricity" between a finger and a metal doorknob in a carpeted room, or when we use instruments and experiments to probe a world we cannot otherwise see.

The Potential To Make Cats and Dogs

This neutralized electrical microworld—the chemical world, the world of *us*—begins with an intermingling of just two particles. Nature mates a single negative charge (an electron) with a single positive charge (a proton), and we have an atom of hydrogen:

hydrogen atom

Next, from two electrons and a nucleus containing two protons, we get an atom of helium:

helium atom

There will be more atoms to come, but even now we are off to a good start. Nearly all of the visible matter in the universe derives entirely from hydrogen and helium, mostly hydrogen.

And with these first two atoms, we already appreciate the subtlety of nature's elemental approach to matter. Small variations in structure (an electron here, a proton there) yield big differences in behavior. Hydrogen and helium, so similar in their basic construction, turn out to be radically different in function. Pairs of hydrogen atoms, on the one hand, readily share their electrons and nuclei to form *molecules* of hydrogen, the next step up in the organizational hierarchy of matter:

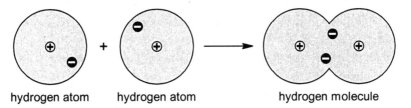

hydrogen atom   hydrogen atom   hydrogen molecule

Hydrogen atoms, moreover, combine not just with themselves, but with dozens of other atoms as well. The helium atom, on the other hand, simply does not react—not with itself, not with hydrogen, not with carbon, not with nitrogen, not with anything. It is a big difference.

The scheme that unfolds is as disarmingly simple as its results are richly complex. Add a proton to a nucleus, an electron outside, and, voilà, a new kind of atom takes the stage: an atom not quite like any other type of atom, an atom that will enter into varying associations with its fellow atoms. Mixing and matching these same few interchangeable parts (electrons and atomic nuclei), nature thus fashions the elements of a chemical universe. The result is a set of building blocks flexible enough to assemble real-world stuff as complex and wondrous as you like, including every one of the aforementioned dogs and cats and barbells and oceans. There is even room for six billion slightly different human beings able to think and feel in no fewer than six billion different ways. At a certain level of resolution, each of them reduces to a pile of electrons and atomic nuclei. Whatever they are and whatever they do, they do so by means of those tiny bits of charge.

Past hydrogen and helium, the buildup continues one atom at a time. Next in line comes lithium, with three electrons balancing a three-proton nucleus. Then comes the fourth element, beryllium. The fifth, boron. The sixth, carbon. Up and up, through uranium and beyond,

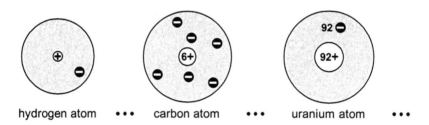

hydrogen atom   • • •   carbon atom   • • •   uranium atom   • • •

each atom adds one electron and one proton to the atom before, until at last (after some 120 elements or more) the accumulations of like charge become insupportable.

With atoms at the ready, nature's work of fashioning complex materials now begins in earnest. Atoms intermingle their nuclei and electrons with those of other atoms, each in a different way, and from these marriages of positive and negative come the innumerably varied molecules and other microstructures that fill the space of an appreciably larger world. From the two atoms in a hydrogen molecule to the thousands and millions of atoms in protein and DNA molecules, nature weaves dazzling complexity from fundamental simplicity. Electrically neutral combinations such as atoms, molecules, clusters of molecules, gases, liquids, crystals, bacteria, tulips, ducks, and elephants all take hold as the attractions and repulsions come into balance. Not *all* the interactions are attractive, because then the structure would collapse; and not *all* the interactions are repulsive, because then the structure would explode; but instead there is a local flattening of the potential, a Jack–Jill valley of stability where neither attractions nor repulsions are dominant:

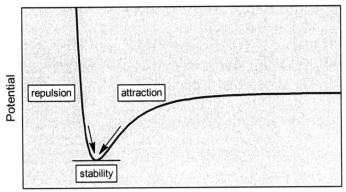

Configuration (distances and angles)

Not too far and not too near, like Jack meeting Jill, a combination of electrons and atomic nuclei settles into an equilibrium when its potential reaches a minimum. Here, where the landscape is locally flat, a small change in position brings neither advantage nor disadvantage; and here, restrained by hills of increasing potential, the particles can stay together for a time. These fragile valleys of stability allow the chemical world to exist.

The Strong Nuclear Force

Give the electric potential its due. It makes atoms from electrons and nuclei. It makes molecules from atoms. It makes oceans from molecules. It makes dogs and cats. One way or another, the interaction between electric charges weaves together all such structures, and the finished products speak both of nature's thrift and of nature's versatility. Still, there must be something more at work. Crowded into the smallest dimensions of the universe, beginning with the nucleus, the electric potential must give way to a stronger force. The influence arising from electric charge cannot hold together the protons in a nucleus.

No. Particles of the same charge have the potential only to repel, never to attract. The closer they are, the more strongly they repel; and packed into an atomic nucleus, practically touching, the ill-matched electrical particles stand closer than the imagination can grasp. At separations of only a millionth of a billionth of a meter, the repulsive force between two protons grows to a staggering amount—some 50 pounds, pushing apart bodies with a mass barely greater than a millionth of a billionth of a billionth of a gram.

The nuclear construction becomes still more dubious once we discover that interspersed among the protons are *neutrons*, bits of mass comparable to protons but with zero electric charge. A nucleus of helium, to take just one example, comes in several alternative forms called isotopes, the most common of which holds two neutrons in addition to its statutory two protons:

helium-4 nucleus

Similar variations appear all throughout the elements. Each nucleus contains a fixed number of protons (one for hydrogen, two for helium, three for lithium, and so on, up through the ranks), blended with a variable number of neutrons. The positive charge remains the same among isotopes of the same element, but the number of neutrons does not.

These neutrons are no passive bystanders. They contribute materially to the structure and strength of a nucleus, and we are forced to ask: How so? By what means? For in a world governed only by mass and electric charge, neutrons would slip by unnoticed. Lacking a net charge, they neither attract nor repel the protons by electrical influence. Nor, as mere specks themselves, can they put enough space between the positive particles to alleviate the enormous repulsion already present. Still more, with only a hairsbreadth greater mass than the less-than-featherweight proton, they are gravitationally challenged as well. What is it then, if not charge or mass, that keeps the neutrons from occupying their own separate worlds? What is it, if not charge or mass, that keeps the protons from flying off to the ends of the universe?

It works, whatever it is, and it works with ferocious strength. The very survival of an atomic nucleus testifies eloquently of an old force subdued and a new, stronger force triumphant. In the face of a tremendous electric repulsion and a paltry gravitational attraction, the neutrons and protons in a nucleus do stay strongly bound—or, to put the matter more emphatically, they stay *very* strongly bound. Nuclei hold their protons and neutrons together far more tightly than they hold their electrons outside. The nucleus holds together with the fearsome energy of the atomic fission bomb and the thermonuclear fusion bomb, energy released for good or ill when protons and neutrons are rearranged to form new nuclei.

We acknowledge this prevailing influence as the "strong" nuclear interaction, the not-electric and not-gravitational potential that allows protons and neutrons to exist close together in atomic nuclei. Mapping it out, we find that the strong interaction ignores electrons entirely but treats protons and neutrons alike. Protons with protons, neutrons with protons, and neutrons with neutrons all interact the same way at the same distances, and they all settle into the same kind of potential well that first trapped Jack and Jill:

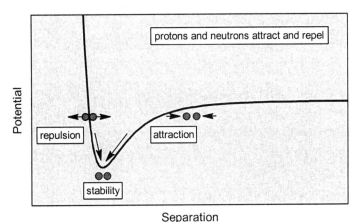

Recognize it as a two-way street. Unlike purely electrical particles (but just like Jack and Jill), protons and neutrons have the potential to go back and forth. They attract, they repel, and ultimately they strike a balance. When they do so, moreover, scarcely any daylight remains in between:

It is a snug fit. Steep walls of rising potential loom on either side, confining the nuclear particles a hundredfold more strongly than the electrical hills carved out by atoms and molecules.

The strong interaction gives protons and neutrons scant room to maneuver. Push two of them together even slightly more,

and the strong interaction will force the particles apart with considerable authority:

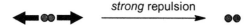

Pull them away a little bit, barely a millionth of a billionth of a meter, and the same interaction will snap them right back:

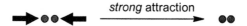

And there, once trapped, protons and neutrons stay trapped: sunk into valleys of nuclear potential, steep and deep. Within those valleys, nature builds nuclei for all the atoms; elsewhere, *outside* those valleys, the electrical world of atoms and molecules begins to fall into place. Outside the nucleus, the strong interaction gives way and the electric charge reasserts itself once again.

Strong it may be, yes, but the strong force overwhelms the electric force only within the close confines of a nucleus. A valley of nuclear potential proves to be as narrow as it is deep, and the strong interaction dies away after the shortest of distances, strong no more. The nuclear potential flattens, making meaningless any distinction between "just out of reach" and "way too far":

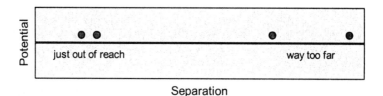

If protons and protons (or neutrons and protons or neutrons and neutrons) are to hang together, then the spacing had better be as close as they can make it. Let them move apart by little more than one diameter, and the nuclear particles break free of the strong interaction. Beyond that threshold, a pinprick in space, the electric potential takes charge of an atomic and molecular world outside the nucleus.

The Weak Nuclear Force

Even inside the nucleus, the strong force does not go unchallenged. The electric interaction may be overmatched, but the protons still keep their positive charges. Every proton continues to repel every other proton, and the electric potential exerts its influence with a force hardly diminished by distance—and not just between neighboring particles, like the strong force, but between all possible pairs. The potential for destruction never disappears.

Some nuclei are stable and last forever, but not all; and when the opposition becomes too much, a nucleus disintegrates. It undergoes radioactive decay, changing either into a different isotope of the same element or into a different element entirely. An electrically unstable nucleus may spew out a proton or neutron to lighten the load,

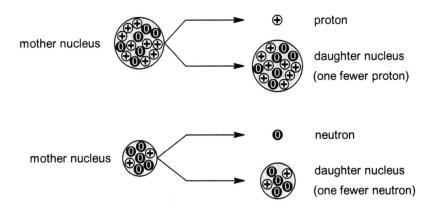

or it may eject a helium-like cluster of two protons and two neutrons (a combination called an alpha particle):

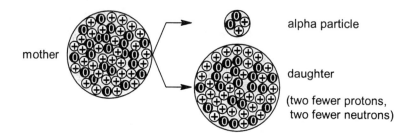

Or it may undergo fission and break into two nuclei of roughly the same size:

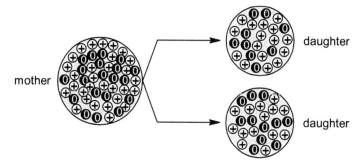

There are other possibilities as well, but the challenge to the strong force is already clear. Nuclear earthquakes can shake protons and neutrons out of their valleys. The battle does not always go to the strong.

Where the force of electric repulsion wins out, the casualty is the nucleus itself. A proton or two comes loose. A neutron or two comes loose. A heavy nucleus breaks apart. In none of these breakups, however, is the integrity of an individual proton or neutron compromised. If a mother nucleus begins with 92 protons and 143 neutrons, then those same 92 protons and 143 neutrons—no more, no less—show up among the daughters. If 84 protons and 126 neutrons go in, then 84 protons and 126 neutrons come out. If there are 70 protons and 85 neutrons at the start, then there are 70 protons and 85 neutrons at the finish. In any contest between the nuclear strong force of attraction and the electric force of repulsion, the protons and neutrons are shuffled about but otherwise unchanged.

So what shall we say, then, to a nuclear event such as this (beta decay),

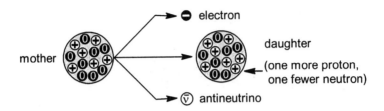

in which a neutron turns into a proton and also shoots out an electron together with an antineutrino? What brings into being, curiously, the stealthy antineutrino, an electrically neutral particle hitherto unsuspected? And from where, we ask, does an *electron* suddenly appear amidst the protons and neutrons, and why is this electron unrestrained by its powerful electric attraction to the protons? If the nuclear strong force does not apply to electrons and antineutrinos (it does not), and if low mass makes gravity irrelevant, and if the electromagnetic force as we initially conceive it cannot be responsible, then what force of nature licenses a nucleus to undergo beta decay?

It is manifestly something new, a fourth fundamental interaction to accompany the three we have already discovered: the "weak" nuclear interaction, arguably the most specialized of all. Epitomized by the transformation of a neutron into a proton,

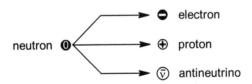

the weak interaction exerts its limited influence over vanishingly small distances, at least a hundred times shorter than those characteristic of the strong interaction. The force it delivers is legitimately called weak as well, amounting to about one millionth the strength of the strong interaction and one ten-thousandth the strength of the electromagnetic interaction. Only gravity, a phenomenally weak force between particles, makes less of an impact in its retail transactions.

Realize right away that a neutron does not hold captive within itself a proton, electron, and antineutrino, languishing there as three intact particles waiting to be freed:

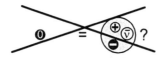

The influence of the weak interaction is more subtle than that, and its subtlety gives us a hint that protons and neutrons might be not quite what they seem. The weak interaction hints of a still finer layer of structure, a substratum built from particles more fundamental than protons and neutrons. It is on the level of those simpler, more fundamental particles—the quarks—that both the weak and strong forces operate more transparently, and it is on that level that we begin to sense the unity of nature's design.

From the Bottom Up: Quarks

From molecules down to atoms, and from atoms down to electrons and nuclei, and from nuclei down to protons and neutrons, we peel away layer after layer from the onion of matter. We descend into ever smaller domains, each simpler and more profound; we discover worlds within worlds. Each world, a foundation for the one above, supplies the building blocks needed to fashion progressively more elaborate combinations. At some point, though, the subdivision must finally end, and the particles supplied must be formed from no others. Everything we have learned about nature, through centuries of experimentation, tells us so.

The quarks are perhaps not the irreducibly simple, indivisible, end-of-the-line building blocks of this ultimate world within, but they are probably only a step or two away. They come in a half-dozen varieties, equipped for all four fundamental interactions. A quark has a *mass*, which ties it to the gravitational interaction. A quark has an *electric charge*, which ties it to the electromagnetic interaction. A quark has a *strong interaction charge*, which ties it to the strong interaction. A quark has a *weak interaction charge*, which ties it to the weak interaction. The strong charges and weak charges do for the strong and weak interactions what the electric charge does for the electromagnetic. They are the sources, the transmitters and receivers of the influence.

By their six "flavors" we know them: the up quark, the down quark, the charm quark, the strange quark, the top quark, the bottom quark. The names are whimsical, but the attributes they represent are not. Quarks affix their signatures to the very real events that take place in particle accelerators, and quarks behave with a consistency that we can mirror in precise mathematical terms. We know quite well what quarks do, even though everyday language fails to embrace their essence. So, with tongue in cheek, we hijack words like flavor and charm to render with ironic familiarity a world utterly alien to the senses.

Of the six flavors, only two—"up" and "down"—play a role in building the protons and neutrons of ordinary matter. An up quark carries an electric charge of $+2/3$, and a down quark carries an electric charge of $-1/3$. Bind together two up quarks and one down quark ($2/3 + 2/3 - 1/3 = 1$), and we have a proton with its net charge of $+1$:

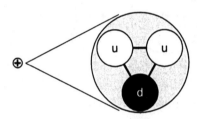

Bind together one up quark and two down quarks ($2/3 - 1/3 - 1/3 = 0$), and we have a neutron. Its three electric charges add to zero:

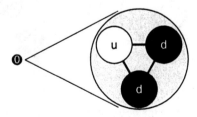

The up quarks and down quarks, dominated by a force arising from their strong interaction charges (not their electric charges), stay tightly confined within the close quarters of a proton or neutron. Lacking the enormous energy needed to break free, each trio of quarks presents itself to the outside world as a single particle, seemingly indivisible. A tangled, intricate web of quark–quark interactions gives the three-in-one particle its unity.

We call these exchanges of influence the "color" interaction, ascribing the purely metaphorical colors of red, green, and blue to the strong interaction charges—in the same spirit that we label an electric charge as positive or negative. And just as protons and electrons combine in equal numbers to form electrically neutral atoms, so do threesomes of red, green, and blue quarks combine analogously to form "colorless" protons and neutrons.

The analogy goes further. When intact atoms or molecules clump into still larger associations,

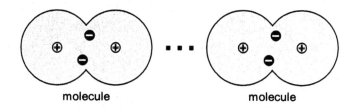

they derive the means ultimately from an electric potential generated by their protons and electrons. When intact protons and neutrons clump into nuclei,

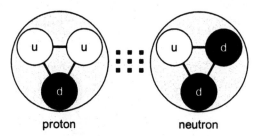

they do so using a potential generated by the strong interaction charges (the color charges) of their internal quarks. The net strong force observed between protons and neutrons (or between protons and protons or between neutrons and neutrons) emerges from the tangle within.

Coming from within, also, is the weak interaction. Not in all nuclei, but certainly in many of them, the weak interaction sometimes subverts the neutrons and protons bound otherwise so strongly. It takes only a change in flavor. The weak force, with weak interaction charges as the source, transforms a down quark into an up quark and hence a neutron into a proton. At the same time, an electron and antineutrino spring loose:

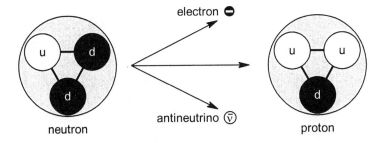

The strong force plays no part here, since neither the electron nor the antineutrino carries a strong interaction charge. Electrically neutral, the antineutrino escapes the electromagnetic force as well.

It is not the strong force, evidently, and not the electromagnetic force, but rather the weak force that allows a neutron to decay into a proton, electron, and antineutrino. The four particles all carry weak interaction charges, and their common endowment makes them all actors in a single play.

Playing the Field

That particles interact, we accept...that the distances between them make a difference, we accept...that a potential for interaction can be mapped experimentally, we accept—all these things we accept as consistent with what we observe, yet there remains the nagging question of *how*. How, to use a metaphor, could two particles ever know the exact distance between them? How could they instantly sense a variation in potential? How could they make contact through empty space?

But consider that space may not be so empty after all. Suppose, instead, that the apparent emptiness has a physicality no less real for being intangible. Suppose that space itself is permeated with force and potential, that space can exchange influence and energy with matter, that space plays the role of a communications network. Suppose that a particle, just by its presence, can erect what we shall call a "field" in the surrounding space: a *field of influence*, a message broadcast to all points beyond. Somehow, let us say, a particle acts as both transmitter and receiver of a signal carried by this postulated field. The particles are the computers. The field is the network that links them.

If so, then the interaction between two particles becomes indirect, not direct. It becomes local, not distant. It becomes delayed, not instantaneous. It becomes physical and realistic, not eerie. Particle 1 broadcasts its signal continually, handing off a message (maybe something like, "my electric charge is +1") first to one messenger, then to a second, then to a third, on and on, one signal after another. The messengers go forth and set up a field of influence throughout space and time, investing in every point a particular value of force, potential, or some other quality. They travel as far as the influence persists, and the strength of the field varies with distance from the source:

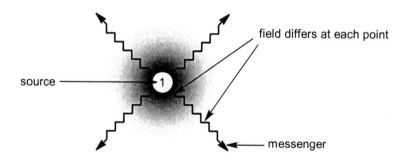

For messengers of the gravitational and electromagnetic fields, the road goes on forever. For messengers that carry the short-range strong and weak forces, the path ends abruptly.

Once erected, a field remains real and physical even when its source particle recedes into the background. The sender of the message may be far away, but the message is in the custody of the field-acting-as-messenger. The information is *here*, locally, at this particular point in space and time. And if a suitable recipient (particle 2) happens to be *here* as well,

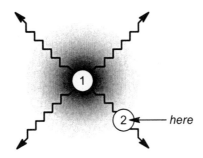

then the field can deliver the message. It transmits the residual influence from particle 1 to particle 2, adjusted for distance and the passage of time.

Make no mistake: the influence is real and resident in the field. We see it in the pattern created when a magnet sits beneath a sheet of iron filings. We see it transformed into a televised image when an antenna snatches an electromagnetic signal out of the air. We see it in the acceleration of a fruit falling from a tree. Everywhere, embedded in fields of all sorts—in the midst of not-so-empty space—we find the potential to move matter and the potential to change matter.

The stage is set. Particles move across fields of influence, giving and taking, broadcasting and receiving. The fields and their particles, together, shape a universe large and small, a universe at once awesomely simple and vexingly complex. Particle 1, over here, affects the condition of particle 2 over there, and particle 2 affects particle 1 in just the same way. No particle fails to give and receive. Bits of stuff draw together and move apart, tugged at by agencies that touch at the very essence of matter: the *gravitational* interaction, arising from mass; the *electromagnetic* interaction, arising from the electric charge; the *strong* interaction, arising from the strong interaction (or "color") charge intrinsic to quarks but not to electrons; the *weak* interaction, arising from the weak interaction charge borne by quarks and electrons alike.

To have, for example, the quality of mass is to assert and also to accept a gravitational influence. A particle with mass moves closer to any other particle likewise endowed, engaging in an ever greater interaction as the distance shrinks and the masses grow. Where gravity is dominant, stars and galaxies form. Waves break on the shore. A gymnast flips in the air and lands on the mat.

To have an electric charge is, analogously, to broadcast and receive an electromagnetic influence and thereby to alter one's disposition in the presence of a second particle. An electrical body draws near to another with a charge of opposite sign and pushes away from one with the same sign, positive or negative. Where the electromagnetic interaction is dominant, atoms and molecules form. Oceans of molecules form. A gymnast forms.

To have any of the other fundamental attributes (such as the color of a quark) is therefore to interact in some other unique and fundamental way, perhaps not yet fully appreciated, but in the end always to do one thing: to communicate. To be a particle of matter is to exchange influence, little more; to be a particle is to move in the field of a neighbor and to move something else in return. Nature's tapestry is forever woven and rewoven in space and time by these tiniest shifts of influence—here and there; yesterday, today, and tomorrow; always with the same few stitches.

For the astrophysicist, the interplay of fields and particles spreads over the whole universe. Here, in the heavens, appears a world dominated by mass, a world stitched together by the gravitational interaction, a very big world.

For the nuclear physicist and the particle physicist, taking a different view, it is a world where mass is so slight that gravity seems to retreat. The strong and weak interactions, not gravity, govern the binding and decay of protons and neutrons inside the nucleus. It is a submicroscopic world, one of the smallest ever visited.

Finally, not least and not most, there is the world of atoms and molecules. The molecular physicist, the solid-state physicist, the chemist, and the biologist all stand between two domains, overseeing a middle ground between the ultrasmall and the ultralarge. Here, neither the strong nor the weak nor the gravitational interaction holds sway. Diligent observers find an electromagnetic world instead, a realm extending from a single atom of hydrogen all the way to solids in bulk. This world, the electromagnetic world, is the one we know most intimately, even if only from a distance—a distance from which the lumpy grain of matter blends deceptively into a continuous fabric, where positive and negative charges blur together into an overall neutrality. It is a world entire, but it is also part of a universe undivided.

All throughout, in worlds within worlds, are the seeds of potential thus sown. It is up to us now to discover exactly how that potential becomes reality. It is up to us, doing our experiments, devising our models, looking and learning, to uncover the rules of the game. It is up to us to illuminate the fields, to draw out the particles, to determine what things undergo change and what things do not. It is up to us to understand the laws of energy and equilibrium, order and disorder, chance and destiny.

3

IN THE EYE OF THE BEHOLDER

The whole universe may be ours to know, but expect to make little progress until we can answer two of the journalist's most basic questions: where and when. Whatever the event—the crash of an asteroid, the ins and outs of an electron, the undulation of a wave, anything at all—we need a way to map its location in space and to sequence its occurrence in time. We need to erect a mental scaffolding of *where* and *when* to support the *what* of the physical world. Underneath it all, we need to devise a frame of reference in which to follow the comings and goings of our particles and fields.

Sometimes, perhaps when we are looking at the Moon, the center of the universe will be the Earth. At other times, as in an atom of hydrogen, the world will collapse into a tiny sphere radiating out from a single proton. For another system we shall have something else, and for still another system something else again, and on and on, on and on. Whatever we agree to, though, our chosen reference frame will not (cannot) be the only one possible. It may be the smartest, simplest, and in many ways the best of all such reference frames, but it is only one choice out of infinity.

Let us accept, then, the first constraint of scientific knowing: *no observer is without bias.* All physical observations, however well intentioned, however painstakingly recorded, however exact they may be, are made from the perspective of a thoroughly arbitrary frame of reference—a convenient reference frame, no doubt, and likely a sensible, mathematically transparent reference frame, but an arbitrary one nonetheless. Like the YOU ARE HERE on a street map, the point of origin is arbitrary. We can shift it. Like the designations north and south or clockwise and counterclockwise, the orientation is arbitrary. We can tilt it. Like meters and miles, the units of measurement are arbitrary. We can change them. We can do all these things, yet the landmarks remain unmoved.

If we are to discern the workings of nature with any objectivity, then clearly we must reconcile the parochial viewpoints of an unlimited number of possible observers. Each explores the same territory with a different map, drawn from a different vantage point. No single one of them has a monopoly on the truth.

Space, after all, is what you make of it. To a sprinter, the Earth is flat and the shortest way from start to finish runs along a straight line. It is a world that begins and ends with one dimension:

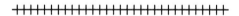

To a sailor, the Earth is round and circumscribed by two dimensions, latitude and longitude:

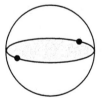

The shortest route between any two points curves along a great circle.

To passengers in a plane, flying ahead smoothly with window shades down and all noise from the outside suppressed, the three-dimensional world inside is at rest. Cups and saucers stay where they are, as surely and steadily as they would on the ground. With no sudden jerk or jolt or rush of scenery to provide a cue, eyes and bodies register no sensation of motion. Sitting feels normal. Standing feels normal. Walking feels normal. If the passengers didn't know better, they might justifiably believe that they were not flying at all. Tell *that*, though, to a traffic controller in the tower, whose radar confirms that an aircraft laden with cups, saucers, and sundry other cargo moves along at an unchanging 500 miles per hour. Motion and rest, we are forced to concede, are matters of opinion, not absolute truths.

Time, too: A particle of some sort (it doesn't matter what) races toward the Earth at tremendous speed, traveling at nearly 186,000 miles per second. It hurtles down from the upper atmosphere in a flash and then mutates into some other form. So claims an observer based on the ground, but to another observer—to one flying alongside the particle at exactly the same speed—the picture is entirely different. "No," reports this equally competent observer, "I agree that your particle eventually disappeared, but in every other respect you are wrong. First, the particle never moved. It remained in one position the whole time. Second, its lifetime was considerably shorter than you report: not *fifty* millionths of a second, but only *one* millionth of a second. Check your ruler and check your clock as well!"

What shall we say? Wherever we turn, two observers present us with two different but seemingly credible points of view. Must one be right and the other wrong? If so, how are we to know? Are we to despair of recording an unbiased description of the world on which all observers can agree?

What a world it would be otherwise, but not so in the only world we know. For here, in the universe large and small, we find that nature enforces everywhere the same fundamental laws: at sea and on shore, in trains and on planes, on Earth and Moon, with us or without us. Nature, blind to human perception, goes about her business independently of any reference frame we elect to establish. Our response, nothing less, must be to understand the tenets of physical law accordingly, in a way worthy of the adjective "universal."

Doing so, here and in chapters to come, we discover that neither time nor space nor motion is absolute. We find hints of conservation and symmetry in the action of mechanical systems. We learn that some things never change, no matter who looks at them, and we learn that some things always change. Our aim is to identify those quantities and events on which we all can agree, and to translate and harmonize those on which we all must disagree.

It begins with that most mundane of tasks, constraining and liberating at the same time: the construction of a reference frame to monitor events in space and time.

A Place for Everything

Space seems to be everywhere, so we take it for granted. But imagine, fantastically, a universe in which space itself were to shrink into nothingness, taking with it also the humble freedom of *place*. Imagine a world in which everything were squeezed into a dot infinitesimally small, a world caught in the grip of an unremitting, all encompassing, irresistibly strong power of attraction. It would be a world in which everything would have the same address, a world in which an object might be anywhere at all...as long as "anywhere" happens to be the one and only point available:

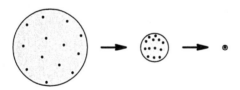

Gone would be the possibility of near and far, of large and small, of bent and straight. Gone would be the possibility of simply being *here* rather than *there*. All such possibilities would vanish into a pinprick space of zero dimensions.

It would be nothing like our present universe, evidently, for in our present universe particles and events do have both the freedom and the room—the space—to establish a sense of place. Two particles, identical in every way, thus become separate and recognizable to an observer with a measuring rod:

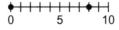

They appear at different positions along a straight line, each assigned a unique number, something like a street address.

Now let that one-dimensional street, with its *one* number per location, expand into a two-dimensional grid:

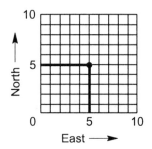

Here an observer needs a pair of numbers, the intersection of two gridlines, to locate any site in this two-dimensional neighborhood. Not one, but *two* signposts point the way: so many steps east or west, so many steps north or south...and there it is, the spot in question.

Crossing at right angles, the two basic directions appear separate and pure to our observer hopscotching east–west and north–south across the grid. A step to the east brings about no displacement north or south; a step to the south brings about no displacement east or west. Neither direction mixes with the other, but the pair of them together cover the two-dimensional space completely. Two perpendicular displacements are all one needs to navigate in two dimensions. Just two elementary steps, no more and no less, label each site with a position shared by no other.

We know, of course, what comes next. A flat surface grows up into a three-dimensional box, and with three directions there will be three numbers to cover the space. One for east–west. A second for north–south. A third for up–down:

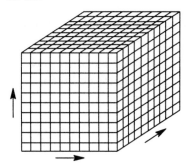

Tell us how many steps to take in each of the three independent directions, and we can find any place at all within the three-dimensional box.

And why stop there? Suppose we have two such boxes, or three, or four, or even an unending row of them, each cut off cleanly from its neighbors. To give every point an unambiguous address, we need only to add a fourth number—a fourth dimension—to the space, a number to tell us in which particular box to step east–west, north–south, up–down:

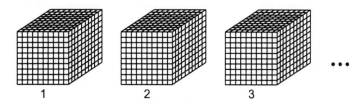

It would be a four-dimensional address in a four-dimensional space, stated perhaps as "building 2, floor 7, corridor 1 east, room 4 north." It would be just a set of four numbers, not the entrance to a mystical world.

Indeed we can conceive of all kinds of space, all kinds of places for things to be and events to occur. We can specify addresses in one dimension, two dimensions, three dimensions, four dimensions and more. There is no limit to either the dimensionality or shape of any space imaginable. Space can be flat, like the lines and planes and boxes sketched above, or it can be curved. It might bulge outward in one place and pinch inward in another. Some spaces, like the inside of a box or the surface of a sphere, are bounded and contained, whereas others are unbounded and never ending. There is a space for everything, and everything has its space.

Yet who is to decide where and how that space shall take shape? Who has the right even to stipulate at what location a space shall begin? Who of all possible observers is privileged to assert that somewhere in the midst of an abstract space, in the midst of a conceptual grid bare of all landmarks, that precisely *here* is the point to be called "zero," the point from which all dimensions originate? You, with your one-dimensional measuring rod, say that particle 1 sits at point zero and particle 2 appears eight positions to the right:

I, with *my* one-dimensional measuring rod and my own idea of where things should start, say that particle 1 appears five positions to the left of zero whereas particle 2 appears three positions to the right:

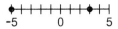

If either of us could prove that *here*, at some one point in a featureless space, we knew exactly where we were—that there was some indisputable, fixed, immovable monument like the throne of Zeus to mark the spot— then maybe one of our notions of zero would have an absolute meaning. But no, neither of us can lay claim to priority of position; neither of us really knows where we are. Your zero is as good as mine.

What brings us together, however, is the realization that these two points are undeniably eight units apart, despite the different labels we apply to the individual positions:

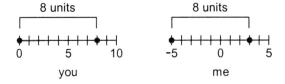

We realize further that our two systems of reckoning are easily reconciled. You can subtract five units from each point in your sequence and thereby reproduce mine,

or I can add five units to each point in mine and reproduce yours:

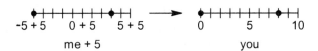

Either way, we translate from one point of view to another and, more encouraging, we share an unvarying, absolute perception of at least one measurable quantity: the distance between two points on a line. It's a start.

Lost in Space

The tale of the measuring rods teaches us that empty space stretches out uniformly in each dimension, everywhere the same, indistinguishable from site to site. We come upon no absolute zero of position, no guiding light to show the way home to every possible observer. All points on the spatial grid are created equal.

The essence of a thing—what it is, how it is put together, what effects it can engender—remains untouched whether that thing exists here or there in an otherwise empty space. Whatever it is, if we pick up the object and translate it somewhere else (carrying along everything that interacts with it), we really do nothing at all. We merely readjust our idea of zero, while preserving intact the distances between individual positions.

Go ahead. Take two particles, drop them into empty space, and superimpose a grid of coordinates on their positions. Pick whatever numbers you like. Since one point of origin is as good as any other, then one spatial grid is also as good as any other. The configuration described by observer 1 (who calls the "east" dimension x and the "north" dimension y) therefore must be identical to the one described by a displaced observer 2 (for whom east is u and north is v):

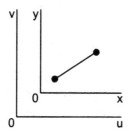

Both observers report the same distance, represented by the connecting line, even though they disagree on the east–west and north–south position of each particle.

Now, plainly, the specific numbers that observer 1 declares for x and y are utterly meaningless, as meaningless as those declared by observer 2 for u and v. Yet this very meaninglessness takes on profound meaning once we begin to discern how particles act and interact on the spatial stage. As observers, we need to formulate mathematical equations to mimic what happens in nature—and if we are to reconcile our infinitely many points of view, those equations must be bulletproof against any shift in spatial perspective. Without knowing a thing about the equation itself, without knowing what symbols will adorn it and what operations it will command, we nevertheless know that we shall have to submit to at least one constraint: that in whatever way our equation incorporates x's or y's or u's or v's (or any other expressions of position), its form must remain unchanged if we were to add 5 to all the x's, or 3 to all the y's, or -22.7 to all the u's, or some other fixed number to any of the coordinate axes. An equation of nature, if it is to meet the first test of a universal law, must withstand a translation of the entire spatial grid. It must be free of all prejudice of position. It must be universally valid even in the face of the arbitrary numbers chosen to set up a spatial frame of reference. It must be "symmetric under translation," indistinguishable before and after a displacement.

Without looking and learning, without experimenting, without testing, we would never know how the Moon orbits the Earth. We would never know how a proton and an electron join to form an atom of hydrogen. We would never know how an atomic nucleus holds together, but we would still know something deeply important about the underlying laws governing all such systems. We would know that these laws, when finally revealed, would have to be translationally symmetric.

It is nature's way. Observers 1 and 2, transcending the labels on their spatial grids, ultimately describe the same orbiting Moon, the same atom of hydrogen, and the same nucleus. They discover that the fundamental forces of nature depend not on arbitrary, meaningless, shifting positions, but rather on a quantity fixed in all spatial reference frames: the distance between points. Nature honors no particular zero of position, and our mathematical descriptions of any structure or event must follow accordingly.

In another kind of universe, the rules might be different. Not, apparently, in ours.

Which Way Is Up?

In the middle of nowhere, adrift in empty space, an unbiased observer has no place to turn—and thus everywhere to turn. Since no single direction is special, then *all* directions are special. With no compass needles to point north, no Sun to rise in the east, and no birds to fly south, one direction becomes as good as another. Nothing of consequence, certainly no law of nature, can depend on what someone calls north, south, east, west, up, down, hither, yon.

Two observers map the locations of two particles, agreeing first that one of the objects appears at point zero. Observer 1 subsequently places the second particle four steps east (x) and three steps north (y). Observer 2, using a rotated grid, places it five steps due east (u). Here is what they see:

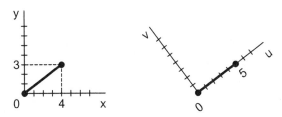

To observer 1, the directions x and y mean pure "east" and pure "north." A step x remains rigorously separate from a step y, whereas x and y together map the space completely. Yet to observer 2, equally convinced of purity and completeness, it is the directions u and v that represent the perfect east and north. "You are wrong," says one to the other, "*your* so-called east is not really east at all. What you have is a contaminated mixture of the *true* east and the *true* north, all jumbled up."

They suffer from, at worst, a simple misunderstanding. The skewed observers soon realize that their differences in position are only apparent, born of an arbitrary definition of north and east. They concede that neither of them is privileged to make such a judgment, and, wonderful to relate, they readily find common ground. Within one system, x and y are pure and complete; within the other system, u and v are no less so. Despite the rotated points of view, each observer measures the same distance between the two particles, an unchangeable five units (look again at the picture). Moreover, knowing the angle between the rotated grids, they develop a formula to convert xy positions into uv positions and uv positions into xy

positions. All it takes is an application of high-school trigonometry, easy if you know how.

In what manner, then, shall we interpret our instinctual notions of north and east and up and down? Doesn't nature show a sense of direction at every turn? Just think of the difference between the way the wind blows and the rain falls, or the difference between climbing up a mountain and climbing down, or the difference between a pendulum clock on its feet and on its side. Think of how a compass needle, aligned with the Earth's magnetic field, points in one seemingly special direction, nowhere else. Consider all of these limitless possibilities (because surely we must), but realize also that in none of them does nature favor a particular direction as absolute. Nowhere is one single direction recognized as intrinsically special by every fair-minded observer, identifiable on its own without some external reference.

For although a compass needle does point in one direction, it is a direction that exists only in relation to something else: the magnetic field of the Earth. Were Earth and compass (and all that goes with them) to be rotated through space, we would still see the needle point in the same direction as the terrestrial field, whatever that direction may be:

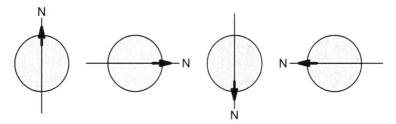

Look beyond the Earth, and there is no longer any up and down. There is no universally established axis to which everything else can be referred.

Nature decrees it; we describe it. A valid expression of physical law, purged of directional bias, may specify orientations only relative to an existing axis. All observers, regardless of their degree of rotation, must describe the same phenomenon in the same way using the same mathematical forms. Their description must be impervious to a simple turning of the spatial grid. Their equations must be symmetric under rotation.

The Clock Has Started

By allowing for space, nature guards against everything being in the same place. By allowing for time, she prevents everything from happening all at once. Just as space creates a perception of "here" and "there," time creates a perception of "before" and "after." We keep track of it with a clock.

Ticktock. A clock cycles from state to state, returning regularly to its starting point. The regularity can come from the swing of a pendulum. It can come from a laser beam bouncing between mirrors. It can come from the daily rotation of the Earth, or the monthly orbit of the Moon about the Earth, or the annual circuit of the Earth about the Sun. A clock can be built from all kinds of processes, but what matters most is that the states recur at evenly spaced intervals. Each cycle must run through the same course of events in exactly the same way.

Ticktock. One cycle. *Ticktock.* Two cycles. *Ticktock.* Three cycles. The uniform march of the cycles, never varying, lays out for us an ordered sequence of standard events:

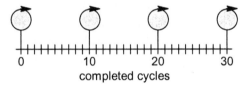

Observing some other, less regular series of events, we then count the number of clock cycles completed from start to finish. That number (the number of uniform ticktocks) gives the duration of the process observed. It tells us the interval elapsed between two arbitrarily labeled points on a scale of time, and its value does not depend on what cycle we choose to call zero. An interval of, say, 10 units remains the same regardless of when the clock starts:

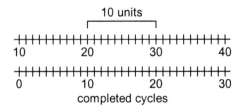

Now if the laws of nature hold constant over time—if the rules and patterns of today are the same as those of yesterday and tomorrow—then a valid physical equation will depend only on elapsed time. It will be an equation free of all references to an absolute zero of time, relying instead on intervals rather than single points. Its mathematical form will be unaffected by any displacement of the temporal axis. It will always be the same law, and from moment to moment an observer will never know what time it *really* is. In such a world, one point on the temporal ladder is as good as another.

But if the physical laws change with time—if the cause that gives rise to an effect today produces a different one tomorrow—then an observer might be able to tell time in an absolute sense. If, for example, the universe is expanding (and compelling astronomical evidence says it is) and if we can project backward to a time when all matter and energy were compressed into a single point (and, again, the available evidence suggests we can), then we shall have discovered an absolute origin of time: a one-of-a-kind instant at which the universal clock started to tick. An observer, measuring the relative positions of the galaxies, could then deduce how much time has flowed since the primordial moment when everything began. In that kind of world, perhaps, the equations of physical law might refer not just to differences in time but to individual points as well.

So what do we have? Do we have a universe in which the rules of the game are fixed, once and for all, or do we have a universe in which they vary from moment to moment? Are Newton's equations of motion valid for all time, or must each generation of physicists discover for themselves how objects move in response to a force? Has light been traveling at 186,000 miles per second since the beginning of time (if indeed there was such a beginning), or is it slowing down or speeding up as we watch?

Nobody knows for sure, but we are beginning to appreciate how the universe presents more than one face. On the largest scales, over unimaginably great intervals in space and time, the laws of nature do seem to show a sense of history. The way things work today, a mere moment in the cosmic narrative, may well be different from the way they worked 14 billion years ago (and different still from the way they may work 14 billion years hence).

Even today, our picture of a world woven together by a gravitational force, an electromagnetic force, a strong force, and a weak force may be incomplete. Astronomers are gathering evidence that an additional fundamental interaction, a repulsive effect opposite to gravity, may be at work over vast distances and possibly changing with time.

On the smallest of scales, too, at a level more basic than that of quark and electron, theoretical physicists are proposing a startling fundamental structure for the microworld: an architecture in which everything is built from tiny, tiny "superstrings" of energy, vibrating in a tightly compressed space of ten dimensions or more, not the three of our everyday experience. Physicists speculate that some of the dimensions, so small as to go unnoticed, may have shrunk to near nothingness since the birth of the universe. If true, then the dimensionality of space itself has a history.

There is another face of the universe, however, a more familiar one not to be forgotten amidst the excitement of new discovery. Observing nature on all but the most extreme scales—from objects the size of quarks and nuclei to stars and galaxies, from intervals in time shorter than a quintillionth of a second and longer than a million years—we find patterns and laws that appear resolutely ahistorical. We see no evidence of an absolute "time zero" in the way that gravity holds the Moon in orbit, or in the way that the electromagnetic interaction creates an atom, or in the way that the strong force binds protons and neutrons, or in the way that the weak force breaks up a neutron. Newton's mechanical equations specify only differences in time, ascribing no absolute meaning to each instant. The same goes for Maxwell's electromagnetic equations. The same goes for Schrödinger's quantum mechanical equations. The same goes for Einstein's equations of relativity. The same goes for so many other equations that describe so much of the physical universe.

For us, open-minded but nonetheless emphasizing this ahistorical aspect of nature, our clocks will tick and tock without reference to any absolute starting point. Our spatial grids, similarly inspired, will make no reference to absolute position or direction. Combining time and space, we shall then have a fully equipped reference frame from which to view a world ever in motion, ever rearranging. And from that vantage point, striving to be objective, our observers will do away next with yet another absolute: the idea of absolute rest.

Never at Rest

Picture a simpler universe, stripped bare of everything except for two generic particles joined by a generic interaction. Choose whatever you wish: Earth, Moon, and gravity; or maybe proton, electron, and the electromagnetic force; or possibly neutron, proton, and the strong force; or, better yet, our old friends particles 1 and 2 with a symbolic line between them. Let one body sit fixed at the zero of an arbitrary spatial grid, around which the other body moves in similarly arbitrary fashion. As a ready example, uncomplicated but in no way special, we can imagine a steady circular motion:

Other than that, nothing: no other particles, no other influence. Just space. Formless, positionless, directionless, empty space.

Now throw a switch and turn off the interaction. In an instant, cleanly and completely, take away whatever influence it is that links particle 1 with particle 2. Break the connection and watch.

Will particle 2 keep moving stubbornly in the same circle at the same speed, or will it stop dead in its tracks? Will it continue round ever more slowly before finally coming to rest? Will it spiral inward and collapse onto particle 1?

Particle 2 suffers none of these fates. Experience and four centuries of observation (beginning with Galileo and Newton) tell us exactly what particle 2 will do. It will fly off in a straight line at a constant speed and thereafter simply keep going. It will obey the law of *inertia*, which demands of objects that they resist change: that they maintain their present condition until *forced* to change. Remove a compelling force and a body continues to do forever what it was doing at the instant it became free.

Particle 2, remember, had been following a curved path, moving a tiny bit in a particular direction before changing course. Bit by bit, pulled along by the interaction, it was making its way around the circle:

Then, abruptly freed of all constraining influence, it flies off in its current direction and moves independently through the featureless void of space:

Particle 2 travels in an unending straight line at a constant speed, while particle 1 remains at rest.

Says who? So says an observer bolted to a coordinate system in which particle 1 stays fixed in place. But to an observer sitting on particle 2, viewing matters from a reference frame no less privileged, the picture appears reversed:

Here it is particle 2 that stays put while particle 1 runs away in the opposite direction. Neither observer can prove convincingly that the other is either moving or standing still, just as passengers traveling smoothly in a plane or train (or, for that matter, merely parked on planet Earth) experience no sensation of motion.

Along comes another observer, flying also in a straight line but with a different speed and direction, and we have yet another point of view. This third observer reports that particles 1 and 2 are both moving in straight lines, although not the same straight lines claimed by the previous two. The speeds appear different, and the directions appear different. Add to that report a fourth and a fifth and a sixth, and we soon realize that all such observers in "inertial reference frames"—all observers whose spatial grid and clocks are moving in a straight line at constant speed—will describe the system in the same essential way, despite manifest differences in certain details. They will agree that a free particle, unconstrained by any coercive force, can do only one of two things: (1) If it happens to be already at rest, it will remain at rest. There is nothing to move it out of its state of inertia. (2) If not already at rest, then a free particle will be coasting in a straight line at unchanging speed. Absent some form of compulsion, it will maintain the same course from one end of the universe to another.

On these two issues the various observers all agree, although admittedly they disagree on everything else. They record disparate values of position, direction, and time, and they even differ on the question of who is moving and who is standing still. The philosophical ideal of absolute rest loses all meaning in a space where "here" is the same as "there." Nature permits no observer to claim, "I know definitely that I am not moving, because I see Zeus sitting absolutely still on his throne—and we all agree that Zeus's throne never budges from its place. It's the rest of you that are in motion, not I." An observer can only say, more modestly, "If you ask *me*, I am at rest. The walls and furniture around me appear to stay in the same place. But that's just my opinion. Yours may be different."

We consider the opinions of not just two or three or even a million fair-minded spectators, but rather an infinity of them. Each inertial observer moves in a different direction at a different speed, and each set of observations is no less valid than any other. The frames are utterly equivalent, and in every one of them a free particle moves in the simplest way possible. It obeys the law of inertia, a consequence of the gray homogeneity of space and time.

Simplicity is good. The freedom to choose an arbitrary reference frame does not obligate us to choose a *stupid* reference frame (like a roller coaster, for instance, where noninteracting particles will appear to be tossed about every which way for no apparent reason). Instead, we simplify our observations wherever possible by specifying inertial systems. We accept a frame as inertial when we see that free particles stay put or move uniformly in straight lines, and we agree to treat observations from all such frames as equally admissible.

Our army of eligible observers still must come to terms with the different numbers they record from their different vantage points. Comparing notes, observers in stationary reference frames find ways to understand each other's displaced or rotated coordinates. So do observers in motion, as we shall see next.

Time, Space, and Motion

Don't think, just because we cannot agree on some absolute, universally acceptable, throne-of-Zeuslike state of rest, that a condition of *local* rest must always be dismissed as meaningless. Global perceptions aside, all kinds of objects and events anchor themselves to a preferred frame of reference: trains on tracks, waves in ponds, winds over prairies, footballers on fields; the list is endless. To recognize, sensibly, that a particular action plays out on a particular stage is not necessarily to be a biased observer. There is a difference between empty space and a space filled with distinguishable landmarks.

A train rides the rails. Relative to those rails, a train stopped at a station is undeniably at rest. So are the passengers waiting on the platform. They can prove it. No observer in any other state of motion sees a stationary track.

A passenger sitting in a railway car, agnostic about whether the train is stopped or moving at constant speed, remains stationary relative to the interior surroundings. An observer in the next seat who sees this particular body stay in one place, *inside the train*, concludes that the passenger is locally at rest. No one looks out a window. If the train is moving uniformly down the track, who in the car is to know?

Back on the platform, an observer records distance and time as a train passes by: 100 feet after one second, 200 feet after two seconds, 300 feet after three seconds...a steady 100 feet covered every second:

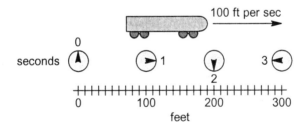

From this ground-based viewpoint, the train and its seated passengers all move forward at 100 feet per second. Inside the train, where the seats remain fixed relative to the walls and floor and ceiling, we know that the same passengers claim not to be moving at all. Their internal positions remain the same while the clock ticks away.

One of the passengers now begins to walk toward the front, and the other passengers measure distance and time: one foot after one second, two feet after two seconds, three feet after three seconds...a steady one foot covered every second, as determined from inside the car. To our observer outside, however, the passenger in the aisle appears to be moving at the much faster rate of 101 feet per second. Relative to the tracks, the new motion registers as an internal walking speed of 1 foot per second *plus* an overall train speed of 100 feet per second. The two speeds add together:

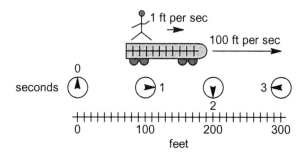

We seem to have just an understandable difference in perception, easy to resolve. A train moves on a track. A body moves inside a train. Inside, an observer measures this-and-this number for the speed of the body in motion. Outside, a different observer adds that same number to the uniform speed of the train. A simple mathematical recipe, cut and dried.

It might be the end of the story, if not for one wrinkle. In a universe without absolute space, without absolute time, without absolute rest, we stumble upon a striking absolute feature in nature's design: the maximum speed at which a signal may pass from here to there. Electromagnetic influence, for one, travels in the vacuum of space at some 186,000 miles per second, perceived as exactly the same number in all inertial reference frames. Even if we reach this maximum speed ourselves, we never catch up to an electromagnetic signal and we never see it at rest. Its speed to us is always a constant 186,000 miles per second, unlike our changeable view of train and passenger. This universal speed limit—and the same unvarying perception of it—holds not just for things electromagnetic ("light"), but for the other fundamental interactions as well.

And that, as Einstein realized, changes everything. Required to give equal opportunity to all inertial observers and also to accept an invariant speed of light, we now need to rethink our view of time as separate from space.

Killing Time

A beam of light cannot be confined to a railroad track, buckled into a seat on a train, or tethered to an astronaut in a space ship. A beam of light propagates as an electromagnetic field, a projection of influence generated by an electric charge. It shows up as a quality impressed onto the void, divorced from the matter that creates it and supported by literally nothing. The electromagnetic field, which we first met in Chapter 2, requires no particular medium for its existence.

No reference frame claims a beam of light as its very own, the way a track claims its train or a railway car claims its passenger. A beam of light does not walk up the aisle, carried along with the car, held fast to the floor. A beam of light, quite the contrary, makes its way through empty space. It moves ahead another 186,000 miles every second, with or without a railway car, with or without a railway track, with or without the ground below or the sky above. It moves ahead, always at 186,000 miles per second, as if nothing else exists.

Fancy this: At rest in the middle of a railway car, a certain particle simultaneously sends out two beams of light. One beam shoots up the aisle, toward the front; the other beam shoots down the aisle, toward the back. After traveling equal distances, the signals register on the two bulkheads as separate flashes of light. Which flash comes first?

For a passenger inside, they arrive simultaneously. Seeing two signals travel equal distances at equal speeds, the train-based observer concludes that the two flashes occur at the same time. The temporal separation between events 1 and 2 is zero:

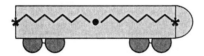

How about for an observer on the ground? To make matters interesting, let's pretend that the train moves much faster than trains usually do, perilously close to 186,000 miles per second. We still expect (most of us, at least) that the flashes will be simultaneous from this viewpoint as well, conditioned as we are to treat time separately from space. Yet our unschooled instincts prove wrong, because the pulse moving backward (to the left) actually hits the detector first when viewed from the ground. Seen from a reference frame in which the train is in rapid motion, the temporal separation becomes nonzero.

The beams, remember, may have been triggered *in* the car, but they are not *of* the car. They adhere to nobody's favorite reference frame, and consequently a ground-based observer sees them move at the same speed as a train-based observer. Inside the train, though, the light-sensitive detectors are secured to the bulkheads with literal nuts and bolts, and therefore an observer outside sees the two detectors as moving separately from the signals. The detector on the left, running toward the left-going signal, lights up first. The detector on the right, running away from the right-going signal, responds some time later:

No other conclusion is possible, given that the observed velocity of light never changes. Only by perceiving time in different ways—by disagreeing on whether the flashes occur at the same instant—are the observers able to agree that the signals traveled at the same speed. The sequence of events thus becomes a matter of opinion, colored by one's state of motion. For an observer inside the train, the events occur simultaneously. For an observer outside the train, they do not.

Back in the everyday world, where trains come nowhere close to the speed of light (not even European express trains), an observer on the ground sees the flashes occur not quite simultaneously, but any difference in time proves vanishingly small. To someone watching a train lumber along at 100 feet per second, the lag occasioned by a light ray moving at 186,000 miles per second is easy to overlook, a change in the umpteenth decimal point. To such observers, thinking practically, the number 186,000 miles per second is tantamount to infinity, and it is precisely this human experience of moving around so *slowly* that shapes our conventional view of time. We naively believe, solely as a result of limited observation, that time ticks the same way no matter how fast or slow something is moving. But to observers in reference frames traveling at velocities close to the speed of light (riding along with highly energetic elementary particles, say), the propagation of signals at finite velocities cannot be ignored. Time, like space, is in the eye of the beholder, and unbiased inertial observers must find a way to come to terms with both.

Relativity and Invariance

Allow a beam of light to bounce back and forth between two mirrors held rigidly in place. Since light moves at the same speed in all inertial reference frames, the relationship between time and distance during each cycle remains constant. We have a simple clock:

With a flash, the beam leaves the first mirror: event 1. After traveling a certain fixed time and distance, it returns home and produces a second flash: event 2. We then ask two inertial observers, one moving uniformly in a straight line relative to the other, to tell us what happens. We want to know the distance in space and time that separates the two events.

Up, down. Up, down. An observer at rest relative to the clock sees the beam remain in one place, just as in the diagram above. The ray of light cycles up and down vertically, but it does not move horizontally. Observer 1 reports the two events as separated only in time, not in space.

Observer 2, for whom the clock moves left to right, sees the beam follow a triangular path. A horizontal component mixes with the vertical, so that the total distance covered by the beam—and hence the time—appears greater than for observer 1:

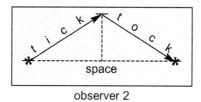

observer 1   observer 2

We convince ourselves, just by looking at the picture, that the diagonal ticktock of observer 2 exceeds the vertical ticktock of observer 1. A clock in motion is perceived as running slower than a clock at rest.

How much slower? The dilation of time depends on the relative speed. Fast motion, close to the velocity of light, stretches out the triangle and delays the ticktock more than slow motion. At any speed, though, observer 2 sees the diagonal stretch in time accompanied also by a horizontal stretch in space, with the exact proportion between time and space constrained by the geometry of a triangle. Given a certain interval in time (the diagonal line below) and a certain interval in space (the horizontal line), the length of the broken vertical line is fixed once and for all. There is only one way to complete the right triangle:

space

Observer 2 notes happily that this vertical leg, a combination of temporal and spatial contributions, is identical to the one recorded by observer 1. Moreover, its length proves to be the same for all other inertial observers at all other speeds:

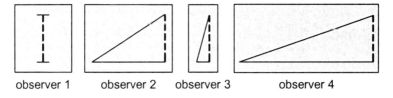

observer 1   observer 2   observer 3   observer 4

With that, our infinitely many observers are able to harmonize their reports. They have a unique number, a mixture of a spatial and a temporal interval that has the same value for all of them. We call that number the invariant (fixed) *space-time interval* between two events. It blends together space and time.

For observers in motion, separate intervals in time and space become as meaningless as the directions east and north in a stationary coordinate system. Time is relative. Space is relative. Their values change according to an observer's state of motion. I say one thing, and you say another. I say that two events occur at the same place an hour apart, and you say that they occur two hours apart with over a billion miles between them. We have no choice but to disagree, since we both observe the electromagnetic field propagating at the same speed.

What unites us is the combined space-time interval, this one fixed quantity common to all points of view. It remains untouched by any perceived blending of space and time, just as a point-to-point distance in space survives an arbitrary rotation of the coordinate grid:

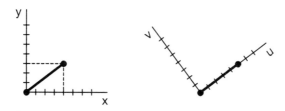

And just as the invariance of distance enables the interconversion of xy's and uv's in space alone, the invariance of the space-time interval makes possible a similar service for space and time together. To unscramble a rotation in space—to express *your* east and north in terms of *my* east and north, and vice versa—all we need to know is the angle between our two coordinate grids. To unscramble an analogous rotation in space-time, all we need to know is the difference in our speeds. Give me that one number, the relative velocity of my reference frame, and I can convert my values of position and time into yours. You can do the same.

With space and time reconciled, we may then work in whatever inertial frames we like, provided that our results (after conversion) turn out to be the same. Specific numbers for position, time, velocity, force, electric field, magnetic field, and a host of other quantities may differ from frame to frame, but the governing relationships may not.

Suppose, for instance, that a reanimated Galileo Galilei observes a passing railway car in which he monitors the system of his choice: billiard balls on a table, an electromagnetic turbine, hydrogen and carbon atoms, a uranium nucleus; anything and everything, no exceptions, no exclusions. Whatever the relevant quantities may be, Galileo sets out to measure their values over space and time. Using ruler, clock, voltmeter, Geiger counter, telescope, or any other such instrument, he determines that quantity A has the value 2 greebles, quantity B has the value 3 thorgs, and quantity C has the value 6 greeblethorgs. From those three numbers and their apparent numerical relationship ($6 = 2 \times 3$), he then goes on to propose a universal law of nature: $C = A \times B$. Does Galileo's inference necessarily carry the force of law?

Looking for a second opinion, we ask Albert Einstein to observe the same system, but from inside the train itself. Einstein, traveling at constant speed relative to the tracks, wakes up in a closed compartment and has no idea whether he is moving or standing still. He is a qualified inertial observer, no better and no worse than Galileo, and he is entitled to his own measurements.

Making those measurements, Einstein reports values of 4 greebles for A, 6 thorgs for B, and 24 greeblethorgs for C. He disagrees with Galileo about the three numbers individually, but he agrees with the governing equation. Einstein, like Galileo, finds that C is equal to A times B, except in Einstein's case the math becomes $24 = 4 \times 6$ rather than $6 = 2 \times 3$. It is a distinction without a difference.

Galileo and Einstein agree on something more important than mere numbers. They agree on how one physical quantity depends on another, no matter how fast or slow an inertial observer may be moving. Even if the train were to whoosh by at nearly 186,000 miles per second (very fast), so that A, B, and C were mixed up considerably in time and space, the two observers would still find a consistent recipe to connect the three quantities. Each set of measurements would make sense within its own frame of reference.

These hypothetical observations thus comply with the "principle of relativity," enunciated first by Galileo himself for low speeds and modified later by Einstein to cover all speeds, high and low. The principle insists that every occurrence unfold in the same fundamental way to every inertial observer, without betraying any uniform linear motion of the reference frame. To be a valid physical law, a descriptive equation must maintain the same form when time, space, and all other susceptible quantities are mixed up. It must respect the relativity of space, time, and motion, and it must respect the invariance of the space-time interval. When it fails to do so, the proposed relationship is rejected on the grounds of motional bias.

If space were absolute...if time were absolute...if motion and rest were absolute...if light emanated only from the throne of Zeus and traveled like a train on a track—if nature's ground rules were something other than what they are, then different observers would have a different story to tell. But that choice is not ours to make.

4

THREE-PART INVENTION

To know the present is to predict the future and retrace the past. Such is the promise of a mechanical universe, the idea that nature might run like a well-oiled machine.

Think of it. Like the cogs and wheels of some precisely tuned mechanism, the parts of a clockwork universe would follow a set course once put into motion. Every move would be predetermined by the one that came before. The state of the machine tomorrow would follow inevitably from its state today, with everything of interest displayed on assorted gauges and readouts. And like all machines, no matter how complicated, its course of operation would ultimately be knowable in full. With enough observation, with enough analysis, with enough familiarity, we might eventually learn what makes everything tick. Omniscience, a bite from the Tree of Knowledge, is the promise offered by a mechanical universe.

For the big things, it is a promise often kept in both letter and spirit. The Moon, put into orbit billions of years ago, travels around the Earth with impressive regularity, never straying far. We calculate, almost exactly, where in its cycle the Moon was a hundred years past and where it will be in a hundred years to come. A ball, also a big thing, rolls predictably down a slope of a certain grade, always passing the same landmarks in the same sequence of times. There are no surprises in such a world. A big world is typically a world governed by the so-called classical mechanics of Isaac Newton, a forthright world in which planets and comets and balls and bullets all follow their set courses. They go where they are pushed and pulled, and we know where to find them.

For the little things, for small particles confined in small spaces, the universe submits to a different kind of government: the shadowy regime of quantum mechanics, under which the promise of omniscience is honored (after a fashion) but devalued at the same time. To be sure, nature does allow us to describe with a set of numbers the mechanical state of an atom, or a nucleus, or a proton, or any other system in the microworld. Nature even allows us to follow the ups and downs of our chosen state as it changes with time, acting in response to one of the fundamental forces. Given knowledge of the present, we can then look ahead to the future and back to the past, but the knowledge granted to us remains soft and fuzzy. It is a knowledge not of certainty, but rather of probability and chance. We say of the Moon orbiting the Earth that it will definitely appear here today and there tomorrow, and we say so (in principle) without a hint of doubt. We say of the electron in hydrogen, more cautiously, that it is *likely* to appear here today and there tomorrow, that it will *probably* appear here today and there tomorrow, that we *bet* it will appear here today and there tomorrow—but, in the end, we cannot be sure. Our knowledge becomes a gambler's knowledge of the odds, no longer certain, no longer a sure thing.

Between certainty and uncertainty, too, there is a third form of governance: a world of chaos, where the promise of mechanics is upheld according to the letter of the law yet mocked in spirit. We find it every-where. We find chaos in the faltering of an arrhythmic heartbeat, in the turbulent flow of white water around a stone, in weather patterns frus-tratingly difficult to predict. We find it again and again in the ordinary occurrences of everyday life, often the richest and most complex to be found. We find how a clockwork mechanical system, following the sim-plest of rules, winds up in an utterly unpredictable final condition—all the while doing nothing fancy, just obeying a straightforward Newtonian directive to go definitely from this state to that. What happens, though, is that the machine is so touchy, so sensitive, so delicate that the slightest alteration in its state today produces a wildly different result tomorrow. Determinism be damned. Unable to pinpoint exactly where the system begins, we forfeit any claim to mechanical omniscience in a chaotic world.

Three regimes. Three machines. Nature presents us with three conceptual machines, each fundamentally different in the information it takes in and the information it puts out. One of them, the Newtonian machine, runs like a perfect clock. We give it the current time, wind it up, and it tells the correct time ever after. The chaotic machine, by contrast, operates like an impossibly finicky clock. If we err only slightly in setting the current time, the inaccuracy builds up so wildly as to become unpredictable. The chaotic clock, responding disproportionately to small deviations, runs ungovernably fast or slow. The quantum machine, finally, runs like a clock that tells time randomly. We never know the time until we ask, and even then it may be twelve o'clock one moment and eleven o'clock the next. Still there is a method in the apparent madness, for we know the odds with some measure of confidence. Inquiring at a particular instant, say, we know that twelve o'clock will come up six tries out of ten—not specifically *which* six tries out of ten, but six tries just the same. It may not be omniscience, but it is not hopeless ignorance either.

We shall devote most of the remaining pages to recognizing how these metaphorical machines differ, and differ they do. Each operates in its own domain. Each responds to influence in its own way. Each establishes its own relationship between cause and effect. Beyond their differences, however, the classical, quantum, and chaotic machines of a mechanical universe share a pattern of operation common to all. The mechanical fates of both Moon and electron, although different in every particular, still follow a set of universal rules: a generic code of conduct good for all kinds of particles influenced by all kinds of interactions. For anybody looking for a thread of unity, it is a good place to begin.

Getting Started

A system starts in one condition. An influence of some sort, conveyed by a field, causes it to change. It finishes in another condition. Whether the changes conform to the laws of classical mechanics or quantum mechanics or whatever other mechanics one cooks up, the job of a curious observer soon becomes clear: to chronicle a moment-by-moment history of the evolving states. The first step is to describe the initial condition.

Picture the system at any instant as represented by a list of numbers, a catalog of instructions needed to realize a given condition. The numbers are like the settings on a machine, and their values change as time goes by. They tell us the whereabouts and current business of the various moving parts. Give us the *mechanical variables*, this master set of numbers, and we can reassemble the machine exactly as it exists at the moment of interest.

The appropriate descriptors will differ from system to system in both number and kind, and we shall depend on observation, experimentation, and analysis to discover what they are. For one system, a full set of specifications might include the instantaneous position and velocity of every particle:

> *Ball 1 is currently at such-and-such a height, at such-and-such a latitude, at such-and-such a longitude, moving at such-and-such a speed in such-and-such a direction. Ball 2 is currently at this-and-that height, at this-and-that latitude, at this-and-that longitude, moving at this-and-that speed in this-and-that direction.*

For another, the numbers might delineate the rise and fall of a wave:

> *The crest rises to such-and-such a height and repeats itself infinitely over such-and-such a distance.*

For yet another, the list might describe the orientations of two magnets:

> *Magnet 1 points north. Magnet 2, separated by such-and-such a distance at such-and-such an angle, points south.*

With no particular system in mind, then, let us imagine the changing values of these mechanical variables simply as entries in an extended table (as conjured up below). The initial condition, the "zero" state in which the machine begins, occupies the first row:

Mechanical Variables

State #	1	2	3	4	5	6	
0	∿	∿	∿	∿	∿	∿	
1	∿	∿	∿	∿	∿	∿	
2	∿	∿	∿	∿	∿	∿	
3	∿	∿	∿	∿	∿	∿	

Running across the top are labels to identify each mechanical variable, and running down the leftmost column are labels to identify a sequence of changing states. Each row contains a complete list of values at the time specified, a snapshot of the machine at one point of its progress. That's all there is. Everything there is to know stands written in the table.

Now suppose, just to concoct a simple example, we study a system described fully by only two mechanical variables. At time 0 (state 0), both variables happen to have the value 0. At time 1, the value 1. At time 2, the value 2. At time 3, the value 3 ... and so on. Entering these numbers in our table, we thus trace the course of the machine from start to finish:

	Mechanical Variables	
State #	1	2
0	0	0
1	1	1
2	2	2
3	3	3
4	4	4
5	5	5

With numbers in hand, we might also choose to display the raw data as a graph—a nicety, perhaps, but it would show everything at once, neatly condensed into a single diagram:

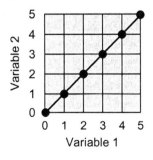

Although graph and table contain exactly the same information, the picture may be better able to convey an overall trend. It is a good device to have.

Why not, then, apply the same pictorial treatment to our generalized mechanical table with its unspecified number of columns? To do so, we picture all the numbers in a particular row as compressed into a single point: a point "plotted" against a set of axes corresponding to the mechanical variables, however many there may be. The axes are not true axes, of course, because spaces with more than three dimensions become awkward to draw, but it hardly matters. The meaning of every point is spread out in the table. There is one axis for each variable (which we shall not even attempt to visualize); there is one point for each of the evolving states; and, point by point, the system traces out a curve as the settings of the machine change with time:

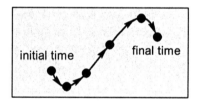

What we have is a symbolic solution to an "equation of motion," a state-to-state mechanical history written into the curve. It tells us where the system once was and where it will someday be. The ghosts of states past, present, and yet-to-come are contained in the points.

And so we ask, What now? Starting out with one set of mechanical variables, where does the machine go next? What path does it take, and what happens along the way?

The answers vary. Planets do one thing; electrons do another; quarks do something else. But one thing they all do is this: *nothing*, unless they are forced into action.

Inertia: Nothing Doing

Without interruptions, time stands still. Without something to make a difference, past and future blend indistinguishably into an unvarying, endless present. Without a spur to action, a mechanical system does nothing new. It persists in its present state. It remains trapped by inertia, nature's inherent resistance to a change of condition.

Inertia. We invoked it in Chapter 3 almost in passing, as a criterion for establishing a simple, commonsensical frame of reference. For Newton, though, inertia was something much more. It was the first law of motion, nature's insistence that a body maintain its current state unless ordered by an external force to do something else.

Say that a spaceship happens to be coasting through a region free of all outside influence, traveling in a straight line at constant speed. There are no gravitational fields, no electromagnetic fields, no disturbing fields of any kind. Inside the craft a small ball is fixed in position by a system of rods, like so:

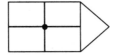

An observer based in the ship, noting first that the ball is locally at rest, then removes the rods and sees a sight conspicuous for what is absent. The ball does absolutely nothing. No force nudges it forward, backward, left, right, up, or down. Nothing attracts it. Nothing repels it. The particle remains rooted in place, floating freely in the empty space inside the craft:

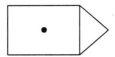

The ball was at rest before, and there it stays. No force disturbs it. The law of inertia applies.

Let there be another observer, an equally qualified *inertial* observer in the sense we understand from Chapter 3: an outside observer moving at a uniform velocity different from that of the first. To this observer 2, the same uninfluenced particle moves in a straight line at constant speed, also forever, also without change. It is not at rest (and never has been), and somehow it persists stubbornly in its motion, suspended inside the traveling spaceship like a pilot fish accompanying a shark. The ball continues to move along with the craft, despite the absence of any physical connection. It is a notable sight, a demonstration of the law of inertia:

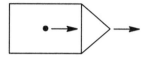

Whatever the free particle was doing last, it keeps doing. No continuing stimulus is needed.

Observers 1 and 2 agree on the elementary fact of inertia. Their different opinions of rest and uniform motion are merely two sides of the same coin, fully equivalent in a universe where space is not absolute. What kind of world would we have, after all, if a free particle could change course without provocation? If, for no apparent reason, a particle were suddenly to speed up or slow down or veer off in a new direction, then empty space would evidently manifest different qualities in different places. The laws of motion would change according to where our system chances to be, either in the white region or the gray:

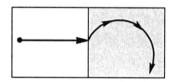

Could it happen? Yes, there is no intrinsic need for the same natural laws to be enforced all throughout the cosmos, independent of position. Just as different nations have different legal systems, the universe might have different local governments as well.

Does it happen? No, remarkably, it does not—at least not in the universe that scientists have explored so far. There appears to be only one system of natural law in the universe as we know it, and inertia is the first article in a global constitution.

Making Change

Variety may be the spice of life, but its role in mechanics is far greater than that of a condiment. It is the essential ingredient, without which nothing can happen. For to break the hold of inertia (to change, to begin to have a history), a mechanical system must be exposed to something other than undifferentiated space and time. If one environment is the same as the next, then there are neither advantages nor disadvantages to being here or there. To do anything new, a system needs the prod of a gravitational field or an electromagnetic field or some other agent able to bring variety to space and time.

In Chapter 2 we spoke of the *potential* to interact, the *potential* to effect change, the *potential* to be different. We spoke of hills and valleys of influence, of attractions and repulsions and ultimate balance. We spoke of stones rolling down slopes and coming to rest in the flat spots of landscapes like the one below:

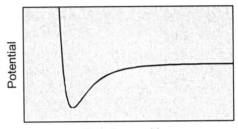

Relative position

Picture again what happens on an ordinary hillside, how a stone picks up speed rapidly down a steep slope and slowly down a gentle slope. How a rapidly rolling stone climbs higher than a slowly rolling stone. How the stone pauses at the top of its climb and then falls back down. Picture how the object trades position for motion, how it moves in response to the curving terrain.

A rolling stone carries energy. It packs a punch. It can hit something else and make it move. A heavy stone hits harder than a light stone, and a fast stone hits harder than a slow stone. Matter in motion can do something with its energy. It can do work.

But if motion spawns energy, then even a stone poised on a hillside, not moving, must possess a kind of energy as well: a latent energy, an energy of position, a "potential" energy. A stationary stone has the potential to move and lacks only the opportunity. Give it that opportunity, and it will indeed start to roll. The body will move in a manner determined by where it is on the hill, faster in some places and slower in others, all the while converting the promissory note of potential energy (the energy of position) into the currency of kinetic energy (the energy of motion). It starts out in one position, with a certain potential energy, and ends up somewhere else. The difference in potential energy is converted into kinetic energy, paid for in full.

A landscape of potential energy—with its peaks and valleys, its sloping hillsides, its mountain passes, its plateaus—thus lends texture to a vacant and homogeneous space. It is a landscape of fields generated by fundamental attributes such as mass and charge, fields in which energy is stored and ready to be dispensed. Any imbalance of potential energy, high in one spot and low in another, gives a particle the wherewithal to make a move. Differences in potential generate the *force*, the push or pull, needed to deflect a system from what would otherwise be an unvarying inertial condition. Differences in potential allow a system to change its mechanical state.

As if on a roller coaster, a particle rides the curve of potential energy. A sharp drop or rise produces a large force and a correspondingly large change in speed. A gentle grade produces a small force. A flat portion, where there is no difference from point to point, does nothing. It all depends on how the potential energy changes at a given position:

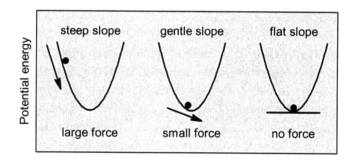

And so we look toward energy, potential and kinetic, as the fuel to power the machine. In some way, according to some prescription, according to some equation of motion that we shall have to discover, it will ultimately be energy that enables a mechanical system to overcome its inertia. It will be the energy latent in position, the potential energy generated by the fundamental interactions, that breaks the uniformity of empty space and creates an impetus for change. Less subtly, too, it will often be the kinetic punch of a moving particle that kicks a system from state to state. In one form or another, though, it will be energy that writes mechanical history. It will be energy that differentiates state 0 from state 1, and it will be energy, the author of change, that gives meaning to time:

The next step is to see what happens en route.

Moving Along

If the first law of mechanics is to remain "as you were," then there must be a second law as well: a law to tell the system what to do when doing nothing is no longer the rule. An equation of motion, with its instructions for turning state 0 into state 1 into state 2, shows the way.

Specify, first, *which* equation of motion. An equation to regulate the deterministic macroworld of stones and planets? Yes? Then apply Newton's second law of motion, as we shall do presently. What about the quantum mechanical microworld of atoms and molecules, where jittery electrons do things quite unlike those done by rolling stones and orbiting planets? Entering that small and special world, beginning in Chapter 7, we shall have recourse to an entirely new plan of action, the Schrödinger equation. Newtonian mechanics does not carry the force of law in a jurisdiction governed by chance and uncertainty. For worlds even smaller, for worlds penetrating into atomic nuclei and subatomic particles themselves, the quantum mechanical equations are different still.

We tackle a multifaceted universe one face at a time, tailoring our models and equations to fit the facts at hand. Whatever mechanical conception proves appropriate, that is the one to use. Discovering worlds within worlds, a practical observer will deal with each realm on its own terms. It is the only sensible approach to take.

To do so, moreover, is not to balkanize the universe, for we shall never lose sight of the hierarchy of matter. From quarks come protons and neutrons. From protons and neutrons come nuclei. From nuclei and electrons come atoms. From atoms come molecules. From atoms and molecules come cells. From cells come organisms. One set of building blocks is assembled into a certain level of structure, which in turn serves as a new set of building blocks for the next level. We ask the same of our mechanical theories: that the various models make smooth transitions from one domain to another.

Take the Moon, for example, a big body presumed to be built from many small particles. If true, then a quantum mechanical equation of motion applied to a huge collection of atoms—pushed to the limit—should be able to describe the lunar orbit using the same mathematical recipe that Newton would. The microworld equation may do its macroworld job inefficiently, perhaps even with indescribable clumsiness and difficulty, but do so it must. Nature passes imperceptibly from small to large, and our descriptive equations have to follow along.

There are other requirements as well, some of which we know from Chapter 3. A valid equation of motion cannot depend on a particular zero point of time or space, nor can it depend on any particular orientation of the coordinate axes. A valid equation of motion must also respect the principle of relativity, offering the same view to all observers in uniform motion. For observers moving slowly, intervals of time will appear to be rigorously separate from intervals in space. Such observers will treat time as one quantity and space as another, never to be mixed. For observers moving exceedingly fast, however, the picture will be different entirely. The demands of Einstein's space-time will come to the fore, and the governing equations will blend together space, time, and all other measurements derived from those primal two.

Beyond the basics—beyond nature's nonnegotiable demand to accommodate observers in reference frames of every possible speed, orientation, and position—the various equations of motion have yet another thread in common. For although each of them acts in a fashion peculiar to its own world, they all enforce the same universal conservation laws. They stipulate that certain quantities taken as a whole (energy, for one) must never change. No matter what mechanical path a system follows, the total amounts always stay constant. The numbers remain fixed.

The conservation laws guarantee an abiding constancy in the midst of change, and they bring together the different faces of the universe in ways that many other laws cannot. They apply equally to all fundamental interactions, to all fields, to all particles, and to all levels of scale. They are rooted most profoundly in the translational, rotational, and temporal symmetry of natural law... but that is a discovery we shall make in due course.

First, before meeting the conservation laws themselves, we look into the workings of an actual equation of motion: Newton's second law, the interrelationship of force, acceleration, mass, and momentum in the Land of the Large and the Slow.

Newtonian Force: A Shock to the System

Since so much of Newton's world moves slowly compared with the speed of light, it appears to be a world in which time is unconnected to space. Our many observers establish spatial reference frames of their own choosing, but their clocks all run at the same rate and yield the same elapsed times. A single clock, shared by everyone, ticks away unperturbed as positions and velocities change variously in all reference frames.

The *position* of a particle: where the body happens to be at a given instant. In a space of one dimension, with all locations falling on a single line, one number will do. Call it x:

In a space of two dimensions, we need two numbers (x and y) to establish not just a distance from the zero point but also a direction:

In three dimensions, there are three independent axes. We need three numbers.

The *velocity* of a particle: by how much and in what direction the position changes from instant to instant. If, for example, the body moves

one meter to the right in one second, then it travels at an average speed
of one meter per second in the direction *x*:

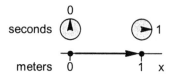

In two dimensions, where the position changes in two directions, there
are two components of velocity. In three dimensions, there are three.

With measuring rod and clock at the ready, we now observe a system
abruptly shocked into action. Our particle, unmolested for a time, is
initially traveling in a straight line at constant speed. It is obeying the law
of inertia, advancing the same fixed distance each instant, when suddenly a
force kicks it from behind. The force persists for (let us say) one second,
by the end of which the particle will have covered more ground than it did
during the second just before. It picks up speed continually, moving faster
and faster as long as the force endures. Pushed forward, the particle
accelerates. It alters its velocity and travels farther:

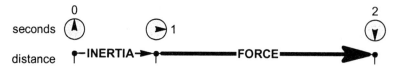

After that, the force goes away and inertia returns. The energized particle
maintains whatever new, higher velocity it finally acquires.

We proceed to imagine another incident, this time with the force coming
from a different direction—from the top down, rather than left to right:

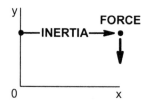

What happens now?

What happens, in the *x* direction, is *nothing*: the force has no
component directed horizontally. It changes nothing. In the horizontal
world, inertia remains in control and the particle maintains its steady

rightward march. Left to right, the freely moving object traverses the same fixed distance every tick of the clock. But look what happens in the up–down y world, a domain completely independent of the left–right world ninety degrees away. The steady force, acting only in the vertical direction, accelerates the body downward while leaving its horizontal progress undisturbed. The line of motion bends into a curve:

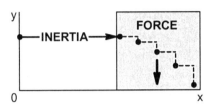

As long as the force persists, the particle travels in both the y direction and the x direction. Its course changes from purely horizontal to a mixture of horizontal and vertical.

Now to observe that a force gives rise to acceleration is one thing, but to detail an exact relationship between those two quantities is quite another. For that, we repeat all our experiments systematically, first with double the force, then with triple the force, then with quadruple the force, on and on, and from that ordered series a connection starts to appear. It is simply this: With twice the applied force, the acceleration is twice as great. With three times the applied force, the acceleration is three times as great. With four times the applied force, the acceleration is four times as great.

We thus have the beginnings of a mechanical model, a direct proportionality between force and acceleration. Subjected to the disturbing influence of a force, a particle changes its state of motion. The speed may change. The direction may change. Both may change. A tailwind makes a body go faster; a headwind makes it go slower. The greater the force, the greater the effect. Whatever the outcome, though, the system is headed for a new mechanical state. In the presence of a force, inertia gives way to acceleration.

But not without a fight. A body resists any attempt to force it out of its inertia into a new state. Matter fights back with its very substance, its mass.

Inertia and Mass

Which needs more force to get off the ground: a firework skyrocket or a rocket to the Moon? Which is easier to move: a single brick or the Great Pyramid of Cheops? Which moves faster when thrown: a five-ounce baseball or a sixteen-pound shot?

It is more a question of quantity than quality: not *what* there is, but rather how much of it is massed together. More stuff, containing more of everything—more electrons, more protons, more neutrons, more particles of all kinds—maintains its inertia and resists a force better than less stuff. "Mass," the amount of stuff doing the resisting, becomes a measure of inertia.

Looking for a link between mass and acceleration, we begin with a single particle and subject it to a uniform force. Let the force come from a coiled spring, or from a gust of wind, or from the blast of a gun, or indeed from any source at all except for mass itself (which produces a gravitational field). Accepting for the moment that one restriction, we impose the force and observe. The particle accelerates, changing its velocity at each instant to an extent commensurate with the force. If the body had been at rest before, then it starts to move abruptly in the direction of the push. After one second of this forced march, our standard particle reaches a certain distance (which we arbitrarily assume to be two meters):

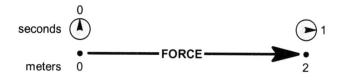

Now glue together two of these standard particles and repeat the experiment with everything else the same: the same force, the same strength, the same duration, the same initial condition. We do so, and we find that the acceleration is half as much as previously. Rather than gaining two meters in one second of acceleration, the pair of standard particles gains only one:

When the mass doubles, the acceleration halves.

The effect is cumulative. With three particles, the acceleration is cut to a third. With four, it falls to a fourth. The acceleration scales inversely with the quantity of matter fighting against the force. Each particle resists individually, and in unity they find strength. The more particles, the more mass. The more mass, the more inertia. The more inertia, the less acceleration.

Putting together our various observations, we arrive eventually at Newton's second law of motion: that force promotes acceleration, that mass (inertia) retards it, and that the resulting acceleration varies as the ratio of force to mass. Double the force? The acceleration doubles. Double the mass? The acceleration halves. Double them both? The acceleration stays the same. It is a tug of war between equally capable opponents, force and mass.

For bodies falling freely in a uniform gravitational field, the contest is forever stalemated. Something called "mass," remember, is the source of gravity, and from Chapter 2 we already know its effect. A particle with twice as much mass transmits and receives twice as much gravitational force. Three times the mass, three times the force. Four times the mass, four times the force; and so it goes, up and down the line: a direct proportionality between force and mass. And as nature would have it, this "gravitational mass"—the attribute of matter that gives rise to a gravitational attraction—turns out to be the same "inertial mass" that gives rise to an intrinsic inertial resistance.

Mass, on the one hand, produces the force of gravity. Force encourages acceleration. Mass, on the other hand, produces inertia. Inertia discourages acceleration, and it does so to exactly the same extent that gravitational force encourages it. Since Newtonian acceleration depends on the quotient force ÷ mass, the mass that contributes to the force of gravity (implicit in the numerator) cancels out the mass that causes inertia (explicit in the denominator). The tension between these two conflicting inclinations, equal but opposite, results finally in a draw. All bodies, regardless of mass, fall with the same acceleration in a uniform gravitational field.

Drop any two objects simultaneously from the same height, and (if no forces other than gravity are present) they will hit the ground at the same time. A cannonball delivers more of a wallop than a feather, but both fall at identical rates, courtesy of the equality between gravitational and inertial mass. Large or small, every object runs through the same series of velocities and covers the same distance in the same time. Whatever extra acceleration a larger body would gain from its mass, it loses from the extra inertia that comes with a bulkier package:

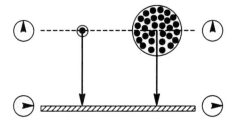

On Earth, the additional forces imposed by air resistance and wind often obscure the ready observation of such wondrous sights. In a vacuum chamber, though, or near the surface of the Moon, or wherever else the air is rare, a feather and a cannonball do fall side by side, never out of step.

Strange? Maybe so. Newtonian mechanics offers no reason for the sources of inertia and gravity to be the same, but they are; and because they are, the second law of motion acts the way it does in a uniform gravitational field.

Leave it to Einstein, with a wholly new world view, to show that the equivalence of gravitational mass and inertial mass is no accident. For that, however, we shall wait until Chapter 5. There is still much to do in Newton's world of force, mass, and acceleration.

Charting a Course

Force. Mass. Acceleration. We supply the values of any two, and Newton's equation of motion promises to supply the third.

The action plays out over some field of force, a field sown with the seeds of change. Over here, a shading of potential energy pushes a susceptible particle into moving at a new speed:

Over there, a wall forces a ball to reverse course:

All throughout the system, wherever there is a potential for change, we assess its strength and direction. Gravitational fields, electromagnetic fields, constraining walls, collisions—whatever the sources may be, our challenge is to map the net force at each point. Measuring the mass of each body, we then use Newton's second law to crank out the acceleration.

The *acceleration* of a particle: by how much and in what direction its velocity changes. We already know (because presumably we can measure) the velocity of our particle just before the force hits. Newton's law, properly outfitted with force and mass, now gives us something we did not know previously: the acceleration, the change in velocity engendered by the force. Added to the old number (the initial velocity), this new number becomes the velocity for the very next instant.

In turn, the new velocity determines how far the particle travels during that same moment. The change in distance, combined with the initial position, tells us finally where the particle ends up. We have a new position and a new velocity, just enough information to turn the crank once again. The job is done.

With that, the equation of motion fulfills its promise. It delivers a new set of mechanical variables in exchange for an old set plus full disclosure of the governing influences. We specify the forces, the masses, and the initial position and velocity of each particle. Newton's second law does the rest:

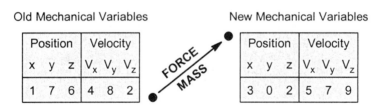

| Old Mechanical Variables | | | | | | | | | New Mechanical Variables | | | | | |

Position			Velocity				Position			Velocity		
x	y	z	V_x	V_y	V_z		x	y	z	V_x	V_y	V_z
1	7	6	4	8	2		3	0	2	5	7	9

There is only one way to go. One specific set of mechanical variables is transformed uniquely into another. There are no second choices, no alternative routes, no detours, no shortcuts, no exceptions. The system has only one possible destination and only one path to get there:

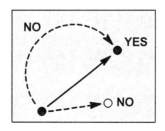

It is Hobson's choice, a choice that is really no choice at all: a fork in the road with all outlets blocked but one.

Free will, incompatible with the second law of motion, disappears in a clockwork universe. In Newton's world, given a certain configuration of force and mass, the end of an affair is predetermined from the beginning. The end of one step becomes the beginning of the next, and from that new beginning there follows a new end, and another after that, and another, and another. State by state, as long as the force persists, a system makes its determined way into the future and remembers the past. Step by step, there is but one possible move to make:

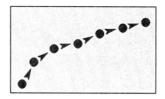

One step or a million, the outcome is never in doubt. A Newtonian observer, describing a single state as it currently exists, becomes simultaneously a witness to history and history in the making.

Constancy and Change

Who would have expected a mere list of positions and velocities to qualify as Knowledge of All Things? Still, in its own way, a Newtonian path is something perfect and complete. Moment by moment, from beginning to end, Newton's equation lays out the whereabouts and current activity of every particle in our chosen system. The second law of motion tells us where the particle was, where it is going, and when it will get there. Regarding that one system, at least, an informed observer becomes All Knowing. The answer to any mechanical question lies buried within the numbers, like a precious metal scattered within an ore.

To extract the nuggets, we look for ways to summarize and condense the raw data. Perhaps there are new combinations of space, time, and mass for us to identify and track, quantities able to reveal the rules of the road with special clarity. Think of the difference, say, between a list that reads "apple, orange, peach, peach, peach, orange, apple, apple, apple," and one that reads "four apples, two oranges, three peaches" or maybe even just "nine pieces of fruit." Dressed in a new suit, the same old information may take on new meaning and inspire new insight.

There are times when Knowledge of All Things does not bring Wisdom. There are times when a complete mechanical history becomes too much to handle, threatening to drown the fundamental simplicity of a process in a sea of numbers. There are times, also, when a configuration of forces is too complicated ever to specify in full, times when we despair of solving Newton's equation of motion.

For times like those, and they are many, we have other arrows in the quiver. The conservation laws are among the sharpest.

Conservation of Momentum

Start with linear momentum, the product of mass and velocity. Of two bodies moving at the same speed, the one with twice the mass carries twice the momentum:

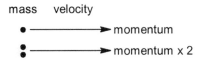

It is this one-two punch of mass and velocity together, not velocity alone, that makes a feather different from a cannonball dropping at the same rate. The bigger they are, the harder they fall.

Now imagine that the velocity and therefore the momentum of a moving object were suddenly to change. Maybe an arrow hits its target, or a ball bounces, or an apple falls from a tree, or a bullet leaves a gun. In each instance the change in velocity (an acceleration) means a change in momentum, and the resulting product of mass and acceleration implies a force. Why? Because Newton's second law connects force directly with mass and acceleration. To say that "force is the product of mass and acceleration" is also to say that "force is the rate at which momentum changes." The bigger the change in momentum, the bigger the force needed to produce it. The *faster* the variation, the bigger the force as well. A swift, strong change in momentum demands a comparatively greater force.

But what if there is no force at all? What if our system, taken as a whole, suffers no push or pull from the outside? What if the system encompasses particle and walls together, and we warrant that the internal world (the solid gray box below) remains uninfluenced by its cross-hatched surroundings?

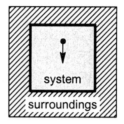

If so, with no force intruding from the outside, then the internal momentum has no way to increase or decrease overall. Where there is no force to effect a change, momentum is conserved. A gain in one place is offset by a loss somewhere else, and the total momentum stays the same. Newton's law demands it, no more and no less.

Do an experiment. Bounce a rubber ball off a hard, flat stretch of ground and watch where it goes. Any child knows what happens. If the ball hits at an angle, then it rebounds at the same angle:

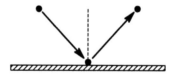

If the ball falls straight down, then it bounces straight up:

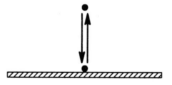

To react in any other way would be to violate the conservation of momentum. It does not happen.

To appreciate why, we do not need to know either the detailed mechanical history of the bouncing ball or the exact tangle of forces present at impact. All we need to know, simplicity itself, is this one bare fact: that the composite system of ball and ground together sustains no force from any agent other than the two parties themselves. Whatever they do during the instant of collision, they do to each other. Between the two of them, there is no change in the total momentum.

The ball hits the ground in a certain direction (let's say straight down), and the ground hits back. The ground exerts a counterforce of the same strength but in the opposite direction, up rather than down:

ball pushes down

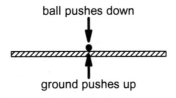

ground pushes up

The ball, pushed by the ground, changes its momentum and moves up at a certain speed. The ground, pushed with equal force by the ball, moves straight down—far, far slower than the ball, of course, since it contains so much more mass, but nevertheless it moves:

ball moves up

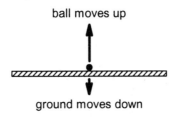

ground moves down

The ground moves just fast enough to absorb *from* the ball the same amount of momentum it gives *to* the ball. One change cancels the other, and the sum is zero: no winner, no loser. The total momentum of the universe remains the same.

Look around. A bullet leaves a gun, and the gun recoils. *Momentum is conserved.* A rocket spews exhaust to the left and moves to the right. *Momentum is conserved.* Billiard balls collide, exchange forces, and move away with new speeds in new directions. *Momentum is conserved.* Newton, observing balancing acts like these everywhere in nature, proposed a third law of motion to codify them: the principle of action and reaction. One body pushes on another, and the second body pushes back in equal measure. Wherever forces are balanced, momentum is conserved. It is the law.

It is the law not just of classical mechanics, but the law of all mechanics. Conservation of momentum applies equally to the fast and the slow, the large and the small, the strong and the weak. Electrons and nuclei, interacting electromagnetically, conserve quantum mechanical momentum when they assemble into atoms. A neutron conserves momentum when it undergoes beta decay, a weak nuclear interaction. Quarks, governed by the strong nuclear interaction, likewise conserve momentum in all their quantum mechanical doings. Nobody has ever found otherwise.

One is justifiably amazed, even awed, at such universality: how a bookkeeping rule taken specifically from Newton's world extends to realms where classical mechanics loses all meaning. Yet the conservation law survives even where the equation of motion does not, and indeed any proper equation of motion (for any process) will demand implicitly that total momentum be conserved throughout the universe. Systems of quarks, electrons, decaying neutrons, and crashing comets evolve in vastly different ways, but they all conserve global momentum from start to finish.

They do so because equations of motion have something special in common, a transcendent quality we first acknowledged in Chapter 3. A valid equation of motion makes no reference to absolute position. Its form is translationally symmetric, unaffected by any displacement of an arbitrary coordinate system. Whether we shift an axis left or right,

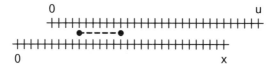

the underlying world of particles and their interactions remains the same. And it is just this symmetry of place that mandates the existence of a steadfast, unchanging, *conserved* quantity intimately related to displacement along a straight line. We recognize that quantity, in retrospect, as the linear momentum.

The proof is abstruse and deep, but the laws of nature come with a fundamental guarantee: that for every manifestation of symmetry in an equation of motion, a mechanical quantity is conserved. Invariance to *anything* ensures the conservation of something else, and that "something" is for us to discover, define, and track. Invariance to absolute position, we assert, ensures the conservation of linear momentum. It is no coincidence. In quantum mechanics as well as classical mechanics, we always associate linear momentum with a shift in position. Whether the equation of motion applies to the Moon in orbit or to an electron in hydrogen, the connection cannot be broken.

A different kind of momentum, angular momentum, arises from rotational motion and is associated with the turning of an axis through an angle. Nature's invariance to absolute orientation—rather than absolute position—thus ensures the conservation of angular momentum:

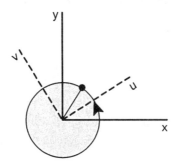

Linear momentum remains constant when no unbalanced *linear* force disturbs a system (something that would push or pull it in a straight line). In the same way, angular momentum remains constant in the absence of any unbalanced *twisting* force (a "torque").

Conservation of Energy

Invariance to temporal displacement, another one of nature's symmetries, necessitates a conservation law of its own: the conservation of energy, a quantity we plausibly expect to generate shifts in time. We have already maintained, in the broadest of terms, that a system needs a variation in energy to progress from one state to the next. Now, with an appreciation of symmetry, we come to realize that the homogeneity of time itself freezes the cosmic pool of energy. The total amount of energy in the universe is forever constant, the same today as yesterday, the same tomorrow as today.

Raise a ball to a certain height, hold it in place for a moment, and then let go. The ball drops straight down, accelerated by the force of the Earth's gravity. Starting from rest, the falling object picks up speed with every passing instant and eventually hits the ground.

Do it again, this time starting from a greater height. The ball follows the same schedule of acceleration, but now it has a longer way down and more time to get there. Gaining new velocity each second, the ball ends its trip moving faster than before:

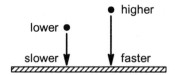

Try a different experiment. Attach a ball to a spring; stretch the spring a certain distance; release. Pulled out of balance, the spring fights to restore its original position. It exerts a force from within and begins to oscillate, alternately compressing and stretching. The ball, accelerated by the force, starts from rest and begins steadily to gain speed. It moves faster and faster until finally passing through a maximum velocity at some point in the cycle. The greater the initial stretch, the greater the maximum velocity:

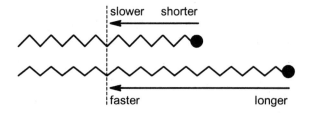

We can do countless experiments of this sort, placing a system at various positions in a field of potential energy. It might be a field of gravitational energy, set up by one or more massive objects (like the Earth, for example). It might be a field of elastic energy arising from a stretched spring, or a field of electromagnetic energy created by electric charges and currents. It might be a field of chemical potential energy, or nuclear potential energy, or some other kind of potential energy, but really the differences are in name only. Regardless of the source, potential energy always amounts to the same thing: a promise of greater or lesser speed *somewhere else* in the field, not here.

Negotiating a difference in potential, a particle has energy either to pay out or to take in. One way to settle the account, as illustrated above, is by a change of speed. Dropping from high potential to low, the body gains kinetic energy. It moves faster. Climbing from low potential energy to high, the same body loses kinetic energy. It moves slower. A twofold change in velocity requires a fourfold change in kinetic energy. A threefold change in velocity requires a ninefold change in kinetic energy. A fourfold change in velocity requires a sixteenfold change in kinetic energy.

To change something's potential is to do *work*, the work of applying force over a distance. We do work on a particle—we change its energy—when we either alter its speed or force it to move between points of different potential. Energy, potential or kinetic, ultimately translates into the ability to make something move.

Like money, energy can be invested in any number of instruments: bank deposits (gravitational energy), stocks (electromagnetic energy), bonds (chemical energy), real estate (nuclear energy), art (radiant energy), jewelry (elastic energy), pork bellies (mass energy), and still other forms of potential energy. Energy can be invested in kinetic instruments, too, put to work in the sundry motions of atoms, turbines, planets, and everything else in a restless universe. Each investment is fully liquid, and funds can be transferred freely from one account to another.

Unlike money, though, energy cannot be created or destroyed. No central bank stands ready to inject new cash into the system, and no authority has the power to retire and demonetize old currency. All the energy that has ever existed in the universe remains with us today. Nobody is making any more of it, and nobody is making any less of it. Energy, strictly conserved, merely flows from one account to another. Not one cent is ever gained or lost.

Any change of energy over here is balanced by a compensating change of energy over there. Potential energy drawn from a field goes into the kinetic energy of a particle in motion. *Nothing is lost.* Particle 1 collides with particle 2, and a fixed amount of energy is redistributed. *Nothing is lost.* Electrical energy becomes mechanical energy; elastic energy becomes gravitational energy; chemical energy becomes electromagnetic energy. One kind of energy becomes another kind of energy, and nothing is ever lost. The symmetry of physical law guarantees it.

Though the Heavens Fall

Pulled down by gravity, a stone falls for a time and finally crashes. Why not the Moon? What keeps the Moon from crashing into the Earth?

For that matter, what keeps the Earth from crashing into the Sun? The force of gravity, always attractive, would have all the planets come ever closer to the Sun (unto a fiery death), yet still the solar system endures.

If what goes up must come down, then why launch a satellite? What keeps the International Space Station aloft, orbiting the Earth in seeming defiance of gravity? Why does Halley's comet visit us every seventy-six years or so, never failing to appear?

It is all the same question, and the broad answer—the first great triumph of classical mechanics—is as breathtakingly simple as its many details are complex. A brief explanation, in the most general terms, will give us a crowning example of the way things work in Newton's world.

The problem is to understand how particle 2, attracted by gravity to particle 1, can become trapped in a closed orbit:

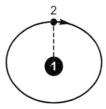

To avoid undue complications, we stipulate that the orbiting body has far less mass than its gravitational partner. If so, then particle 1 remains effectively stationary and our attention turns justifiably to the motion of particle 2 alone. For even though both bodies attract each other with the same force, the more massive of the two undergoes proportionally less acceleration. It appears to remain in place.

We begin by conceding that Chicken Little was right, to a point. The sky, and everything in it, is undoubtedly falling. The Moon is falling toward the Earth; the Earth is falling toward the Sun; and yes, if Earth and Moon were doing nothing else but falling, then sooner or later their mechanical histories would come to a crashing end:

But no, the end is not near, because there is more than one dimension to the motion. Particle 2 is also moving perpendicular to the force of gravity, running away from the attraction while simultaneously submitting to it.

Picture what might have happened once upon a time, when particle 2 was traveling independently through space, carried along solely by inertia. Imagine the body moving uniformly in a straight line, unaffected by anything else, until one day it comes under the influence of particle 1. Deflected by gravity, particle 2 then begins to fall toward particle 1, all the while maintaining its inertial motion perpendicular to the force. The result is a change in course. Rather than getting to point A, particle 2 falls instead to a point B below the original line of motion:

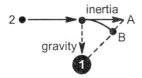

The trajectory, shaped now by movement in two independent directions, starts to curve.

What happens next is determined by how much kinetic energy particle 2 brings to its meeting with particle 1. If the motion perpendicular to gravity is too slow, then the falling body soon runs out of room to fall. Its curved path terminates with a crash landing on the surface of particle 1, like a cannonball fired from atop a mountain:

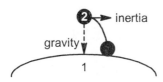

But if particle 2 moves fast enough, then it can slip over the horizon and continue to fall. It falls on a straight line toward the center of particle 1 ("down," if you like) while persisting in its current inertial motion perpendicular to the force:

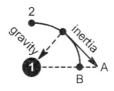

Once again, particle 2 finds itself at a point B below the point A to which inertia alone would have carried it. The Moon, for instance, falls roughly a twentieth of an inch every second, a tiny bit compared with an inertial run of about forty thousand inches during the same time—but enough to maintain a stable orbit.

And so it goes, step by step. Particle 2, carried along by sufficient energy, orbits round and round, trapped in perpetual free fall, never crashing. With gravity always pulling the body inward (toward particle 1) and inertia always pushing it off on a tangent, the path taken is usually an ellipse, a distorted circle:

Details of the orbit, including the exact shape, are determined by the initial speed, direction, and position of particle 2 as it enters the gravitational field of particle 1.

Now those details make for a complicated mathematical solution, but from the geometry alone we can already infer a critical feature: the variation of speed at different points along an elliptical orbit. Separation between the two bodies waxes and wanes during a cycle (look at the diagram above), and the gravitational influence follows suit. Stronger up close and weaker far away, the force responds fourfold to a twofold change in separation. At twice the distance, the gravitational force drops to a fourth. At three times the distance, it drops to a ninth. At four times the distance, it drops to a sixteenth. This same relationship, called the "inverse square law," applies to electrostatic force as well. It is a statute of some importance, one that helps shape the universe on scales ranging from atoms to galaxies.

Approaching particle 1 at different distances, particle 2 thus moves faster and slower in response to a variable force. Close by, under a stronger pull, the orbiting body speeds up. Its kinetic energy rises, and its potential energy falls. Far away, under a weaker pull, the tables are turned. Velocity and kinetic energy go down, and potential energy goes up. The particle completes less of its circuit during each second of travel:

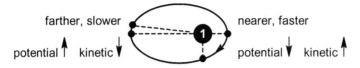

farther, slower   nearer, faster

potential ↑ kinetic ↓   potential ↓ kinetic ↑

Yet the more things change, the more they stay the same. Any temporary gain or loss of kinetic energy is offset by an opposite change in potential energy. The total amount of energy, kinetic and potential together, remains the same no matter how fast or slow the body moves in the gravitational field. Motion is purchased at the expense of position, and both energy and angular momentum are conserved.

The story has no surprise ending. From the moment that particle 2 comes under the influence of particle 1, the clockwork machinery of a classical universe determines what the orbit will be. The rest is history, and in a classical universe those who remember the past are doomed to repeat it. Play the game as often as you like—have particle 2 shoot by particle 1 in the same direction, at the same distance, with the same speed—and the resulting orbit will always be the same. Newton's laws of motion make it a foregone conclusion.

5

MASS AS A MEDIUM

Early in the twentieth century, a landmark case came before the Supreme Court of the Universe. The facts of the matter were as follows:

1. A generic particle, forced to move at increasingly high speeds, failed to conserve momentum. During collisions with walls and other particles, the sum of all individual momenta (mass × velocity) was supposed to remain constant. It did not.

2. Subjected to an ever stronger force, the particle refused to reach or exceed a finite velocity designated as c (300,000,000 meters per second; equivalently, 186,000 miles per second). The body was legally obligated, however, to accelerate in direct proportion to the force impressed, no matter how much speed was needed to do so. It did not comply.

Alleging violations of the second law of motion, I. Newton (Plaintiff) brought suit against A. Einstein (Defendant), claiming that Defendant incited a natural system to violate the laws of nature. Defendant responded by asserting that Newton's laws of motion, although an excellent approximation for the low-speed world in which they were deduced, did not apply to bodies moving at high speeds. Defendant further asserted that Newton's laws were framed under a principle of relativity utterly inappropriate, again, for rapidly moving frames of reference.

The Court ruled unanimously in favor of Einstein. As a first step, they threw out Newton's second law under all circumstances where speeds become comparable with the velocity of light. Next, they authorized Einstein, his heirs, successors, and assigns to develop a revised classical mechanics suitable for inertial observers moving at any speed.

In issuing their opinion, though, the Court upheld Newton's laws at low speeds, citing especially the transcendent value of the conservation laws. They noted that any reasonable observer would accept Newton's mechanics not only as correct for low-speed systems (because they produce highly accurate results) but also far simpler to apply than Einstein's reformulation (which, after a lot more work, would ultimately give the same answers).

Accordingly, the Court ordered Einstein to construct a new, broader mechanics in which:

1. All inertial observers shall agree that light travels through empty space at the same speed c, come what may: a fixed, invariant, constant, unchanging, *absolute* 300 million meters per second. No exceptions.

2. No event of any kind (no ticking clock, no barking dog, no falling Moon) shall ever prove that a particular inertial frame is in a state of absolute motion or absolute rest.

3. All inertial observers, no matter how fast their relative motion, shall perceive the same mechanical laws. Their equations shall always share the same mathematical form, even though individual measurements of space, time, velocity, and other quantities may vary. Observers shall be able to reconcile any such differences and convert values from one frame to another.

4. When redefined to satisfy the aforementioned conditions, quantities such as momentum and energy shall be strictly conserved in all reference frames.

5. The new equations and definitions shall turn smoothly into Newton's expressions at sufficiently low velocities.

Thus empowered, Einstein went to work. First, applying his special theory of relativity (Chapter 3) to Newtonian mechanics (Chapter 4), he discovered something nobody bargained for: that mass is a medium in which to store energy, and—same thing—that energy can be congealed in the form of mass. This equating of mass and energy, the iconic $E = mc^2$, changed forever the way we think of matter itself.

It was a priceless gift for the intellect. $E = mc^2$ gave us matter and antimatter. $E = mc^2$ gave us the annihilation of mass and its afterlife as energy. $E = mc^2$ gave us the reincarnation of mass from pure energy, as if by magic. $E = mc^2$ gave us nuclear power, and it also gave us, inevitably, the nuclear bomb.

Second, creating the general theory of relativity, Einstein abolished the "force" of gravity and discovered yet another role for mass: as the medium by which space-time is made to curve. It was one of the Biggest Ideas of all time.

General relativity shattered any prior interpretations of cosmic structure on a large scale. Before general relativity, space-time was a passive stage on which things happened. After general relativity, space-time is simultaneously director and actor. The local contours of space-time, sculpted by mass, both make things happen and respond to whatever happenings take place.

General relativity gave us a wholly new picture of gravity, and, no less, it gave us a radically new (and still evolving) way to view the universe. It gave us ripples in space-time. It gave us black holes and wormholes. It gave us the Big Bang.

From Clocks and Rulers to $E = mc^2$

If you want to believe in something, believe in this: Interactions do not occur instantaneously. It takes time for a signal to go from here to there. The speed of light is finite and fixed, the same value for all observers. It is a large number, but not infinity.

If the rules were otherwise, then time would never get mixed up with space. A clock in motion would tick at the same rate as a clock standing still, with none of the s-t-r-e-t-c-h-e-d ticktocks that we saw at the end of

Chapter 3. What happens simultaneously in one reference frame would happen simultaneously in every other, and all observers would agree on the same sequence of events. Time would remain above the fray, flowing at its own pace, unperturbed by motion or distance. As observers, we would simply set our clocks to Cosmic Mean Time and worry no more.

And (behold!) that is just what we do in any application of mechanics to things in the Land of the Slow—to things like planets in orbit, or projectiles in free fall, or billiard balls on a table, or even helium atoms rattling around a container. For how different from infinity, really, is the speed of light (300 million meters per second) compared with particle speeds ranging from a few meters per second to a few thousand meters per second?

No, in the Land of the Slow be glad that we can use Newton's mechanics and Galileo's principle of relativity. Simple. Accurate. Sensible. All observers of that world see a spatial grid laid out in three dimensions, with everything marching to the steady beat of the same drum. A single clock (to measure time) stands aloof from three rulers (to measure space), and the clock in the dashboard ticks away without consulting the speedometer.

Life is different in the Land of the Fast, but babies learn to walk before learning to run. Before setting foot in Einstein's mechanical universe of four dimensions, where clocks become rulers and rulers become clocks, we must first map the terrain more closely in Newton's universe of three.

Getting Around in Three Dimensions

Scalars and vectors…we met them in Chapter 3, without naming names. We return now for a fresh look, the first step on the road to understanding the space-time connection between energy and mass.

A *scalar*, like a rung on a ladder or a point on a scale, has size but not direction. Temperature is a scalar. Distance is a scalar. Time is a scalar. Scalars are simple quantities, impervious to rotations, and just a single number (call it *t*, in honor of "time") tells us all there is to know:

 t

Time, distance, and temperature remain the same regardless of how a spatial grid may be oriented.

Another kind of quantity, called a *vector*, has both size and direction: a position vector, for example, specified by a distance (50 miles) and a directional bearing (north northeast). We build a position vector from components in independent directions, and we need as many numbers as there are perpendicular directions in space.

To keep the diagrams simple, let's agree to draw in just two dimensions (x, y) while remembering that a complete space contains three (x, y, z):

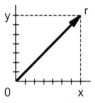

Now look at what we have. The x component changes nothing in the y or z directions, but together the components combine to produce a scalar magnitude (the length of the arrow, r) and an orientation (the arrow's relationship to the axes). It is a sight already familiar to us from the trajectories of Chapter 4, where we saw how inertia in one direction carries a particle independently of the force in a perpendicular direction.

Note here especially—because down the road, in four-dimensional space-time, there will be a subtle difference—that each perpendicular component contributes positively to the total length. Neither x nor y diminishes the effect of the other as the vector comes together:

The geometry of a right triangle does not allow otherwise.

Our next step is to recall how a vector responds to any rotation of its coordinate system. Some things change. Some things remain the same. The length, a scalar quantity, stays fixed while the old vector components (x and y) turn into new components along new axes (u and v). The new components become mixtures of the old components, and the relationship between new and old depends on the degree of rotation:

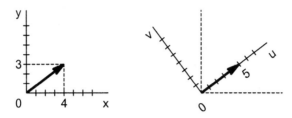

Applying the rules of trigonometry, we can then convert the old components into the new. In three dimensions, the general recipe would amount to this:

1. u will be some combination of x, y, and z that involves the angles of rotation.
2. v will be some other combination of x, y, and z involving the same angles.
3. w will be yet another combination of x, y, and z involving the same angles again.

One, two, three. By building a vector from perpendicular components, we guarantee ourselves a quantity valid in any rotated coordinate system. Each observer records an invariant length, and everybody understands how to convert one's own variable components into everybody else's. Problem solved.

Need a rotation-proof velocity? Build it up from three perpendicular components (V_x, V_y, V_z). Do so, and the magnitude of your velocity (a directionless speed of 50 miles per hour, say) will remain the same under any arbitrary rotation of your axes. The components will change, to be sure, but they will change in the same predictable way as the coordinates of a rotated position vector:

1. V_u will be some combination of V_x, V_y, and V_z that involves the angles of rotation.
2. V_v will be some other combination of V_x, V_y, and V_z involving the same angles.
3. V_w will be yet another combination of V_x, V_y, and V_z involving the same angles again.

Need a rotation-proof acceleration? Construct a vector. A rotation-proof linear momentum? Angular momentum? Electric field? Construct a vector. A rotation-proof directed quantity of any kind? Construct a vector.

Write an equation in which vectors combine only with vectors (or scalars combine only with scalars), and rest assured. The relationship will be unaffected by any tilt of the axes, no questions asked.

Getting Ahead in Three Dimensions

Now that harmony reigns among stationary (but differently oriented) reference frames, we go on to consider Newtonian systems in uniform motion. Our conditions—straight line, constant speed—are simply those of an ordinary inertial observer, except for this one restriction: let the relative motion be slow. Let it be slow enough to ensure that all observers mark time with the same clock.

We return to our train from Chapter 3, moving along once again at a constant 100 feet per second. Observer 1, a passenger, measures distance relative to the seats. Observer 2, standing on the platform, measures distance relative to the tracks. The train passes by at its smooth, steady speed, advancing the same 100 feet each second:

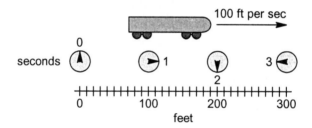

Observer 1, sitting in seat zero, claims to remain stationary all the while the clock ticks. "My location never changes," says observer 1. "Relative to this railway car, both my position and velocity continue to be zero. What do you see, observer 2?"

Observer 2 sees a different picture. To observer 2, the train adds a fixed 100 feet of distance each second. To observer 2, the position of "seat zero" appears 100 feet down the tracks after one second, 200 feet down the tracks after two seconds, and 300 feet down the tracks after three. To observer 2, the velocity of observer 1 (sitting in seat zero) is a constant 100 feet per second.

Observer 1 then agrees to walk forward at a rate of one foot per second, and they both record time (t), position (x), and velocity (V) while this new motion takes place. Remember that velocity measures the change in position between one time and the next, and now look at what they perceive:

Observer 1 (Inside)			Observer 2 (Outside)		
t	x	V	t	x	V
0 s	0 ft		0 s	0 ft	
1	1	1 ft / s	1	101	101 ft / s
2	2	1	2	202	101
3	3	1	3	303	101

That's enough, finally, for the two of them to deduce a pattern and for observer 2 to tell observer 1:

1. I measure the same elapsed times as you do. What you call "one second," I call one second as well. We have no argument.

2. My reading of your position, however, is consistently greater than your reading by a certain fixed amount: the elapsed time multiplied by the speed of your train (100 feet per second, relative to my tracks).

3. Since we disagree on your position, we must agree to disagree on your internal walking speed. My reading here is equal to your reading (1 foot per second) *plus* the constant speed of your train (100 feet per second). You're moving faster than you think.

   Moreover, since we define momentum as (mass) × (velocity), my understanding of your momentum differs by a constant as well: the product of your mass and your train's constant speed.

4. I measure the same acceleration as you do (in this case, zero). Since our speeds differ from second to second only by a constant amount, we are both sure to observe the same change in velocity.

Our observers realize further that the Newtonian laws of motion apply equally well in either inertial reference frame. Although they measure different numbers for position, velocity, momenta, and energy, the two observers do perceive consistent relationships among the various quantities. Any change in momentum, for example, appears identical to both—simply because the velocities in question always differ by a constant amount (see point 3 above). They find the same equation linking force, mass, and acceleration, and they also find the same conservation laws.

It was Galileo who established these rules for relative motion and thus laid the foundations of Newton's world. A simple world it is, too: a world in which time is time and space is space, and never the twain shall meet. Scalars go with scalars. Vectors go with vectors.

Give them an angle, and our observers know how to reexpress position, velocity, momentum, acceleration, or any other vector in a rotated reference frame. Give them a relative velocity, and they can do the same for a system in uniform motion.

With everybody looking at one clock, the rules of mechanics are clear to all. A change in linear momentum (mass×velocity) produces a force. A force produces an acceleration. Momentum is conserved. Energy is conserved.

And along comes Einstein, to crank up the speed and introduce time to space. We sketched some of the consequences at the end of Chapter 3, and now, with a revised mechanics to understand, we come back for more.

Four of a Kind: Space-Time

Observer 1 mounts a light-clock on a rocket (or, more commonly, on a high-speed particle like a muon) and sets out to measure the time:

A simple device, the bouncing beam proves to be a model of mechanical reliability. Outside, the surroundings whoosh by at a steady 10,000 meters per second, or 100,000 meters per second, or 1,000,000 meters per second, or 10,000,000 or 100,000,000 or even close to 300,000,000 meters per second. Inside, within the reference frame of observer 1, space stands still. The beam goes up and the beam comes down, never moving off its spot. Ticktock. It is a coming and going that takes place wholly in time, with no change in horizontal position. From cycle to cycle, the beam completes its circuit in the same time.

Carried along with a moving reference frame, the clock monitored by observer 1 enjoys a measure of privilege. It is *primus inter pares*, first among equals. For although all inertial reference frames are created equal, in no other system will the ticktock be all time and no space. In no other system will the beam of light travel a shorter path. In no other system will the time be less. Observer 1, with a certain pride of place, calls it the "proper time"—the time recorded by a clock locally at rest. It becomes an invariant interval between a tick and a tock in space-time, a single number (a scalar) on which all other inertial observers can agree.

Everybody certainly agrees that light travels at the same fixed velocity, and everybody therefore agrees that the distance traversed by the beam is equivalent to an interval of time. Recall, then (from Chapter 3), how things must look to observer 2, who sees rocket and clock pass by at high speed:

 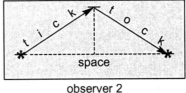

observer 1   observer 2

Observer 2, measuring an elongated path traced out by the rocket's motion, views the two events (the beam's departure and its eventual return) as occurring in both space and time. "My dear observer 1, how can you possibly say that the beam did not advance horizontally? I saw it clearly moving from left to right, and the time required for that trip was longer than you say!"

We know, of course, that our observers eventually realize that motion causes intervals of time and space to mix together. Just as *xy* space is rotated into *uv* space,

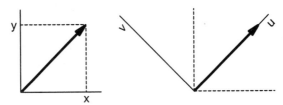

so, too, are space and time able to reflect varying points of view. The spatial and temporal components of one space-time frame "rotate" into another space-time frame, to an extent determined by the relative velocity:

observer 1   observer 2

Some things always change. Some things never change. Just as observers in tilted reference frames all measure the same invariant interval in space (a fixed *distance*), observers in moving reference frames all measure the same invariant interval in space-time: a fixed *proper time* for the clock, proportional to the dashed vertical lines in the diagram above. The greater the velocity, the more that time mixes with space; but through it all, the space-time interval remains the same. If not, then observers 1 and 2 would disagree on the speed of light.

There is one twist, and it is a twist that leaves space and time with a remnant of difference, even in a space-time universe. Whereas two spatial components reinforce each other to create an invariant distance, the spatial component in an invariant space-time interval diminishes the temporal component. Two comparable triangles suggest why:

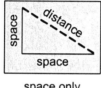   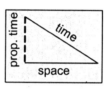

space only   space-time

For a mixing in space alone, the invariant quantity shows up as the longest side of a right triangle. Each of the perpendicular sides contributes to something bigger than itself, and thus the combination must somehow involve a "plus" sign. For a mixing in space-time, the invariant quantity is one of the two shorter sides. The contributions oppose each other. A "minus" sign enters somewhere into the mix, and it becomes a mark of separation.

Observer 2, representing all inertial observers for whom the clock appears to be in motion, finally works out a reconciliation with observer 1. And if the terms of that agreement sound familiar (see below), it is because they are: the language is analogous to what we used a few pages above to describe vectors in space alone. Observer 2 says:

1. My time coordinate (t') will be some combination of your time (t) and your three spatial coordinates x, y, and z. The specific mixture will depend on our relative velocity.

2. My spatial coordinate u will be another combination of your four coordinates t, x, y, and z. The mixture will be determined, again, by our relative velocity.

3. Similarly, my second spatial coordinate, v, will also be a mixture of your t, x, y, and z.

4. Ditto for my third spatial coordinate, *w*. Each of us, with four space-time coordinates, will then group our particular numbers into four-dimensional vectors. Our vectors will have different components but exactly the same magnitude (if by "magnitude" we understand a quantity that will somehow include a minus sign).

One, two, three, four. By building a vector from four independent components, we guarantee ourselves a quantity valid in any inertial reference frame. Observer 2, using a cut-and-dried mathematical recipe, is able to convert four components measured in a personal reference frame into four new components in any other. Everybody's vector has the same "length," and everybody is happy.

Need a directed something-or-other in space-time, the equivalent of an ordinary vector like velocity or momentum? Do this: Find three objects that combine like a vector in three-dimensional space (a set analogous to *x*, *y*, *z*) and put them together with a fourth object (analogous to *t*). Not all combinations will work, but if you choose the appropriate objects, then this gang of four (the components A_t, A_x, A_y, A_z) will become a four-dimensional vector in space-time. The A vector will do everything with A_t, A_x, A_y, and A_z that the space-time position vector does with its components *t*, *x*, *y*, and *z*. The mixing instructions for A_t, A_x, A_y, A_z will be the same as those for *t*, *x*, *y*, *z*, and the invariant space-time interval (like the length of a vector in space alone) will be the same in all reference frames. The space-time association of A_t, A_x, A_y, A_z will be just as acceptable to observers 1 and 2 as the space-time association of *t*, *x*, *y*, z.

Associations of space and time, like subsidiaries of a parent corporation, function together without entirely sacrificing their individual identities. The closeness of the association between temporal components (*t* or A_t) and spatial components (*x*, *y*, *z* or A_x, A_y, A_z) depends on the relative speed between two reference frames. Approaching 300 million meters per second, the merger is virtually complete. The Space Company and The Time Company become fully integrated as Space-Time, Inc., and all intervals involving space and time are mixed with abandon. At low speeds, however, the clock and ruler part company, and the incorporation of space into time shrinks to nothing. All the temporal components crystallize out of the mixture, and each trio of spatial components becomes an ordinary three-dimensional vector rotating only in space. Einstein's world freezes into Galileo's world.

Take velocity, for instance, a directed quantity that tells us how far an object travels during a given interval of time. In Galileo's world, with its one universal clock, a velocity vector has just three components: one for each direction in space. Since nobody ever disputes the time elapsed, different observers need only to reconcile their different views of position.

We already know what to expect here, both from our earlier arguments and from the experience of everyday life. Inside the train, a passenger walks up the aisle at 1 foot per second. To an observer at the station, the same passenger moves at 101 feet per second, the sum of two velocities: a local walking speed (1 foot per second) plus the speed of the passing train (100 feet per second). The addition of velocities suggests that time ticks at an identical rate for both observers, and nobody is surprised. In Galileo's world, material objects move slowly and light moves fast—indeed so fast that the invariant speed of light, c, might just as well be infinity.

Not in Einstein's world. In the Land of the Fast, an assertion such as "the object traverses 100 million meters every second, and therefore its velocity is 100 million meters per second" is fraught with controversy. Both "100 million meters" and "per second" are matters of opinion, and different observers will volunteer different distances and different times. The only common ground is the absolute speed of light, which at 300 million meters per second no longer even seems particularly fast.

What are conscientious observers to do? They need to register both a change in position and a change in time, but they need to do so in a way that can be understood and accepted in all inertial reference frames. A four-dimensional vector, with its mix-and-match components and invariant length, resolves all doubts. First, to establish a change in position, observers 1 and 2 rely on four coordinates already certified as relativistically sound: t, x, y, z, the makings of a position vector in space-time. Second, for the accompanying change in time, they use a quantity that has the same value in all frames: the proper time, an invariant interval in space-time. Then, introducing a temporal component of velocity (V_t) to three spatial components (V_x, V_y, V_z), they put together a four-dimensional vector immune to motional bias. The four components of velocity scramble among themselves in just the same way as t, x, y, z.

Making that connection, our observers build a bridge from the Land of the Slow to the Land of the Fast. At low speeds, velocities expressed as four-dimensional vectors do appear to add together; and if there are any deviations, they are too slight to notice. No reasonable observer, one suspects, will complain about the difference between 101 feet per second and 100.9999999999999...9 feet per second. Time is divorced from space, and Galileo's principle of relativity survives the challenge of space-time.

At higher speeds, though, the rule of addition breaks down. Say a rocket passes overhead at ½ c. Observer 1, riding inside, launches a particle and measures its local velocity as ¼ c. Observer 2, looking from the outside in, then reports that the particle's velocity is noticeably less than ¾ c, which would have been the result of a simple sum. No other perception is possible if observers 1 and 2 are to measure an invariant speed of light.

At the very high end, finally, there is another limit: the speed of light itself, c, one of the great constants of the universe. Let the rocket shoot ahead at nearly light speed, and let observer 1 fire off a particle also at nearly light speed. What do our two observers see now?

This time they see almost the same thing. They each record a particle velocity of just under 300 million meters per second, with barely a hair's difference between them. The moving object, pushed to extremes, approaches but does not exceed the fixed speed of light:

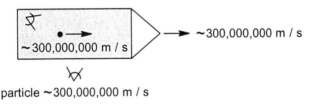

particle ~300,000,000 m / s

It is the Land of the Absolute Fastest, the end of the line. Neither trains nor planes nor rockets nor muons nor light waves travel faster than c. Doing so would require all the energy in the universe, and then some.

To understand why, we now need to see what becomes of momentum and energy in Einstein's world.

Conservation of Momentum-Energy

As velocity goes, so goes the rest of mechanics. For if nature imposes a universal speed limit, then Newton's second law must ultimately fail. If no object may be pushed beyond c, then the proportionality of force and acceleration becomes useless at high velocity. We have to abandon it.

The news gets worse, too, because the Newtonian conservation laws also start to fail. Momentum, notably, is no longer conserved at high speeds—not if we continue to define momentum as (mass)×(velocity), and not if we insist that all inertial observers measure the same finite speed of light. The mixing of space and time affects the perception of velocity, and it necessarily affects the perception of momentum as well.

Conservation of momentum, however, is something too good *not* to be true. It strikes us as a law too regular, too predictable, too strictly enforced to be merely an accident of a world in slow motion, a fluke. It tells us something infallibly correct about the "before" and "after" of a system, even when we know nothing about the detailed forces that control the motion in between. Conservation of momentum is a pattern that asserts itself in every corner of Newton's and Galileo's world. It is not a mere numbers game. It is not a sleight of hand.

Conservation of momentum is a trusted friend in the Land of the Slow, and observers in the Land of the Fast are glad to make its acquaintance as well. A revised form of momentum, conserved in all inertial reference frames, comes as part of a four-dimensional vector that also includes a conserved energy in the bargain.

Our instructions are straightforward. By multiplying the mass of a particle by the four components of its relativistic velocity (treated in the section just above), we put together what we need: an observer-proof momentum and energy, good for reference frames in all states of inertial motion. The three spatial components of this "momentum-energy" vector correspond to momentum, and the lone temporal component corresponds to energy. It is the same connection we asserted back in Chapter 4, part of the foundation on which the conservation laws rest. Momentum is associated with displacement in space, energy with displacement in time.

The momentum-energy vector tells us, for any relative velocity, the values reported by an observer 2 who views observer 1's reference frame passing by at constant speed. Observer 2 takes, for example, a component of momentum measured by observer 1 and multiplies this number by another number called "gamma." Gamma depends on the relative velocity of the two reference frames, and it has a different value for each of the infinitely many possible choices. It appears everywhere in the theory of relativity, and it shapes the landscape of the Land of the Fast.

Let's take a look at that shape. For convenience we express the relative velocity as a fraction of the speed of light. A stationary frame has a relative velocity of 0. A frame moving at 150 million meters per second (½ c) has a relative velocity of 0.5. A frame moving at the speed of light itself has a relative velocity of 1. Here it is, then, gamma:

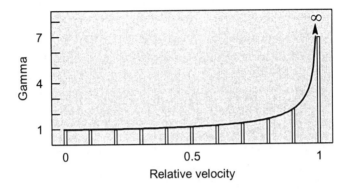

Gamma has the value 1 when the relative velocity is zero, and it effectively stays that way over a broad range of speeds. Under these conditions, toward the left-hand portion of the curve, we amble along in the world of Galileo and Newton, the Land of the Slow. We find a world untouched by a relativistic correction factor that differs scarcely from 1—which is to say, no correction at all. Observer 1 says that the quantity (mass)×(velocity) is conserved, and observer 2 agrees. Galileo's relativity and Newton's mechanics, although subordinate to Einstein's broader enactments, exercise their full authority. Observers in motion reconcile their different views of velocity by applying the sum rule, and all is well.

Even at one-tenth the speed of light (fast!), the correction factor amounts to only 1.005, an increase of a half percent. At two-tenths the speed of light, gamma is still just 1.02. But at half the speed of light, gamma becomes 1.15; and at three-quarters the speed of light, it becomes 1.5. Einstein's relativity begins to assert itself, and momentum starts to looks nothing like Newton's uncorrected product of mass and velocity. No, in the Land of the Fast, the conserved quantity is not (mass)×(velocity), but rather (gamma)×(mass)×(velocity). And as velocity climbs toward its ultimate limit, c, gamma shoots up rapidly to infinity: from a value of 7 at a relative velocity of 0.99...to 70 at 0.9999...to 7000 at 0.99999999, ever upward. It is a limit impossible to attain, a limit that would require a particle to draw upon an infinite amount of energy and momentum.

Meanwhile, whether moving fast or slow, everybody can now agree on three points concerning mass, momentum, and energy. Each of these arguments, traced back to its source, derives from the constancy of c and the absence of any absolute reference frame:

1. Momentum and energy, yoked together as a four-dimensional vector, enjoy a relationship analogous to the one that connects space and time. Compare: Just as different observers report different components of space and time, they also report different components of momentum and energy. Rising above these differences, though, is a polestar by which all inertial navigators set their course: the magnitude (or length) of the four-dimensional vector. It is a scalar, a single number, and it has the same value in all reference frames.

   That's what vectors do. They preserve their length when their components are scrambled. For a clock in space-time, the invariant interval—the magnitude of the four-dimensional position vector—is the so-called proper time. Observers manipulate space and time in a certain way, and everyone comes up with the same number. For a momentum-energy vector, the invariant magnitude is proportional to the *mass*. Observers manipulate momentum and energy in a certain way, and they all come up with this same mass.

2. Momentum, the spatial contribution to momentum-energy, is conserved in all reference frames. The three components of momentum constitute a self-contained three-dimensional vector with its own fixed magnitude.

   Different observers, their views affected by relative motion, report different numbers for the total momentum of a system. And so what if they do? It is enough that they agree on the phenomenon of conservation—that whatever their individual numbers may be, the values remain the same before, during, and after the events in question.

3. Wherever and whenever momentum is conserved, the energy must be conserved as well. If not, the magnitude of the four-dimensional vector would not remain fixed.

Linked together by an invariant mass (perceived as the same in all frames), momentum and energy thus become partners in a broader conservation law. In a space-time universe, you can't conserve one without the other.

Energy and Mass

We come at last to $E = mc^2$, the equivalence of energy and mass. "Energy," says this deceptively simple equation, can take the form of mass. If we choose our reference frame so that the total momentum is zero, then our momentum-energy vector will be *all energy*. Energy, and energy alone, will give that vector its magnitude, a number proportional to mass.

Pick up a handful of stuff, anything you like. Multiply its mass by c^2 (the speed of light, times itself), and *presto!* The number that falls out is an energy, a genuine energy that can be tapped to do work. It is energy bundled into mass, energy that is intrinsic to any material particle.

On the one hand, they appear to be an odd couple. How can energy, we wonder, possibly be equivalent to mass? How can energy—this ethereal, invisible presence, this intangible spur to action—be equated with mass itself, the very embodiment of substance and tangibility?

On the other hand, $E = mc^2$ surfaces almost as an afterthought, a necessary but incidental consequence of space-time. Once the spatial momentum vector takes the form it does, the fate of its temporal associate is sealed. Because if absolute time is abolished, then the temporal component of a four-dimensional momentum vector can be only one thing: the product of mass and the temporal component of four-dimensional velocity. When all the mathematics is done, finally, we are left with the quantity (gamma)×(mass)×(c^2).

What is it? It appears on the scene with units of (mass)×(velocity)^2, just like energy. It is a single number, just like energy. It is associated with time, just like energy. It is indeed the total energy of a free particle, and it increases with velocity in the same gamma-like way that momentum does:

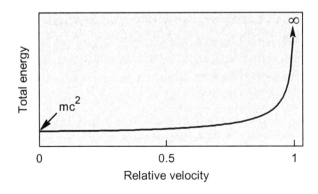

Look at the left-hand edge of the curve, where the relative velocity is zero and gamma is one. Here, in a reference frame *at rest*, a stationary particle has zero momentum and zero kinetic energy (no energy of motion), yet its total energy is not zero. It has an energy equal to mc^2: a "rest energy," an energy of standing still, an inherent energy of being. It is the energy conferred on a piece of matter not by motion, not by a collision, not by a gravitational field, not by an electromagnetic field—not by anything else but mass itself. This internal energy mc^2, the same number in all inertial frames, is the amount that nature deposits in the bank when a particle is born.

The totality of energy in the universe remains the same, despite any local ups and downs. There is a debit for every credit, a loss for every gain. If, say, a particle wants to lose internal energy, then it must literally give something of itself: a portion of its mass. The mass of the particle goes down, and the energy released flows into a different account.

Picture a field of potential energy contoured like a valley between two hills. On one of the hills, poised at the top, a particle sits motionless. No force pushes it to the left; no force pushes it to the right. The potential is flat at the summit, and there the particle sits. It remains at rest, doing nothing, until suddenly a slight nudge from the outside gets the ball rolling:

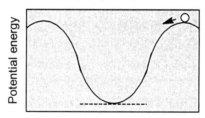

Falling from high potential to low, the particle picks up kinetic energy during the descent. Now faster, now slower, it rolls to the bottom of the hill and begins to climb up the other side. Kinetic energy carries the particle up the hill, higher and higher,

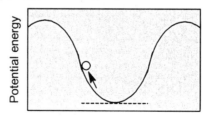

and here is the question: How high does it climb? Will the particle become trapped in the valley, or will it climb all the way up the second hill?

If the body is to be confined to the valley, then it must lose some energy. Like a rambunctious child, it must quiet down. For if the particle takes all the kinetic energy the field can give—if all it does is trade potential energy for kinetic energy, dollar for dollar—then the transaction amounts to a wash. Whatever kinetic energy the particle gains going downhill is just enough to carry it uphill, back to the summit once again. Drained finally of kinetic energy, the object returns to its original condition of rest: a point of maximum potential energy, undiminished by the expense of motion.

Suppose, though, that the particle does not keep every bit of energy the field bestows. It has a mass, remember, a built-in store of energy, an endowment that depends neither on its own motion nor on the influence of any external agent. And by giving away some of that internal energy now, by *losing mass*, the particle avoids climbing all the way up to the top. It settles in the valley, finding a new stability with less mass and less rest energy than before. Somebody else pockets the difference.

Another example: A certain kind of uranium nucleus breaks into pieces, and when the dust settles, the shards contain less mass and less rest energy than the original atom. Is energy lost? No, the difference goes into the kinetic energy of the fragments, which fly off at high speed to deliver explosive power and heat to the surroundings. Nucleus by nucleus, the redeployed mass seems tiny enough (smaller than a trillionth of a trillionth of a gram), but every little bit adds up. Put enough little bits together, and the result is a nuclear fission bomb. It is a convincing demonstration of how a morally indifferent universe shuffles around mass to enforce the conservation of energy.

Spectacular, yes; special, no. Nature exacts the same price—a local change in mass—whenever TNT explodes, whenever methane burns, whenever atoms and molecules hang together or come apart in a field of potential energy. Nuclear energy is no different from chemical energy, and the mass lost or gained by a nucleus is no different from the mass lost or gained by a molecule. If there is a distinction, it is one of extent and not kind. Atoms and molecules, held together far less strongly than nuclei, exchange energy and mass in much smaller doses. They need to put up proportionally less mass than nuclei (millions of times less) to balance accounts.

$E = mc^2$, we see, is no rare occurrence. Wherever particles interact, look for the possible interconversion of mass and rest energy: in the formation of a molecule, in the decay of a nucleus, in the cauldron of the Sun (where four million tons of matter are converted into 100 billion billion kilowatt-hours of energy every second). The business of $E = mc^2$ is an everyday affair, a routine transaction in the global ebb and flow of energy.

Space-Time and Gravity

Dorothy awakens with a start and sees it happen. A wobbly shelf, way up high, suddenly comes loose; and from that very instant, with all support gone, the bowling ball and golf ball have a mechanical destiny to fulfill. The crash that follows is only a matter of time.

Dorothy watches as the two bodies fall together, accelerating in lockstep the entire trip. Side by side, they drop a total of 144 feet from ceiling to floor in three seconds (it is a very tall ceiling, apparently). Dorothy notes that the balls cover 16 feet in the first second, an additional 48 feet in the next second, and finally 80 feet more in the third. Represented horizontally to save space on the page, the schedule of distance and time is exactly what one expects for a body falling near the surface of the Earth:

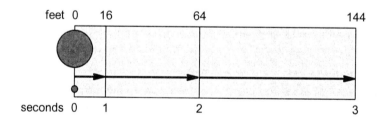

Nothing is out of the ordinary here, but Dorothy wakes up slightly disoriented after her long sleep. Still a bit foggy, she stares openmouthed as the two balls hit the floor at the same time.

"That's funny," wonders Dorothy, "the bowling ball is so much more massive than the golf ball, yet both descend at the same rate of acceleration (if I correct for air resistance, of course). How odd!"

Happily, her head soon clears. "Oh, yes," exclaims Dorothy with not a little satisfaction, "I read about that in Chapter 4. It's like a tug of war, with mass pulling on both ends of the rope. All objects, large mass or small, fall at the same constant acceleration in a uniform gravitational field. The mass of the Earth attracts the mass of the falling body, but the mass of the falling body fights back.

"I remember it all very clearly now. The 'gravitational mass' of a body, which makes the thing susceptible to gravity, contributes to an attractive force. *The more mass, the more force. The more force, the more acceleration.* Yet that same mass, acting in the capacity of 'inertial mass,' gives the body inertia and makes it resist any attempted acceleration, measure for measure. *The more mass, the more inertia. The more inertia, the less acceleration.* It turns out—just because inertial mass and gravitational mass happen to be one and the same—that everything falls with the same constant acceleration. A coincidence, I suppose, but that's gravity for you."

Imagine Dorothy's utter shock now when she looks through the skylight and sees only the black nothingness of outer space. The Earth is gone. The Sun is gone. The Moon, the planets, the stars are gone. Not a particle of mass remains to create a gravitational field. Everything is simply gone away.

Nevertheless, there is gravity. Objects fall to the ground at the same constant rate of acceleration as on Earth. Dorothy and her bed remain rooted to the ground, just like on Earth. Everything in the house feels normal, just like on Earth.

And then, peering into the blackness outside, Dorothy sees a truly bloodcurdling sight. Some sinister *being*, riding what appears to be a jet-powered broomstick, is pulling the house through space by means of a hook and chain. Second by second, the broomrider accelerates the house at the same rate that a body would fall if it were in the Earth's gravitational field. The house advances 16 feet in the first second, another 48 feet in the next second, 80 more feet in the third. The house hurtles through space, ever faster, steadily picking up velocity—just like a body falling to Earth.

"Well," sighs Dorothy, "I guess I'm not on Earth anymore. Must have happened while I was sleeping, and it sure is a shame not to have all that other mass around. But maybe I shouldn't complain, come to think of it, because I can't tell the difference. After all, *I still have gravity.*"

The Principle of Equivalence

What our observer calls "gravity" is really just the effect of accelerated motion. In the vacuum of space, far from any large mass, no "force" of gravity pulls the two balls down from ceiling to floor. Instead, it is the floor of the accelerating house that moves upward and intercepts whatever objects lie in the way:

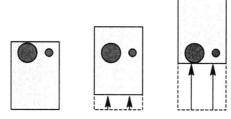

To an observer standing on that floor, the balls of different masses appear to fall side by side, as if accelerating downward in a gravitational field. But to an observer outside, who sees the floor rush up while the balls stay put, the view is different and perhaps more transparent. The masses of the two objects, large or small, clearly have nothing to do with the rising of the floor. The outside observer has no notion of a stay-at-home, force-fighting *inertial mass* in conflict with a footloose, force-compliant *gravitational mass*. The outside observer invokes no force at all.

Gravity? Gravity can be switched off just as easily as it can be switched on, if what we mean by gravity is merely a consequence of accelerated motion. Inside a freely falling elevator, for example, the unfortunate passengers have no sense of gravity after the cable is cut. Both the reference frame (the elevator) and its contents fall to Earth with the same acceleration: 16 feet in the first second, 48 feet more in the next second, 80 feet more in the third; all the way down they go, faster and faster until the ground intervenes. The elevator picks up speed at each instant, and so do the passengers. Everything falls independently at the same constant rate of acceleration.

An external observer sees elevator and passengers descending side by side, in free fall, each body undergoing the same gravitational acceleration. Inside the elevator, however, the passengers claim to be at rest. They float freely inside the cabin. They feel no attraction to the walls, ceiling, or floor. They see all the fixtures remain in place. "Gravity? *What* gravity? *What* acceleration?" says one passenger to the other. "We are free of all force. We are locally at rest in an inertial reference frame, motionless, and we are not going to change our current state unless and until something happ...."

Einstein gave it the status of a founding principle, the cornerstone of his general theory of relativity: that all uniformly accelerated reference frames are fully equivalent; that none is more privileged than any other; that observers inside a closed compartment cannot distinguish gravity or the lack thereof from ordinary accelerated motion. The principle of equivalence is tantamount to saying that gravity and inertia are the same thing, and it solves, in one stroke, the mysterious equality of gravitational mass and inertial mass. It does away with the two masses as separate entities, uniting them instead into just plain *mass*.

The special theory of relativity (what we have called simply relativity until now) gives equal weight to all inertial reference frames. It asserts that observers moving at constant relative velocity must view all physical phenomena in essentially the same way. Nobody can tell the difference between standing still and moving in a straight line at constant speed. Everyone must understand and accept the viewpoint of everyone else.

The general theory extends this equality of observation to reference frames undergoing uniform acceleration. It insists that observers in such frames must view all physical phenomena in the same way, and it leads inevitably to a rewriting of the laws of gravity. It must. Because if nobody can tell the difference between uniform acceleration and a uniform gravitational field, then what actually is gravity?

How, for instance, are we to construe the passage of a light ray through an accelerating reference frame? The ray, moving in a space devoid of all mass, follows a straight line. It enters a window of our accelerating compartment,

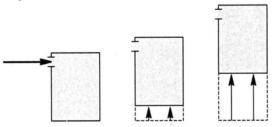

continues for a time, and then strikes a detector at the far end. Suppose that the acceleration upward is contrived to simulate the gravitational pull of the Earth. What happens to the light ray?

It depends on one's point of view. Observer 1, standing outside, sees a beam traveling straight through a rising compartment. The movement of the floor has nothing to do with the movement of the beam:

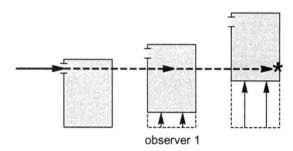

observer 1

"The beam persists in a straight line," reports observer 1. "No external agent deflects it from its path."

Observer 2, planted inside the accelerated reference frame, claims just the opposite. Observer 2 sees the beam fall to the floor in a parabolic arc, just like any projectile launched horizontally in the Earth's gravitational field. It falls 16 feet in the first second (during which it travels 186,000 miles horizontally!), an additional 48 feet in the next second (plus another 186,000 miles), and so on. The drawing is wildly exaggerated, but the path of the beam appears to curve:

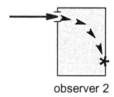

observer 2

"I am obviously standing in a gravitational field," says observer 2, "since my feet remain firmly on the floor, and since all objects—including this beam of light—fall with the same constant acceleration."

"But it can't be gravity," argues observer 1, "because you are nowhere near any *mass*. I assure you that it is only your compartment undergoing the acceleration, not the beam. The beam travels straight through."

"Say what you will," replies observer 2, "but everything I observe in here bears the mark of terrestrial gravity. I can do no experiment nor make any observation that convinces me otherwise."

To settle the argument, they turn to an observer 3 who follows the path of a light ray as it passes by a suitably massive object (the Earth, say). This third observer, reporting from an unaccelerated reference frame, concludes that light indeed does alter its path in the presence of mass. Observer 3 sees the same bending as observer 2, exaggerated once again for emphasis:

observer 3

The deviation near the Earth is small, but it is real; and near the Sun and other large stars it becomes big enough to measure with confidence.

All observers must now come to terms. They must agree that the path of *anything* (including a light ray) bends when viewed from an accelerated reference frame. They must agree that all such points of view enjoy equal status, and they must agree that the effects of acceleration and gravity are indistinguishable.

Let them first apply the principle of equivalence to the gravitation of light. Observer 3 (whose observation rests upon a large chunk of mass) is no more privileged than observer 2 (whose observation arises solely from acceleration). According to the principle of equivalence, though, both observers voice a legitimate opinion. Both claim to witness gravity in action, and so they ask: If gravity arises from the mutual attraction of two masses, drawn together by a force, then how is light—which is pure electromagnetic energy—able to gravitate? Does a beam of light derive an effective mass from $E = mc^2$, the equivalence of energy and mass?

Yes it does, according to special relativity, but Einstein found a broader, more satisfying approach: the general theory of relativity, under which the *force* of gravity is abolished entirely. Under special relativity, unaccelerated observers renounce all claims to absolute velocity. Under general relativity, accelerated observers renounce all claims to absolute acceleration. They renounce any hope of distinguishing constant gravity from constant acceleration.

Any kind of motion, we know, will alter an observer's perception of space and time, but accelerated motion goes a step beyond. Accelerated motion changes one's very perception of geometry, one's perception of shape and proportion. Things may appear contracted in one direction of motion and not in another. Spatial relationships—temporal relationships as well—may be distorted. Flat surfaces may appear curved. Clocks may appear to run slower. And remember: whatever bending and warping the one observer attributes to acceleration, another observer (asserting the principle of equivalence) will attribute equally well to gravitation, to *mass*.

Mass takes center stage once again, but this time in a dramatically different role, one that is about to overturn our whole idea of the nature and geometry of space.

Mass and the Curvature of Space-Time

In Newton's world, space is eternal and unchangeable: a place simply for particles to be and forces to act, a stage that remains unaltered while the actors come and go. It is an empty platform, adaptable to any script, but it is a platform that plays no active role in the dramas that unfold. Particles collide, fields rise and fall, energy and momenta are exchanged, yet the geometric construct of space remains untouched and unnoticed. For director, cast, and audience, the play is the thing. They pay little attention to the stage itself.

We picture the Newtonian stage as a void stretching out endlessly in three perpendicular directions, an emptiness waiting to be filled. We lay it out with straight lines,

and we apply the rules of Euclidean geometry. Rules such as: *Parallel lines never meet.* And: *The three angles in a triangle add up to 180 degrees.* And: *The ratio of the circumference of a circle to its diameter is equal to π (3.14159...).* They are the rules of a flat space,

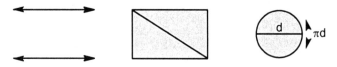

and a *flat* space, exactly that, befits the rectilinear inertia prescribed by

Newton's first law. For a particle moving freely in such a space,

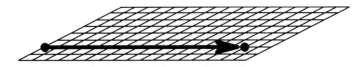

the natural order of things is to stay the course: to keep doing whatever it was doing last, to maintain a constant speed, to continue in a straight line. To go straight is to do what comes naturally. To go straight is simply to follow the gridlines of a flat space. And if a Newtonian force comes along to deflect a particle from the straight and narrow, then so be it. Although the trajectory of the body may bend, the framework of space remains flat:

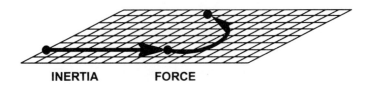

INERTIA   **FORCE**

The inertial reference frame plays host to the action but does not take part. Space and time remain the same before, during, and after the events.

Now is this long-standing picture of static space and time appropriate for an accelerated reference frame, and thus for the description of gravity? Einstein says no. *No* to space and time as separate entities. *No* to space as everywhere flat. *No* to space as eternal and unchanging. *No* to space as merely a mental construct, a latticework of imaginary gridlines. Space-time emerges from general relativity as something tangible, something literally to be grasped and molded. Space-time stops being the passive host and becomes very much an active player. The presence of nearby mass—and its twin, energy—warps the fabric of space-time. *Mass* takes a flat sheet (our two-dimensional proxy for a four-dimensional world we are unable to draw),

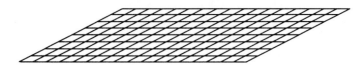

and puts bends and kinks into it, like a gymnast bouncing on a trampoline.

Over here, a configuration of mass makes space-time look like a sphere. Over there, a saddle. Somewhere else, maybe even a sombrero:

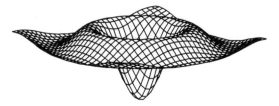

The principle of equivalence demands it. Since spatial and temporal relations appear warped to an accelerated observer, then they must also appear warped to an observer in a gravitational field. You ascribe the warping to motion and I ascribe it to mass, but our views are indistinguishable. We are both right.

According to general relativity, mass exerts its influence not by force but by inertia. The Sun does not use its mass to attract the Earth by force, to *pull* the Earth closer, to deflect the Earth from its "natural" inertial trajectory, a straight line. No, the Sun uses its mass instead to channel out a curved path for the Earth and thereby change flat space-time into curved space-time. Like a railroad worker throwing a switch, the Sun redirects the inertial motion of the Earth onto a different track.

There is no compulsion. The Earth continues to enjoy force-free, uncoerced inertial motion, but now it is inertial motion along a curved path. A beam of light, no different, switches onto a curved track when it flies too close to the Sun. Gliding by, the light ray traces out for us the gridlines of a curved space-time. It does what comes naturally. It obeys the law of inertia over a terrain no longer flat, a terrain peppered locally with mass. Coasting along, it takes the straightest path possible in a four-dimensional world where Euclid's straight lines have been bent out of shape.

Getting Straight

Gravity is geometry, Einstein tells us, and it is specifically the geometry of a curved four-dimensional space—no ordinary four-dimensional space, either, but rather space-time. All the rules are different.

Right from the start, even when we measure the length of a vector in flat space-time (an unaccelerated reference frame), we find a difference that goes beyond just the number of dimensions. In plain space, the components of a vector all contribute constructively to its magnitude. In space-time, by contrast, there is a minus sign buried within. The picture is already familiar to us from special relativity, and it means that curved space-time will never be quite the same as curved space alone:

          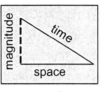

space only   space-time

Aware of that fundamental difference, we shall nonetheless use an ordinary sphere (a concession to our three-dimensional minds) in what is yet to come. Let it serve as a suggestion, however limited and imperfect, of the way geometry might shape up in curved space-time.

The mathematical problem here is to bend four-dimensional space-time in a way that accounts for a local distribution and flow of mass. It is a fearsome challenge, one that taxed Einstein for many years, but what shines forth at the end is beautiful for its simple elegance. After all the equations are written, the theory asks and answers this basic question:

> QUESTION: If the most natural motion in the world is inertial motion (stay the course! steady as she goes! move in a straight line!), then what does inertial motion mean in a neighborhood where all available roads are curved?

> ANSWER: Do the best you can. Follow the straightest path possible under the circumstances.

Start out somewhere on the Equator, sight due north, and take a small step: one meter, one foot, one inch, one millimeter, one micron, one nanometer; whatever you like, but *small*. Make the step small enough for the direction of travel to appear straight, as if the Earth were flat:

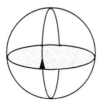

It is an illusion with which we are all familiar.

Now take another small step, again sighting straight ahead. Not to the east, not to the west, but due north:

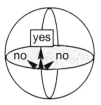

Do it again, and again, and again. Take a step straight ahead each time, and then, after a long series of small steps, stand far away and look at the path traced out. It lies along a great circle, a curve that passes through opposite poles:

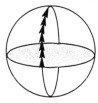

We call this special path, the straightest possible on a curved surface, a "geodesic." It is just one example of many, but it illustrates for us the general meaning of straightness in a bent environment. The message is simple: if you want to go straight, follow a geodesic.

Two surveyors, believing the Earth to be flat, soon discover that the old rules of geometry no longer apply on a sphere, where the straightest line possible is a geodesic. They learn, for example, that the three angles in a triangle may add up to more than 180 degrees. They learn that the ratio of a circle's circumference to its diameter may be less than π. They learn, if they travel far enough, that parallel lines move ever closer and eventually cross. Starting out in the same direction and continuing straight ahead in *flat* space, their paths would never meet:

On the surface of a sphere, though, they meet at a pole:

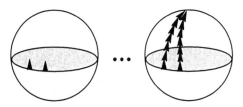

Each surveyor sets forth in the same direction, due north. Each surveyor, looking straight ahead, does nothing but follow a geodesic, step by step. Neither consults the other. Neither adjusts to the other. They move independently, guided only by the geometry of their surroundings, and still they draw inexorably near. An observer from above might well invoke a force of attraction. The surveyors know better.

Surveyors of the cosmos, guided by Einstein's general theory of relativity, now know better as well.

From Falling Apples to Black Holes

Finally, lest we forget: *Newton was right*. What Newton revealed about the mechanics of heaven and Earth—everything from a cannonball in flight to the orbit of a faraway planet—is correct, both logically and mathematically. His laws of gravitation keep the Moon in the sky. They tell us the shape, size, and velocity of the orbit. They tell us when and where to expect the arrival of a meteor. They tell us the path of every projectile falling to Earth. They do all these things, and they do so with formidable precision. The numbers are impeccable.

What Newton's laws fail to do, however, is explain their own success. Newton calls upon an otherworldly gravitational force able to sense inter-particle distances, but he remains silent on the origin of this mysterious stranger. Although he gives us a recipe that is demonstrably correct, we are left in the predicament of schoolchildren learning arithmetic by rote. We wonder why the recipe takes the form it does.

We wonder, too, how Newton's gravitational field can satisfy even the most basic demand of special relativity: that a signal may not be transmitted instantaneously, that a certain time must elapse for information to pass from here to there. Yet a Newtonian force field makes so such allowance. Newton's model makes no mention of a lag between the transmission of a force by one mass and its reception by another.

Einstein's theory steps into the void. General relativity, by connecting mass with the curvature of space-time, provides a mechanism and a justification for gravitation. Every clump of mass, everywhere in the universe, has a contribution to make. Mass projects gravitational influence by the distortion it creates, strongest near the source and tapering off in the distance. Think of how a body makes a depression in a mattress, how the wrinkles smooth out toward the edges, how the curvature diminishes with distance. Think of how a small object might roll, unforced, into the hollow at the edge of the recumbent mass. Think of how a restless sleeper, tossing and turning, continually shapes and reshapes the geometry of the mattress. Conjure up these images, for they convey something of the meaning behind the recondite mathematics of general relativity.

The general theory, with its curved space-time, explains precisely why gravitational effects grow weaker with distance and stronger with mass. It shows how a piece of mass, moving from point to point, continually shapes and reshapes the space-time around it, carving out new furrows while coasting through the furrows of everything else. It demonstrates how a mass in motion sends out waves of gravitational influence (ripples in space-time), and it shows that these weak disturbances travel at the fast but finite speed of light, c. General relativity thus gives us a way, on a grand scale, to understand why Newton's numbers are correct. Einstein's space-time becomes a flexible scaffolding on which the universe erects gravitational structures that run the gamut in size and strength. General relativity takes on everything from Newton's falling apple and falling Moon to the astrophysicist's neutron stars and black holes.

Newton was right, says Einstein, because space-time stretches out nearly flat in places where mass is scarce and gravitational fields are weak—as it does right here, in our own solar system, no great concentration of mass by cosmic standards. And where the curvature of space-time is sufficiently gentle, the complicated equations of general relativity reduce to the much simpler equations of Newtonian mechanics. There are a few small deviations that only general relativity can explain (tiny deviations, really), but still we apply Newton's laws locally with full confidence. After all, they took us to the Moon and back.

But there is more to the universe than Earth, Moon, and Sun, so look deeper and see where general relativity wields its power unchallenged. Go beyond our gravitationally temperate neighbors, and look to the extremes: to the fantastically dense concentration of mass in a collapsing star, say, where the curvature of space-time becomes anything but gentle. Look back to the birth of the universe, the Big Bang, when all the matter and energy there ever was (and perhaps ever will be) appears to have emanated from a single point. Or if you want to confront the ultimate in present-day material excess, look into a black hole, an object in which so much mass is packed so densely into so little space—imagine ten Suns compressed into a sphere with a radius less than twenty miles—that the local curvature becomes infinite.

Be careful, though. Whatever falls into a black hole, stays there. Nothing, not even a beam of light, can escape from its infinitely warped gravitational labyrinth.

6

TAKING CHARGE

It happens in an instant: a streak in the sky, a sudden illumination, then darkness once more. But in that flash of lightning, so brief, we peek into a world usually unseen, a world thrown abruptly out of balance and just as abruptly restored. It is the electrical world, the world of electrons and nuclei, the arena of atoms and molecules, light, heat, and sound. It is the world of our senses, a world where outward calm belies the tension within.

Gravity has no authority here. Atoms come together and go about their business solely on the basis of electric charge, not mass. In the gravitational macroworld, where *mass* reigns supreme, forces are small and distances are large—but the collective masses are enormous, and everything pulls together. A single electron, proton, atom, or molecule counts for little in the gravitational interplay of Sun, Earth, and Moon. It is a speck of mass with less power than a foot soldier in an army of millions. Only a numberless host of such particles, massed together, can muster an influence sufficient to shape the world at large.

Not in the electrical microworld. In the electrical microworld, where charge displaces mass, forces are large (*how large? at least a billion billion billion billion times stronger than gravity, and up to a million*

132

times greater than that) and distances are small (*how small? a few hundred trillionths of a meter, or less*). One particle faces off against another, and all transactions are retail. Single electrons and nuclei, tiny as they are, wield their potent electric charges in a world balanced by opposing forces. There is a positive for every negative, and opposites rarely stay separated for long. When they do part company—as they might, say, when huge numbers of electrons are stripped from molecules of H_2O in a cloud—nature hurls a thunderbolt and rights the wrong. Negative charges and positive charges, driven apart by accident, come together intimately once again. It is the way of the electrical world, the world of charge. Opposites attract, likes repel, and an overall neutrality holds the forces in check.

Nature endows the electron with an electric charge, and that fixed negative charge (symbol: $-e$) stays with the particle wherever it goes, inalienable and inviolable. Like mass, the electric charge is one of a handful of truly fundamental attributes that make an electron an electron. Whether the electron remains bound inside a water molecule, whether it flits from nucleus to nucleus in a copper wire, whether it materializes from the beta decay of a neutron, whether it flies through a television tube— during all its comings and goings, whatever else happens, the electron retains the same electric charge. The value $-e$ is a number that can be measured in the laboratory to great accuracy, and it is a number on which all observers can agree. The fundamental charge of an electron is not open for debate. It stands forever as an invariant property of the particle.

So, too, for a proton. So, too, for a neutron. So, too, for any particle that behaves as if it were an integral, noncomposite lump of matter. A proton carries a positive charge of $+e$, and this elementary charge remains $+e$ (exactly opposite to that of an electron) in any associations the proton makes or conditions it endures. A neutron, with zero electric charge, never grows the least bit positive nor the least bit negative, come what may. *All* fundamental particles preserve their electric charges intact, as long as they live, and the grand total of all the charges in the universe never varies: zero. There are as many elementary positive charges as elementary negative charges, and nature insists that things remain as they are. Electric charge, like energy and momentum, is strictly conserved.

Electric charge cannot exist without matter, and matter cannot exist without electric charge. If we are to understand the inner workings of matter (and especially the small world of atoms and molecules), then we must first begin to understand the electric charge and its influence.

The Two Faces of Electric Charge

In this corner, static electricity: the spark (*ouch*) that sometime flies between finger and doorknob. In that corner, magnetism: the trusty realignment of a compass needle. Two different forces?

No. The difference is only superficial, because both interactions arise from a single source, the electric charge. An electrostatic field comes about from charges at rest, and a magnetic field comes about from charges in motion. And since motion and rest are merely matters of opinion, who is to say which is which?

Standing Still: The Electrostatic Interaction

We pick up where we left off in Chapter 2, with the recognition that charged particles have the potential to interact. Opposites attract,

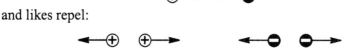

and likes repel:

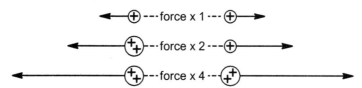

The resulting force acts along the line joining the two centers.

The bigger the charge, the bigger the electrostatic force. Double the charge of one particle, and the force doubles in direct proportion. Triple the charge, and the force triples. Double both of the charges, and the force quadruples:

The electrostatic force between two particles thus depends on the product of the two charges. It is the same pattern we observe for Newtonian gravity, with mass replaced by charge.

Acting particle to particle, the electrostatic force wells up from the electric charge of every proton and every electron in the universe. Its effect is cumulative. No matter how complex the structure, we always count so many electrons (a whole number) and so many atomic nuclei (another whole number); and out of this multitude, two by two, every pair of particles has its say:

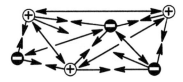

Each interaction adds to the total.

We do another experiment. Keeping the two charges constant, we double the distance between them and measure the force. It falls to a fourth:

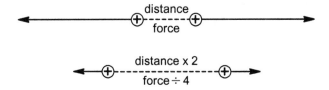

At triple the distance, the force drops to a ninth. At quadruple the distance, it drops to a sixteenth. The electrostatic force, specified in full by *Coulomb's law*, depends on the inverse square of the distance.

So does Newton's force of gravitation. So does the intensity of a light bulb. So does the sound of a bell. An inverse square law always develops when a given influence radiates outward from a central source, symmetrically in all directions. The broadcast influence might be radiation from the Sun, or gravity from the Sun, or an electrostatic force from a charged particle, or any number of things. Let the specific source here be a positive charge, our present concern, and think of a single particle—all alone in a vacuum—as throwing up a field of electric force around itself. With nothing else to make a difference, all directions in space become equivalent. The influence emanates with equal strength all around:

Now introduce into this one-particle universe an Eve to interact with Adam; bring in, slowly and from a distance, a second particle with a minimal positive charge of its own. Place charge 2 at some distance r from the center of charge 1, and measure the force:

What do we observe?

We observe a repulsive *push* delivered outward from charge 1 to charge 2 in the direction of the arrow. But since no one direction commands special privilege, we also observe exactly the same push anywhere along a sphere of radius r:

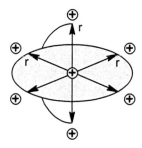

Realize, though, that the central source has only so much force to give: that it radiates a fixed amount at each instant, an amount proportional to the value of its birthright charge. Whatever force may be available, large or small, the full amount must flow out through the imaginary sphere surrounding the source.

If so, then the jolt imparted to a particle on the surface of that sphere must always be less than the maximum—less, because most of the influence misses the target. Nearly all the force shoots out along every radius except the one where particle 2 happens to be. What remains for particle 2 is only the small proportion it claims of the total surface area, a single arrow out of the spherical quiver.

And how much is that? We know from geometry that a sphere has area $4\pi r^2$ (proportional to the square of its radius), and so we conclude that the force must decrease just as much as the number r^2 increases. The dilution shows up in the wider spacing between "lines of force" at greater radii:

large force   small force

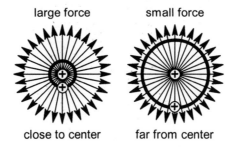

close to center   far from center

Observe: The particle close to the center sits in a region where the force is more concentrated and therefore stronger. Chalk it up to geometry. Radiating outward, showing no preference for any particular angle, the force field simply has less of a distance to spread out before encountering the nearby test particle. The lines piercing the inner ring are still closely spaced, diverged only slightly from the source, and a correspondingly large force flows out through each element. Farther away, however, the density of lines becomes progressively thinner and the influence falls away. The same total amount of force flows through a larger area.

Magnetism: Poles Apart

Bar magnets may offer less of a spectacle than a lightning strike, but they offer no less of a window into the (usually) hidden world of charge. There must be something, after all, something *inside* a magnet that makes its north and south poles attract and repel:

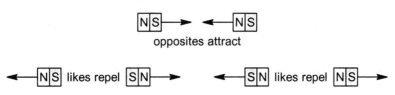

There must be a field of some sort, too, since both the strength and direction of the force vary at different points outside the magnet. We know, because we can see the effects of a force and even feel its pushes and pulls. We can hold a child's magnet under a sheet of paper, sprinkle a powder of iron filings on top, and see traces of the magnetic field right before our eyes: lines of force, mapped out by the filings as they twist and turn into position. We observe a similar pattern for a compass needle moved around the globe, and it looks roughly like this:

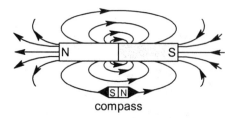

compass

Magnetized iron filings, compass needles, and other such objects orient themselves along the lines of force in the directions shown. North seeks south, and south seeks north, and eventually everything comes into alignment. Where the field lines are densely spaced, close to the source, the force is large. Where they are thinly spaced, far away, the force is small.

We see the pattern again and again. These lines are the signature of a *magnetic dipole* ("two poles"), and the dipolar field is no mere curiosity. It is something new, something without an exact analog in the world of stationary charge. Comparing a magnetic dipole with its electric counterpart,

electric dipole   magnetic dipole

we soon discover that the magnetic poles differ fundamentally from the oppositely charged ends of an electric dipole. Whereas an electric dipole can be broken into two isolated charges (monopoles) with fully separated fields,

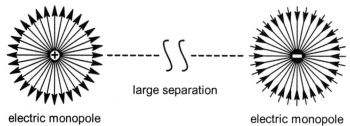

large separation

electric monopole   electric monopole

a magnetic dipole cannot. A magnetic dipole cannot be pulled apart into isolated north and south monopoles, with each source producing a field of its own:

magnetic monopole   magnetic monopole

No, the magnetic dipole represents a bar magnet in its most basic, irreducible form: a north pole and a south pole, indivisible, an inseparable duet. Cut a bar magnet in half, and the thing regenerates as if it were an earthworm. One intact magnet becomes two intact magnets, each enjoying an undivided set of north–south poles. They produce the same dipolar patterns as before:

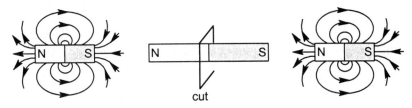

cut

We cut again and again, and all we do is create more magnets. We create magnet after magnet, each one smaller than its predecessor, until finally we come down to the last magnetic dipole still standing. It is a single atom of iron.

And there, hidden within that one atom of iron, we ultimately find the true source of all magnetism: a charged particle executing some kind of motion. We find an electron, in particular, doing something other than just standing still. To stand still is to render oneself immune to magnetism. A stationary charge neither produces nor responds to a magnetic field.

To move beyond a purely electrostatic existence, to become a producer and consumer of magnetism, a charged particle must get going. It must move from point to point as an electric current.

Magnetism and Electric Current

Bring a current-carrying wire near a compass, and watch what happens: the needle deflects. It is responding to a magnetic field.

Pass a wire through a sheet of paper, sprinkle iron filings all around, and turn on the current. The filings arrange themselves into a pattern of concentric circles:

They are responding to a magnetic field.

Run a current through a circular loop, and the iron filings reproduce the dipole pattern of a bar magnet:

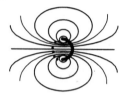

A clockwise circulation causes the field lines to point in one direction, and a counterclockwise circulation causes them to point in the opposite direction:

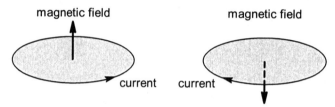

It is magnetism, pure and simple. Whatever a bar magnet does, a loop of current does just as well. The bits of iron are responding to a magnetic field, a field produced by the movement of electric charge.

Set a charged particle in motion, and it will generate a magnetic field. A single electron in hydrogen, a swarm of electrons in a circuit, a beam of positive ions in a particle accelerator—whatever the particles may be, if they have electric charge and they *move*, then a magnetic field will accompany the motion. The field will depend on where the particles are headed and how fast they are going. It will depend on the geometry of the circuit. It will depend on the amount of current passing through. It will depend on whether a charged particle is confined to an atom or whether it flows through a wire. It will depend on all of these things, and sometimes in a complicated way, but by observation and experimentation we learn how to sort everything out. We learn how to predict the strength and direction of the magnetic field at all points in space.

What is it, then, this magnetic field that acts so mysteriously, this invisible force that can move pieces of iron from behind a curtain? What does it do? What influence does it convey, and to whom does its authority apply?

Its influence extends, first of all, only to an electrically charged particle in motion. The movement of one charge gives birth to a magnetic field, and the movement of another charge causes that field to make itself known. Standing stock still, at rest in a magnetic field even of immense strength, a charged particle suffers no magnetic force at all. Only when the object *moves* through a magnetic field does it acquire a new potential to change its mechanical state: speed, direction, energy, momentum, angular momentum, and all the rest.

Start with a charged particle moving freely over large stretches of space, all by itself, not part of an atom, not subject to any external influences. Unconfined and unencumbered, the free particle (even one as small as an electron) belongs really to the Land of the Large, where the classical mechanics of Chapter 4 holds sway. Left undisturbed, the body will keep going in the same direction at the same speed. Hit by a force, it will accelerate. It will change speed or direction or both. Such is the law of the land.

Now, names aside, a force is a force is a force, and Newton's second law makes no distinction among a gravitational force, an electrostatic force, a magnetic force, or a ratseldorfian force. Newton's second law says only this: "Tell me where you are and how fast you are going *right now*, and then tell me what force you are experiencing *right now*, and I will tell you exactly where you will be and how fast you will be going the very next instant." The question becomes merely one of identifying how much force is present and in what direction it points, never mind the source.

Let our free Newtonian particle pass through a magnetic field, and suddenly it is no longer free. The body in motion sustains a well-defined, measurable magnetic force, a force that depends on three things: (1) the magnitude and sign of the particle's charge (q, for generality), (2) the speed and direction of the particle (its velocity), and (3) the strength and direction of the magnetic field. The force created will point in one direction for a positive charge, and in the opposite direction for a negative charge. Everything comes together in the following way:

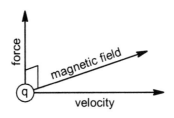

The force delivered is proportional to the charge, velocity, and magnetic field. Increase or decrease any of these quantities, and for a given geometry the magnetic force goes up or down in direct proportion.

Magnetic force has additional geometric peculiarities, though, certain details that set it apart from electrostatic force. First, the force acts perpendicularly both to a particle's line of travel and to the magnetic field. The resulting push or pull changes the direction of motion. Second, the amount of force depends on the angle between the original velocity and the magnetic field. A particle moving parallel to the field endures no force at all,

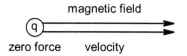

whereas a particle moving at right angles to the field receives the maximum dose, a strength equal to (charge) × (velocity) × (magnetic field):

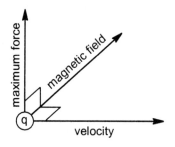

For angles in between, the strength falls somewhere in the middle.

The rules for charge, motion, and magnetic field will be different for electrons bound into atoms, because electrons bound into atoms reside in the Land of the Small, where quantum mechanics controls the action. When we come to that land, beginning in the next chapter, we shall realize that such particles no longer follow the beaten paths of Newtonian mechanics. We shall be forced to concede that classical ideas of force and acceleration lose their original meaning, and we shall realize that electrons in atoms do not circulate like current in a wire loop. But we shall also appreciate that an atomic electron certainly does move in some fashion, in a way perhaps *analogous* to current in a loop, and we shall observe that an atomic electron does produce a magnetic dipole as a result of that motion. And for some electrons, in some atoms, under the right conditions, those quantum mechanical magnetic dipoles all point in the same direction:

That kind of cooperation, understood in its proper context, gives a bar magnet the power to move iron.

A Stitch in Time: Electromagnetism

Not even identical twins always play together, and neither do electric and magnetic fields. If one is to affect the other—if there is to be an interplay between a magnetic field and an electric field—then it is not enough just to maintain a steady course. Something must change. The electric field and the magnetic field must vary with time; the landscape of influence must be reshaped moment to moment. To hold steady is to disunite the two fields and deny them their shared origins.

Think of what it means to exert a steady electric influence, always the same, never varying. We begin with an arbitrary collection of charged particles, and we suppose further that a particular inertial observer sees the distribution as static. All the particles remain in place, unmoved by the passage of time:

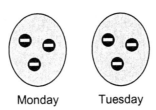

Monday   Tuesday

If so, then the electric field surrounding that static distribution is static as well. The force it delivers to a charge stays the same from one instant to the next. Point to point, all through space, the electric force may vary in strength and direction (it is a vector), but it holds steady in time. If a test charge is pulled with a certain magnitude in a certain direction at a certain place, then the effect will be the same at whatever time we choose to observe it:

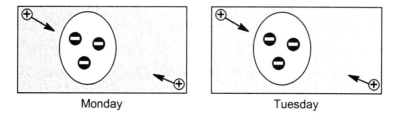

Monday   Tuesday

Nothing more can be done. A stationary distribution of electric charges produces a stationary electric field and nothing else. There is no accompanying magnetic field.

Likewise, a steady distribution of moving charges (a constant electric current) produces a steady magnetic field. If the same quantity of charge passes through the same area during the same interval of time,

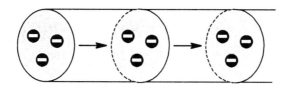

then the magnetic field holds constant as long as the steady current flows. Acting only on particles in motion, such a force does nothing to a particle at rest. A constant magnetic field cannot induce a stationary particle to get going.

A *constant* magnetic field, yes, but what about an inconstant magnetic field? Do something different now. Turn the current off and on. Increase the current. Decrease the current. Jiggle an electric circuit near a wire, or jiggle a wire near an electric circuit. Wave a bar magnet near a coil. Wave a coil near a bar magnet. Do anything at all to make a magnetic field vary with time, and you will simultaneously awaken an electric field. Like twins separated at birth and later reunited, an electric field and a magnetic field will discover their common origins and act as one. Electricity will emerge from magnetism.

To see it with our own eyes, we simply move a magnet in and out of a wire loop. An electric field develops. The field creates a difference in potential energy across the loop, high in one place and low in another, and the charged particles are thus induced to move. They fall from high potential energy to low, and the movement registers as an electric current:

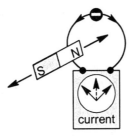

The electric current, in turn, generates a magnetic field of its own in opposition to the very change that created it.

We do not even need a wire. We do not need an actual flow of charged particles. We do not need an electric circuit to convince ourselves that a changing magnetic field carries an electric field in its wake. All we need is a magnetic field fluctuating in empty space. The fields exist independently of the charged particles that create them and the charged particles that respond to them.

Fair is fair. If a changing magnetic field gives rise to an electric field, then does a changing electric field also give rise to a magnetic field? Indeed it does, and we expect no less in light of the commonality between all things electric and all things magnetic, not to mention the relativity of motion. Both influences originate from the same source, the electric charge, and both work together naturally as one: as a unified electromagnetic field, able to deliver energy to a charged particle in any state of motion. Only when the fields remain static, when time stands still, do we speak of an electric influence as unconnected to a magnetic influence. Allow one of the fields to vary with time, and an electromagnetic union is sure to take place.

Riding the Wave

It was James Clerk Maxwell, a giant of nineteenth-century science, who brokered the merger of classical electricity and magnetism into a single corporate body. Interpreting the work of others and adding a crucial finishing touch of his own, Maxwell put the electric and magnetic fields on an equal footing, the only proper relationship for two phenomena that are really one.

Doing so, he unexpectedly laid the foundations of technological society with its panoply of electromagnetic invention: televisions, computers, mobile telephones, satellite communications, electric transmission lines, and everything else that requires the transport of electromagnetic influence through space and time. Go back to the beginning and see. It was James Clerk Maxwell, solving a set of abstruse mathematical equations, who first brought to light the *electromagnetic wave*: nature's designated emissary in a world born of electric charge.

Blueprint for an Electromagnetic World

With four brutally concise equations, Maxwell summarized all there is to know about the electric and magnetic fields of a deterministic, Newtonian-style universe. It is not the quantum universe, granted, but it is a realm of great scope and significance nonetheless.

The arm of Maxwellian electromagnetic law is impressively long. Its four equations apply all throughout the classical world, governing everything from radio telescopes to hair-thin optical fibers to the magnets used in medical imaging technology and more. Much more. The equations are exact and complete, and they deal with all possible incarnations of the electromagnetic field: under *static* conditions (where nothing changes with time), under *dynamic* conditions (where everything changes with time), in a *vacuum* (where the fields fill up "empty" space with a self-sustaining force and energy), in a *material medium* (where the fields interact with the charged particles of atoms they encounter).

What Newton's three laws of motion do for classical mechanics, Maxwell's four equations do for electric and magnetic fields. They provide an unerring record of what once was and what someday will be. Given an initial specification, the equations tell us exactly how the values of the fields change from one instant to the next. They brook no argument and recognize no uncertainty. Future and past follow inevitably from the present.

The mathematical expressions are brief enough to fit on a tee shirt (a novelty occasionally seen on college campuses), but their full elaboration fills thick textbooks and provides the substance for years of advanced courses. In the exacting language of vector calculus, Maxwell's theory lays out nature's complete specifications for the classical electromagnetic field. The four equations cover the territory from alpha to omega, and we have already touched upon the content of each one:

1. Gauss's law for electricity, derived from Coulomb's inverse square law: the way in which a configuration of charged particles gives rise to an electric field.

2. Gauss's law for magnetism: the assertion that isolated magnetic charges (monopoles) do not exist.

3. Faraday's law of electromagnetic induction: the way in which a changing magnetic field gives rise to an electric field.

4. Maxwell–Ampère's law for the magnetic field: the relationship between electric current and a magnetic field (Ampère's law), and also the induction of a magnetic field by a changing electric field (Maxwell's displacement current).

It is specifically the last two, the laws of Faraday and Maxwell–Ampère, that create the possibility of electromagnetic interplay and consequently the possibility of electromagnetic history. For by introducing time and change to the equations, these two laws enable us to distinguish "before" from "after" in the evolution of electric and magnetic fields.

Were it not for the laws of mutual induction, the electric field and the magnetic field would remain forever separate. Neither would be able to act on the other, even if the distribution of charges and currents were to vary with time. Nature chooses instead, however, to yoke together the two fields: to render them thoroughly interdependent, inseparable, a concerted application of the same fundamental influence. They act in tandem, with unbroken symmetry of cause and effect:

A varying magnetic field induces a tag-along electric field, and *a varying electric field induces a tag-along magnetic field.*

The faster a magnetic field changes with time, the stronger is the electric field that it induces. *The faster an electric field changes with time, the stronger is the magnetic field that it induces.*

The induced electric field develops perpendicular to the changing magnetic field. *The induced magnetic field develops perpendicular to the changing electric field.*

Interchange "electric" and "magnetic" in each statement, and the meaning remains the same.

Maxwell's equations also reflect another kind of electric–magnetic symmetry, the relativity of rest and uniform motion. I say a charge is at rest. You say it moves at constant velocity. I claim that my stationary charge produces an electric field and nothing more. You claim that the same charge (it's in motion, right?) produces not just an electric field, but a magnetic field as well. I make my observations from a perfectly valid inertial frame, and so do you—and immediately we know, convinced that nature treats all such reference frames alike, that we must both be correct.

We are. Using Maxwell's equations and their solutions, we reconcile our observations of the electric and magnetic fields. We realize that our state of motion affects the way we perceive a charge at rest and a charge in motion. We put together four-dimensional vectors in space-time, just like those described in Chapter 5, to account for our varying perceptions. We agree, in the end, that observers in different inertial frames see different mixtures of electric and magnetic fields, but the differences are of little import. They point instead to an essential unity.

Published four decades before Einstein's theory of relativity, Maxwell's equations are relativistically correct as they stand. Unlike Newtonian mechanics, Maxwellian electrodynamics needs no modification to accommodate an invariant speed of light. It is already built in. An observer-proof speed of signal transmission (c) emerges directly from the mathematics, and the equations show further that c has the invariant value 300 million meters per second: the speed of light in vacuum.

Maxwell's equations, by coupling a changing electric field with a changing magnetic field, thus give physical meaning to the injunction LET THERE BE LIGHT (or, more generally, LET THERE BE ELECTROMAGNETIC WAVES). To call an electromagnetic wave into being, though, we need first to break the inertial monotony of a charge at rest or a charge in uniform motion. We need an accelerated charge, a charge that changes its velocity. We need to shake up the source.

Making Waves

Somewhere in the world, a charged particle appears to be at rest. It sits motionless, always in the same place, filling the surrounding space with a constant electrostatic field. The field is real. It has physical meaning. It exists in space-time, and it lends substance to what would otherwise be sheer emptiness. Permeated by electric influence, pregnant with force and potential, the vacuum takes on a material aspect as tangible as the water in a pond. And like the glassy waters of a pond at rest, the field of a stationary charge remains still. The lines of force hold steady, unchanged in space and time:

Another inertial observer will see a different kind of pond, but it will be a quiet pond just the same. It will be a mixture of constant electric and magnetic fields, and our varying perspectives are no cause for alarm. Maxwell's equations, certified as relativistically correct, give us the license to choose any inertial frame we wish.

Imagine now that our glassy pond of electric field is suddenly disturbed, as if someone were to drop a stone into its metaphorical waters. Just as we see ripples in a real pond, traveling outward in all directions,

we see ripples in the electric field as well. Let's compare. In one pond, the disturbance of a certain material (water) produces an oscillating wave that goes forth as a regularly repeating cycle of crest and trough:

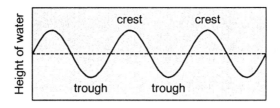

In another pond, the disturbance of a different kind of material (a field) also produces a wave that traverses space and time. Here, however, the wave traces out not the ups and downs of a wall of water, but rather the ups and downs of an electric field:

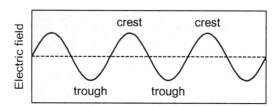

What causes such a disturbance? From where does it come? We look to Maxwell's equations for an explanation, and there we find one. In a body of water, a wave comes from a stone breaking the surface. In an electric field, a wave comes from a charged particle undergoing an accelerated motion. It is the law of the classical world, validated every time we turn on a radio.

Any kind of acceleration will provoke a disturbance in the surrounding field and cause a charged particle to radiate energy. The object might speed up or slow down, traveling all the while in a straight line. It might follow a curve. It might go into a circular orbit, or an elliptical orbit, or a hyperbolic orbit, or a parabolic orbit. It might bounce up and down in regular fashion, as if attached to a vibrating spring:

Pushed and pulled, up and down, back and forth, a charged particle oscillating in this springy way could well produce the sinusoidal wave illustrated above—but it is not our place here to work out the details. The exact pattern of radiation depends on the acceleration that drives the source, and we trust in Maxwell's equations to supply the correct description. Experience teaches that our trust is well founded.

This much we know for certain. Electric energy will be radiated in some way. There will be a changing electric field, and there will also be something more. There will be, in addition to an electric field, a magnetic field that tags along. The fourth of Maxwell's equations comes into play, the law that provides for Maxwell's very own "displacement current," namely: *a changing electric field induces a magnetic field at right angles*.

The effect is reciprocated. A varying magnetic field induces an electric field, and a varying electric field induces a magnetic field. Not one field, then, but two (an electric field and a magnetic field) radiate outward from the accelerated source. They work as a team. They create each other, and they support each other. The electric field drives the magnetic field, and the magnetic field drives the electric field. The oscillating fields move together in lockstep, mutually perpendicular, always at the invariant speed c (in a vacuum). They carry their combined electric and magnetic energy straight ahead:

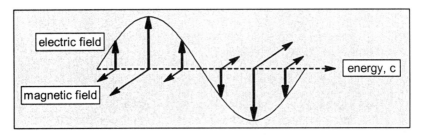

They are *waves*. They are electromagnetic waves, ripples of influence propagating through a unified electromagnetic field at the speed of light. They convey the message of a changing field to points far beyond the source, and they take on a life of their own. Once launched, an electromagnetic wave keeps going. It continues to exist even after the source that created it is long gone.

Waves in general, and electromagnetic waves in particular—in the classical world, the world of Newton and Maxwell, a wave is something special, something altogether different from a particle. We need to know what it is. We need to know what it does.

A Wave Is ...

Suppose a colleague from the planet Kveldar asks us to characterize an electromagnetic wave (or any other kind of wave, for that matter). The description has to be unambiguous and exact. What can we say?

We begin with the material that does the waving. In a water wave, it is water. In a sound wave, it is air. In an electromagnetic wave, it is an electric field and a magnetic field, those twin agencies that exert forces on electrically charged particles.

Next, using the appropriate mathematical formula, we identify some basic waveform that repeats from cycle to cycle. Perhaps it will be a square wave,

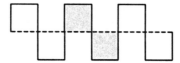

or a sawtooth wave,

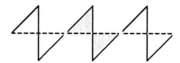

or any number of other simple patterns, but most often it will be a pure sine wave that commands our attention:

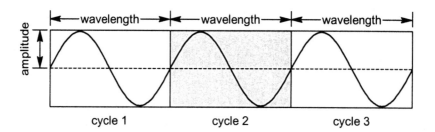

They are ubiquitous. Sines and cosines, elementary building blocks of even the most complex waves, crop up all around us: when a spring stretches and contracts...when a body moves in a circle at constant speed...when a violinist plucks a string...when a radio transmitter broadcasts an AM signal...when heat flows between surfaces at different temperatures...when many, many other natural processes occur. The primitive sine curve is a recurring motif in vibratory phenomena, and we shall use it almost exclusively to illustrate our waves. Whatever we say about a sine wave, in particular, will be valid for any wave in general.

A series of numbers tells the tale. The *amplitude* (see the diagram above) specifies the excursion of the disturbance from its baseline to either a crest or trough. The *wavelength* specifies the distance between any two equivalent points on successive cycles. The *period* specifies the time required to go through one cycle and begin anew, and the *frequency* (the reciprocal of the period) specifies the number of cycles completed during a given time.

Fast? Slow? Think about the meaning of wavelength and frequency, and piece together the *speed* of a wave, the rate at which the disturbance moves forward. It must be equal to (wavelength)×(frequency), the distance covered by one cycle (the wavelength) multiplied by the number of cycles per unit of time (the frequency). Give us any two out of three—speed, wavelength, frequency—and we automatically know the third. The lower the frequency, the longer the wavelength must be to maintain the same speed.

An example: A radio station broadcasts electromagnetic radiation with a wavelength of 3 meters per cycle. The waves oscillate at a frequency of 100 million cycles per second, and thus they are able to travel at the prescribed speed of 300 million meters per second (*c*):

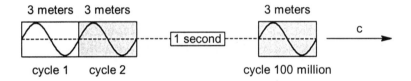

Break it down: (1) The waveform rises, falls, and comes back to its starting point over a distance of 3 meters. (2) Every second, another 100 million of these 3-meter cycles are generated and sent on their way. (3) The advancing wave covers 3 × 100 million meters per second.

Another example: A lamp emits blue-green light (official name: visible electromagnetic radiation) with wavelength equal to 500 billionths of a meter. The waves oscillate at a frequency of 600 trillion cycles per second, and they travel at the corresponding speed of 300 million meters per second:

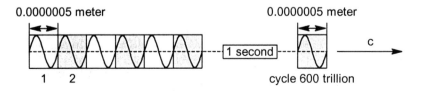

Visible light, we see, evidently vibrates millions of times more rapidly than radio waves (it compresses far more cycles into the same space and time), but both disturbances advance at the same speed. They should, too, since they both arise in the same general way: from charged particles undergoing varying degrees of accelerated motion. They are members of the same family, a family known as the "electromagnetic spectrum," and each member has its characteristic range of wavelength and frequency.

At one end of the spectrum, with long wavelengths and low frequencies, we find radio waves. At the other end, with short wavelengths and high frequencies, we find gamma rays. In between, from low frequency to high, we run continuously through the gradations of electromagnetic radiation: microwave, infrared, visible, ultraviolet, X ray. Together, oscillating over an infinite range of frequencies and wavelengths, the various waves represent the limitless possibilities of the electromagnetic spectrum. They are all different, yet they are all the same.

A few more numbers now and we are done. The *polarization* of a wave specifies the direction in which the disturbance bobs to and fro. For an electromagnetic wave, the electric and magnetic fields oscillate at right angles to the direction of travel. Looking into an oncoming beam, we see an electric field in a particular direction (up, down, up, down) and a magnetic field (left, right, left, right) advancing toward us at a speed of 300 million meters per second:

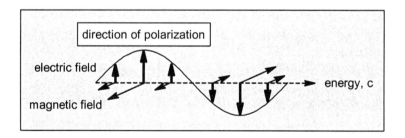

The combined electromagnetic energy, transported perpendicular to the oscillating fields, hits us straight on. There is an electric contribution to the classical energy, and there is a magnetic contribution. Each component is proportional to the square of the respective field, and the total energy is the sum of the two.

The *phase*, finally, specifies where in its cycle a wave of a certain frequency first begins to vibrate: a phase of 90 degrees (out of 360) for a quarter-cycle, 180 degrees for a half-cycle, 270 degrees for three-quarters of a cycle, 360 degrees for a full cycle:

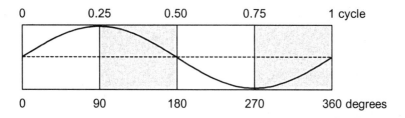

Two waves coming from the same source are likely to have the same amplitude, the same wavelength, the same frequency, the same speed, but not necessarily the same phase. It is no small distinction. Phase is what makes a wave a wave, and a variance in phase makes a difference. A wave, spread out all over space and possessed of a phase, can do things that a classical particle cannot.

... What a Wave Does

If two sister waves have the same phase—if they differ by zero degrees (zero cycles) or 360 degrees (one cycle) or 720 degrees (two cycles) or indeed any number of full cycles—then they are alike in every way, virtually identical. They run through their paces in exactly the same sequence, and we describe them as being "in phase." If they chance to meet, they reinforce each other point by point. Undergoing "constructive interference," they produce a new waveform with twice the amplitude:

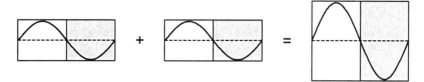

It can happen.

But if the same two waves have different phases—if they got started at different times, at different points in their cycles—then they behave thereafter like self-destructive enemies. A phase difference of, say, half a cycle (any odd multiple of 180 degrees) produces a pair of opposites. When one is up, the other is down; and when one is down, the other is up. Any meeting of the two proves fatal. They undergo "destructive interference" and cancel out:

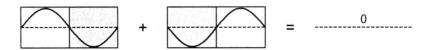

Waves can do that.

Particles cannot. Particles bounce around and exchange energy, but classical particles do not undergo interference. They do not share the same space at the same time, whether constructively or destructively. Forced into overly close quarters, particles suffer repulsion and move apart. Waves, by contrast, occupy a common space as joint tenants. Waves augment and diminish each other as they interfere, but they do not exclude one another.

Sometimes, like ships passing in the night, waves cross paths in just one location and move on. They meet. They interfere. They keep going:

Other times, though, two waves find themselves in joint possession of a more extended region of space: a length of string, perhaps, or a drumhead, or the cavity inside a microwave oven, or the gap between two mirrors, or, in general, any place of *confinement*, any place with no way in and no way out. And there, trapped inside, traveling waves can coexist only when they stand fast—when identical waves going in opposite directions mutually interfere to create "standing" modes of vibration.

It happens whenever a violin string vibrates, whenever an organ pipe resonates, whenever a ray of light bounces between parallel mirrors.

Pinned by two walls, a wave traveling to the right interferes with its reflection traveling to the left:

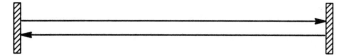

Back and forth go the two disturbances, adding here and subtracting there, until finally they are forced to accept a limited set of possibilities: a discrete series of standing waves, the only configurations able to withstand confinement. Not *all* frequencies and wavelengths, but only a select few can be supported between the walls. Because unless the displacement falls to zero at each barrier, the running waves run into each other in all the wrong places and eventually they disappear. But by fitting themselves precisely to the space—by squeezing in one, two, three half-cycles or more—they are able to stand in place. The combined disturbance vibrates up and down, with no net movement to either side. Like this, for one half-cycle:

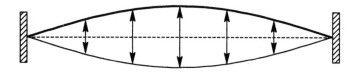

Like this, for two half-cycles:

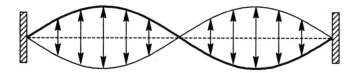

Like this, for three half-cycles:

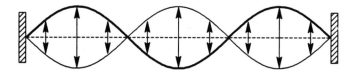

And thus the count continues, one by one, ever upward: four half-cycles, five half-cycles, six, and higher. With each new mode of vibration comes one more "node" (where the displacement drops to zero) and one more maximum. The nodes arise from destructive interference, and the maxima arise from constructive interference:

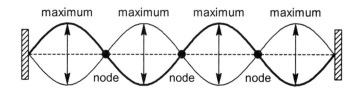

It is one of nature's most general rules for the propagation of waves: *confinement leads to quantization*. A continuous set of options (be my guest—vibrate at any frequency you like, no limits) is winnowed to just a discrete few (confine yourself to a whole number of half-cycles, nothing in between). Quantization inevitably results when two waves extend over the same restricted space, interfering constructively and destructively in different places. Waves can do so. Particles cannot.

A classical particle, remember, can be in only one place at one time. A wave can be in many places at once. An ideal particle is a localized pinpoint of mass, energy, and charge, whereas a pure wave (a single wave with a single frequency) is a diffuse and delocalized disturbance. It exists everywhere at the same time:

A single wave, its frequency fixed, cannot be pinned down in any one location. Its whereabouts remain completely undetermined.

A *mixture* of pure waves, however, takes on a localized aspect that becomes increasingly particle-like as more and more different frequencies join together. Occupying the same territory, interfering constructively in some places and destructively in others, the independent components combine to produce a narrower spread in space. Mix together enough pure sine and cosine waves in varying blends,

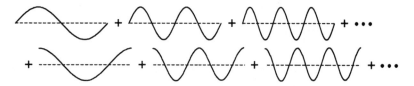

and localized concentrations of amplitude begin to appear. We may see

something like a "ringing" pattern,

or a standing wave, or a square pulse, or a spike,

or something else entirely. Anything is possible. It all depends on how many pure waves contribute to the "superposition" (the mixture) and in what proportions. The greater the spread in frequencies, the narrower is the spread in space.

Take your pick. Either: I want to know exactly *where* my wave can be found, without having to settle for a wishy-washy "everywhere" that really tells me nothing about its location. Put together so much of this frequency and so much of that frequency and so much of however many other frequencies it takes to make the combination large only in some specified region. Or: I want to know exactly *what* kind of wave I have, without having to settle for a wishy-washy "every kind under the Sun." Give me a disturbance that is simple and pure, a single component that vibrates at a single frequency. Take your pick, one or the other, because you cannot have both in equal measure. To narrow down the position is to spread out the range of frequency and wavelength. Waves are like that. Certainty about one thing means uncertainty about another.

Superposition and uncertainty, phase and interference, confinement and quantization—these are the properties that distinguish waves from particles in the classical world, the joint empire of Maxwell and Newton. There are waves and there are particles, and every subject of the empire knows where it is going and where it has been. It is a world of comforting regularity, where radio signals move from transmitter to receiver with no less determination than the Moon in its orbit. Maxwell's equations of electromagnetism and Newton's laws of motion set the course for waves and particles alike in the Land of the Large.

Basking in the light of electromagnetism and mechanics, we are tempted now to take one more bite of the classical apple: to go from macroscopic to microscopic, from Earth–Moon to proton–electron. If Newton's laws enable us to plot the course of the Moon, we ask, then why should they not do the same for the electron in hydrogen? Why not combine Newton's second law (which relates force, mass, and acceleration) with Maxwell's equations (which determine the electromagnetic forces arising from electric charges and currents)?

"Yes," says Newton, "why not? A hydrogen atom looks just like Earth and Moon, or Earth and Sun, or, in fact, any particle attracted to any other particle by a central force. The attractive force accelerates particle 2 toward particle 1 (that's my second law, incidentally), while inertia (my first law) keeps particle 2 moving away at right angles. With proper balance between the competing pulls, a stable orbit can be maintained:

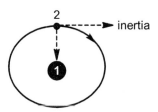

"Just substitute an inverse-square electromagnetic force for an inverse-square gravitational force, and the problem is otherwise the same as it was at the end of Chapter 4."

"No," says Maxwell. "Sorry, but an electron cannot persist in orbit around a proton. It's impossible. Moving in a curved path, this charged particle would undergo acceleration and therefore radiate energy:

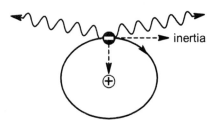

"Steadily losing energy, the orbiting electron would slow down and eventually fall into the proton (something like the way you fall off your bicycle when you stop pedaling). No, Isaac, the two of us can explain many things, but not the existence of atoms."

No. It will not be enough merely to tinker with the classical laws of mechanics and electromagnetism. We need different laws for a different world, the quantum world: the Land of the Small, the Land of the Few, the Land of the Discontinuous. We need to explain what Newton and Maxwell cannot.

Turning the page, descending into a new world, we shall learn to expect the unexpected. We shall discover, first of all, that waves and particles no longer split into separate camps. We shall discover instead that quantum particles, like waves, have phases; that quantum particles, like waves, undergo interference; that quantum particles, like waves, submit to localization and quantization; that quantum particles, like waves, are subject to superposition and uncertainty.

It is time to think small.

7

NEVER CERTAIN

Think small, because there is a big difference between large and small, a difference more telling in quality than in quantity. It is the difference between the observer and the observed, the difference between the act of measurement and the measurement itself. It is the difference between the comprehensive, certain knowledge of classical mechanics ("yes, definitely, I know for sure") and the limited, probabilistic knowledge of quantum mechanics ("most likely it's this, but it could also be that").

An observer of the classical macroworld has no doubts. The classical observer, dealing always with objects of sufficient mass, confidently expects to measure a body's position without affecting its momentum. Can an astronomer, after all, change the course of the Moon simply by looking passively through a telescope? Can a batter straighten out a curveball by mere observation? Can a driver alter the position of a car just by glancing at the speedometer?

For observers of the quantum microworld, however, there are no truly passive measurements. A quantum observer can never be an uninvolved bystander, a neutral spectator able to look but not touch. Any attempt to fix the location of an electron (even merely to shine upon it the weakest light imaginable) will irrevocably alter its instantaneous speed and direction. Whatever the lightweight particle was doing before, it is doing no longer. To probe a system is necessarily to disturb it.

The disturbance occurs randomly and uncontrollably, a result not of imperfect technology but rather of a nonnegotiable "uncertainty principle" that lays down the law of the microscopic world. Uncertainty (or indeterminacy) means that specific pairs of physical quantities cannot be determined simultaneously to unlimited accuracy. *Any* measurement of position, even the slightest touch, renders uncertain the linear momentum and velocity in the same direction. *Any* measurement of angular momentum renders uncertain the angle of rotation about the corresponding axis. *Any* measurement of energy renders uncertain the lifetime of the state under observation. These special pairings, remember—linear momentum and translation, angular momentum and rotation, energy and time—derive from the symmetry of physical law (as we noted in Chapter 4), and they remain harnessed together in both classical and quantum mechanics. The more accurately we determine one member of the pair, the less accurately we can determine its associate at the same instant. There are limits to measurement. There are limits to knowledge.

The boundaries are always present, but in the macroscopic world of Newtonian particles and Maxwellian fields they escape notice. Who would argue, for example, with the classical assertion that the mass of an object can be changed by imperceptibly tiny increments? Just imagine using an eyedropper to add water to the Atlantic Ocean. We take turns, you and I, each trying to outdo the other in the delicacy of our drops, and after a while we realize that the game has no real winner. However small a droplet you add, I always manage to produce another one just a little bit smaller. You, in turn, follow with one smaller than that; and I again, and you again, and I again, until at last, after uncountably many subdivisions, we reach the limit of fineness: a single molecule of H_2O, an infinitesimal drop in the proverbial bucket. With a mass of 30 trillionths of a trillionth of a gram, it counts for practically nothing at all.

Look at how tiny the number is (0.000000000000000000000003 g), and realize that already we find a trillion trillion H_2O molecules (1 followed by 24 zeros) in a single ounce of water, let alone the Atlantic

Ocean. And since there are a million trillion H_2O molecules even in a millionth of an ounce, and a thousand trillion H_2O molecules in a billionth of an ounce, and a trillion H_2O molecules in a trillionth of an ounce—a very small bucket, by any measure—we sensibly conclude that mass varies continuously in the classical world. Mass can always be made a little bit smaller, a little bit closer to zero. Nature seems to dish out mass in infinitesimally small amounts.

Charge, too. Energy, too. Momentum, too. In the macroscopic world, wherever Newton's laws and Maxwell's equations are enforced, nature's face appears smooth and continuous, with no perceptible grain. Only by interrogating the small world, the microscopic world, do we discover that there are indeed limits to subdivision. Narrowing our focus, shrinking the bucket ever more, we discover eventually that the grain of nature is not infinitesimally fine. We discover that neither matter nor energy nor momentum nor anything else in nature can be sliced so thin as to fade away on a microscopic scale. Every transaction is processed in small but nonzero amounts (called "quanta," the low-value coin of the realm), and the inevitable little bumps and bruises stand out when the affected system is small itself. For the Moon, the appearance of certainty proves to be only a justifiable illusion created by the minuscule disturbance of a huge body. For an electron, the more fundamental laws of quantum mechanics come to light. Uncertainty cannot be ignored.

The classical laws, Newton's for particles and Maxwell's for electromagnetic fields, rest on a foundation of certainty and continuity: the take-it-for-granted certainty that (if only we are clever enough, gentle enough, careful enough) we can measure anything at all to arbitrary accuracy. It is a question only of engineering, a technical challenge that is ours to solve. Nature, macroscopically smooth and continuous, places no obstacles in our classical path.

The quantum laws, which give the lie to Newton and Maxwell, are built instead on a foundation of uncertainty and grainy discontinuity: the understanding that accuracy in measurement is constrained ultimately not by technological difficulties, but by the nature of things themselves. Try as we might, we cannot make change for a quantum. We cannot make the grain of nature finer than it is.

Uncertainty and discontinuity become the basis of a jittery microworld of molecules, atoms, electrons, nuclei, and things smaller still. The very notion of a mechanical path, the sum of all classical knowledge, becomes meaningless under the indeterminate regime of quantum mechanics. How, for instance, can we speak of a well-defined, exact sequence of position and velocity (as we did in Chapter 4) when the two quantities are not knowable at the same time? We cannot. Our only recourse is to abandon the classical equation of motion, which asks in vain for information that nature refuses to give: the initial values of incompatible quantities such as position and velocity, measured simultaneously to unlimited accuracy. In a quantum universe, we cannot do it. We need new laws for a new world.

Expect a world nothing at all like the orderly classical universe, where particles are particles and waves are waves (and everyone knows the difference, just by looking). No. In the chancy world of quantum mechanics, things are not always what they seem. Particles act like waves; waves act like particles; and the old distinctions—even the distinction between light and matter—begin to blur into a new kind of reality.

At the Barricades: Classical Light and Matter

In a macroscopic universe, matter appears to be one thing and light another; and nobody, not even the most unschooled observer, is likely to confuse the two. The differences are plain to see. Lumps of matter, everything from dust motes to billiard balls to the moons of Jupiter, represent the literal embodiment of substance. To be a classical particle is to be something tangible and massy, something that can be grasped with the fingers. X rays and moonbeams, by contrast, slip through the fingers like shades and spectres, ever elusive. They are electromagnetic waves, ethereal and intangible, and they behave in ways that particles do not.

A particle delivers its energy in a localized package, all in one place, all at one time. A wave carries its energy across a broad front. Approaching a fork in the road (two holes in a barrier, let's say), a classical particle takes one path or the other, not both at once:

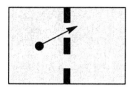 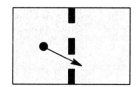

A wave passes through the two openings at the same time:

The particle, undivided, emerges intact on the other side. The wave, its front interrupted, regenerates anew at each hole.

Look closely now, because what happens at the barrier—at the two holes, and then beyond—makes a particle a particle and a wave a wave. The openings may arise from breakwaters in a harbor, or holes in a wall, or slits in a screen, or regularly spaced atoms in a crystal, or any number of possible arrangements. Whatever form they take, though, these apertures will be our windows on the world of light and matter. We shall shoot at them with particles and wash over them with waves. We shall see what goes in and what comes out, and with such observations we shall begin to make out the difference: the difference between a classical macroworld (in which some things are particles and some things are waves) and a quantum microworld (in which everything is both particle and wave together).

First, the classical.

Particles: Shooting the Gap

Start by blocking off one of the two holes, leaving just a single aperture open to a spray of macroscopic particles—shotgun pellets, maybe, or marbles, or anything else of that sort. It doesn't matter exactly what they are. Imagine only that we shoot these generic particles randomly through a range of angles, and afterward we register the arrival of each shot on the far side:

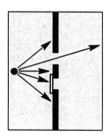

Ping! A particle hits the wall. *Ping!* Another. The gun fires in random directions, but Newton's laws determine exactly what happens every step of the way. One by one, many of the particles pass straight through the hole and follow a direct line to the wall beyond. Others, however, arrive somewhat off center, owing either to their original trajectory or to a deflection. Small deviations prove more likely than large deviations, and eventually the pattern of hits becomes clear. Greatest in the center, it falls off symmetrically to either side:

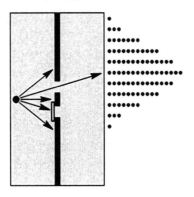

And if now we open both holes and fire away, then the same kind of pattern takes shape across from each opening. Particles passing through the first hole produce one profile, and particles passing through the second hole produce a mirror image of the first:

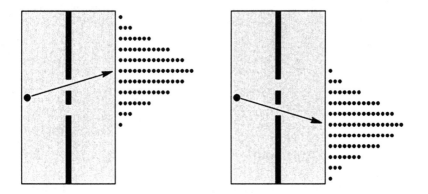

Each particle thus goes its own way, emerging independently from one hole or the other. Since the bypassed hole has no effect whatsoever on the subsequent trajectory, it is as if the particle faces just a single opening, not a pair. The distribution overall becomes simply the sum of two unrelated patterns, with a single peak growing up in between:

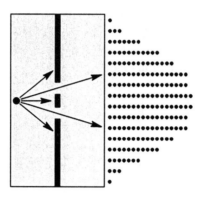

With that, we have the signature of a classical particle. Unlike a wave, a classical particle does not undergo interference. It is never a house divided. It can be in only one place at one time, and it cannot split its energy between two distant sites.

Waves: Slipping Through

They approach like marchers in a parade, row after row, with everyone cycling through the same sequence of synchronized steps. They are waves,

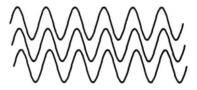

and we already know (from Chapter 6) some of the things that waves can do. We know that a pure wave has a frequency, a wavelength, an amplitude, a phase. We know that the energy of a wave varies with the square of its amplitude, which means that a wave with twice the amplitude carries four times the amount of energy:

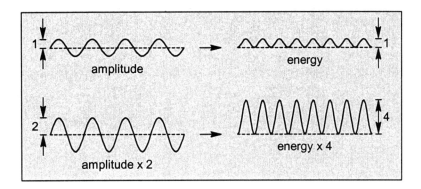

We know as well, within the realm of classical law, that the variation with amplitude is continuous: that any change in amplitude and energy can always be made a little bit smaller, a little bit closer to zero.

We know that pure waves (water waves, sound waves, electromagnetic waves, waves of all kinds) are diffuse rather than compact. They spread out. They work their way around obstacles. Arriving at a barrier broken by openings of a certain size and spacing, a wave bends around the edges. The approaching wavefront reemerges as a pair of daughter waves radiating outward in all directions. It is a process called "diffraction,"

and the circular ripples start out with the same phase at each hole. Like a row of marchers crossing a line with their legs all in the same position, they share a common point of departure. It might be a crest or a trough or a point anywhere else in the cycle, but it will be the *same* point—and the memory of this common phase will determine what happens to the diffracted waves wherever they cross paths:

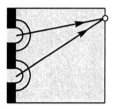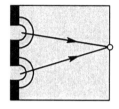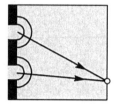

The possibilities are endless. Separated initially in space, the daughter waves travel different distances before arriving at an eventual rendezvous. And as siblings often do, they find themselves sometimes in harmony and sometimes in conflict. If each of the wavefronts advances a whole number of cycles (arbitrary example: 1000 and 1001),

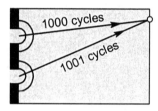

then they meet up *in phase*: crest to crest, trough to trough. They interfere constructively at that point, creating a new disturbance with twice the amplitude and four times the energy:

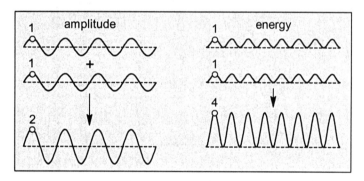

It is as if 1 + 1 = 4. Traveling alone, free from interference, each wave would deliver one unit of energy at its crest or trough. Interfering constructively, however, the waves combine to produce four units of energy. They reinforce each other, and the whole becomes greater than the sum of its parts.

But we have only part of the story so far, because somewhere else along the line there is sure to be an encounter where $1 + 1 = 0$—where one of the wavefronts travels a *half-cycle* more than the other (or three half-cycles, or five, or any odd number of them):

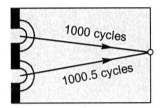

Meeting crest to trough, 180 degrees out of phase, here the waves interfere destructively and suffer complete annihilation. The combined amplitude is zero. The combined energy is zero:

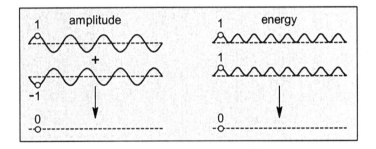

The whole becomes less than the sum of its parts, and in this way the interfering waves comply fully with the conservation laws. Whatever energy is gained from a crest-to-crest meeting in one place (where $1 + 1 = 4$) is lost in a crest-to-trough meeting in another place (where $1 + 1 = 0$). The average energy remains two units, no more and no less than what the waves carry separately. The total energy, although shuffled from place to place, remains constant. It always does.

Up and down the line, then, the diffracted waves interfere constructively and destructively. In some spots, the combined energy rises to a peak. In other spots it falls to zero. At points in between, where the destructive interference is not total, the energy runs through a continuous range of intermediate values. The full pattern, coming together from all possible encounters, develops into an array of alternating maxima and minima:

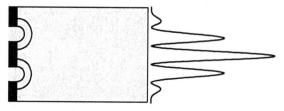

It is the diffraction signature of a wave, a signature manifestly different from that of a classical particle. Count the ways: A wave goes through both openings at once. A particle goes through only one. A wave has a phase. A particle does not. A wave undergoes interference. A particle does not. A wave, unlike a classical particle, makes more of itself in some places and less of itself in others. The energy profile it produces (as in the diagram above) is not simply the sum of the two one-hole patterns shown below:

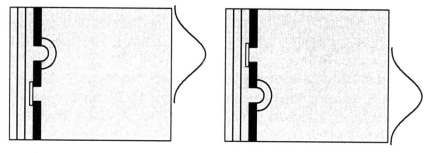

A diffuse wave negotiates the openings in a way that a grainy particle does not.

Or does it? Matter, we know, appears smooth and continuous on a macroscopic level, but is actually coarse and grainy on a microscopic level. Might not light—an electromagnetic wave—prove to be grainy as well, if only we look closely enough?

Waves of Grain

It is not merely a metaphor. To make a photograph is literally to "paint with light": to capture, on the blank canvas of a film negative (or some other suitable surface) an image of the visible electromagnetic radiation emanating from an object. Any scene will do, but we take particular interest now in one of light diffracting through two openings in a screen, as discussed just above. For here is surely an image, painted with an electromagnetic brush, that testifies persuasively to the wavelike character of light. The effects of interference, preserved on film as light and shadow, argue that electromagnetic radiation propagates macroscopically as a wave, smoothly and continuously with no discernible grain:

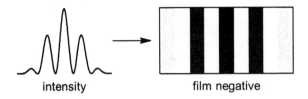

intensity   film negative

Still, it is a picture painted on a grainy canvas. Whether formed by the chemical granules of conventional film or by the electronic array of a digital camera, the light-sensitive surface gives only the illusion of continuity. Think of a photographic medium as a matrix of individual dots waiting to be filled in,

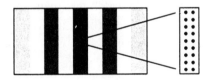

and think of the image produced as akin to a pointillistic painting:

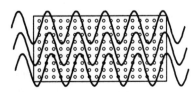

Smooth from a distance, the picture looks coarse and grainy under magnification.

Now ask: Is the image painted with a broad brush, one that sweeps over the whole canvas with a single stroke, touching all the dots at once? If so, the light would spread out and bathe the receptive surface everywhere at the same time, like a wave:

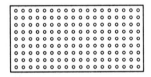

Or are the dots painted one by one, perhaps randomly, as if with a fine brush? If so, each dab of light would impart a particle-like *ping* to just a single point:

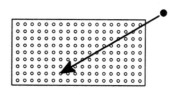

Which is it—particle or wave, coarse or smooth, random or systematic? Is nature's electromagnetic field continuous at all levels of observation, or does the field turn out ultimately to be as grainy as the charged particles that create it?

There is only one way to find out. We need to look and learn. We need to do experiments.

Double Exposure

Any photographer will tell you: light is a wave, and it paints a picture with a broad brush. The proof? Expose a piece of film to steadily increasing amounts of light, and watch the image take shape. It develops uniformly over the entire surface, filling in gradually as the exposure time becomes progressively longer:

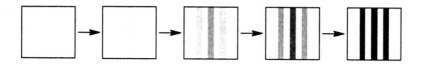

If seeing is believing, then at first sight we have reason to believe that light differs fundamentally from matter. Matter is grainy, but electromagnetic waves appear to be smooth.

Just to be sure, though, we make another series of exposures, this time with exceedingly feeble light. We do so, and (*surprise!*) the intermediate snapshots paint a different picture entirely: a picture not of waves washing uniformly over the surface but of bullets spraying randomly in all directions. The grains of the film light up individually, here and there, until at long last—after a sufficiently long time and many, many individual hits—the image takes its final form:

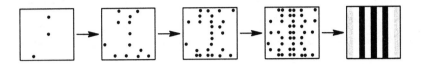

Appearances deceive, we realize, and we learn not to trust our eyes. At low resolution, viewed from a distance, the electromagnetic field appears to be a wave. At high resolution, viewed up close, the field shows itself to be composed of discrete lumps, like molecules in the ocean or grains of sand on the beach.

They are "photons," the fundamental particles of the electromagnetic field. They carry energy. They carry linear momentum. They carry angular momentum. Look at them one by one, and they deliver their goods as a localized package, like a particle.

But an individual photon also has a frequency, like a wave. An individual photon has a wavelength. An individual photon has a phase. Put together enough individual photons, and they behave like a macroscopic electromagnetic wave. It all depends on one's point of view.

The Photon: Energy and Momentum

The energy of a classical wave depends on the square of its amplitude. To make a wave more energetic, all we have to do is increase the amplitude. This wave, for example,

carries more energy than this one:

The energy and momentum of a single photon, however, vary not with amplitude but rather with frequency. A photon with a high frequency (and correspondingly short wavelength) delivers greater energy and momentum than a photon with a low frequency:

low energy ... low momentum   high energy ... high momentum

low frequency ... long wavelength   high frequency ... short wavelength

The amount of momentum or energy delivered, moreover, comes in a measured dose: a fixed quantity, always the same, one quantum per photon.

And how much is that? To gauge the size of a quantum, we have recourse to one of nature's fundamental constants, an invariant quantity called "Planck's constant" (designated as h). An imposingly small number, Planck's constant sets a numerical standard for the quantum universe. It has units of (energy)×(time)—or, equivalently, (momentum)×(length)—and it establishes the size of each lump of electromagnetic energy, linear momentum, and angular momentum thrown about the microworld:

A quantum of energy: h multiplied by frequency

A quantum of linear momentum: h divided by wavelength

A quantum of angular momentum: h divided by 2π

Lumps. Localized bundles of electromagnetic influence. A certain shade of violet light, for example, reaches our eyes as tiny packets of energy, each photon carrying 500 sextillionths (0.0000000000000000005) of a joule. Not 50 sextillionths of a joule. Not 5 sextillionths of a joule. Not 0.5 sextillionth of a joule. No, if a field of that particular color is to grow or shrink by just a little bit more energy, then that little bit must be 500 sextillionths of a joule (one photon), or 1000 sextillionths (two photons), or 1500 sextillionths (three photons), or any other multiple of a whole quantum. "Exact change only," says nature, "we accept nothing but integral photons."

For red light it will be a quantum of a different size (and different still for infrared light and for ultraviolet light and for X rays and for gamma rays), but it will be a fixed quantum just the same. The universe doles out electromagnetic energy one indivisible lump—one photon—at a time, and the size of that lump depends only on the frequency.

The more lumps, the more energy. To add photons is to increase the total energy of an electromagnetic field, to build up the amplitude of an electromagnetic wave without changing its frequency. In systems interacting with only a handful of photons, each one stands out as an individual granule of momentum and energy. But if we address a sufficiently large crowd (10 quintillion violet photons, say, to give a macroscopic energy of 5 joules), then we fail to notice any gaps in the distribution of energy. There *are* gaps, strictly speaking, but one photon more or less out of 10 quintillion is akin to one molecule more or less in the Atlantic Ocean. The difference is imperceptible even to the most discriminating observer, and the microscopically grainy electromagnetic field blurs into a macroscopically continuous wave.

A Matter of Chance

From grain to wave, then, a macroscopic electromagnetic field builds upon the contributions of individual photons, and with this new appreciation we return once more to our telltale diffraction experiment. Consider again what does *not* happen. Just as a child does not behave like a miniature adult, neither does a single photon produce a miniature (but fully formed) diffraction pattern in one go. The characteristic array of light and dark bands does *not* blossom all at once, springing up weakly with the first photon and growing progressively stronger thereafter:

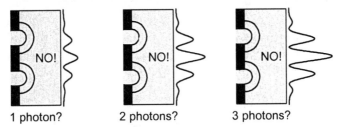

1 photon?   2 photons?   3 photons?

Instead, the two-hole interference pattern takes shape ping by ping, photon by photon, and the maxima and minima become apparent only as more and more photons make their mark:

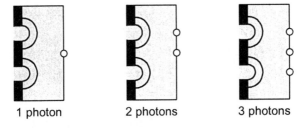

1 photon   2 photons   3 photons

The process unfolds randomly, too, with the outcome of any given event utterly unpredictable. *Ping!* A photon hits the film. Where? We cannot say for sure until it actually happens. We can only lay down the odds—that 60% of the time the impact will be in this place, and 30% of the time it will be in that place, and 9% of the time over here, and 1% of the time over there—and in the long run, after many hits, our predictions will be astonishingly accurate, a triumph of statistics. To determine the fate of a single photon beforehand, though, is not for us to say. Like a gambler rolling the dice, an observer of the quantum world knows only the long-term probability of occurrence. The when-and-where of a specific event is never certain.

It is the way business is done in a quantum universe. Not just photons, but electrons and protons and neutrons and all the other particles of the microworld fall victim both to wavelike interference and to the capricious rule of probability. For no sooner do we discover that light, an electromagnetic wave, is more tangible than we think, we also discover that matter is less tangible than we think. Electrons and their particulate cousins do not behave deterministically like little bullets, always going where they are told and never interfering. A beam of electrons, for instance, diffracts through a microscopic lattice of atoms in a crystal (the practical realization of our two-hole barrier) and produces an interference pattern analogous to that of a beam of photons. "Impossible!" says Isaac Newton, "Massy particles cannot do that. A particle, large or small, must go through either one hole or the other. There is no mechanism for solid particles to create an interference pattern."

Nevertheless, they do. One by one, the electrons arrive on the far side as undivided, localized, particulate *lumps*; and one by one, the lumps pile up randomly in a way that can only be ascribed to interference—to interference generated by a single particle emerging (somehow!) from two holes at once, as if it were a wave:

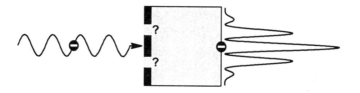

Were an incoming electron to go through one opening or the other (not both of them together), then we would observe the sum of two independent one-hole patterns:

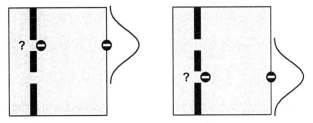

But we do not. Electron by electron, the two-hole pattern develops haphazardly but unmistakably. The particle, endowed with an apparent wavelength and phase, interacts with both openings simultaneously and undergoes interference...with itself.

To a classical observer, nothing could be stranger: particles that behave like waves, waves that behave like particles, particles that follow no set course. A complete breakdown of deterministic law and order.

To a quantum observer, it is simply a fact of life in a world where small effects loom large and the outcome is always in doubt. A world with laws of a different kind.

Which Way Did It Go?

Now you see it, now you don't. Suppose that we set out deliberately to track the electron as it interacts with the barrier, unsatisfied as we are with the notion of a particle going through both holes at once. Let's take a look. Our plan is to shed a little light on the matter (just enough light to see what happens, but with minimal disturbance), and the tool of choice will be a single photon. A one-photon bulb is the dimmest illumination possible in a grainy microworld.

Our lamp has an effective range of approximately one wavelength (called λ, Greek "lambda," for brevity),

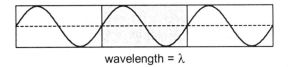

wavelength = λ

and so we shall need to make λ small enough to establish whether the electron comes out of the first hole or the second hole. To hit an electron with a photon, though, is to jolt it with a dose of momentum, an amount equal to Planck's constant divided by the wavelength:

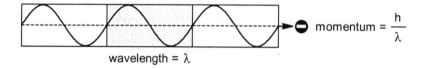

wavelength = λ

momentum = $\dfrac{h}{\lambda}$

And now, like one billiard ball being struck by another, the electron suffers an unavoidable change in momentum. The very act of illuminating the particle's position to within a distance λ causes the momentum to be indeterminate by h/λ.

We have a choice to make. On the one hand, a short wavelength enables us to pin down the electron's location with great precision. The ensuing change in momentum, however, degrades our ability to determine the original velocity. We learn *where* the particle is, but not what it is doing. On the other hand, a long wavelength spares the momentum but leaves us in the dark concerning the whereabouts of the electron. Somewhere in the middle, we find a compromise: a tradeoff where neither the uncertainty in position (λ) nor the uncertainty in momentum (h/λ) is infinitely large or infinitesimally small.

It is called Heisenberg's uncertainty principle, and it merits the status of a fundamental law. The more precisely one measures position, the greater is the uncertainty in momentum. The more precisely one measures momentum, the greater is the uncertainty in position. Multiplied together, the uncertainties in position and momentum yield (in the best of circumstances, where any disturbance is minimal) a number roughly the size of Planck's constant, as suggested by our heuristic example involving λ and h/λ.

We can do no better than that. The minimum product of the two uncertainties, h, is small but not zero, and no amount of technological wizardry can ever make it less.

Constrained by the uncertainty principle, we now carry out our plan to see explicitly what happens to each electron at the two-hole barrier. We turn on the lamp and watch, and indeed—waves or no waves—the particles do seem to hold together as indivisible lumps. Viewed in the light of our dim bulb, each electron comes through one hole or the other before registering its arrival on the film. An electron here and an electron there, one by one, and eventually a fully developed pattern becomes clear: a pattern different, though, from what was recorded earlier with the lights off. This time, having forced each illuminated electron to stand up and be counted ("I am going through hole 1 or hole 2"), we no longer observe the interference characteristic of both holes at once. What appears instead is the independent sum of two one-hole patterns, consistent with electrons that behave like bullets rather than waves:

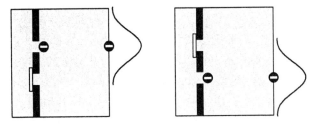

All the electrons originating definitely from hole 1 (we know, because we see them do so) act as if hole 2 is blocked. Likewise, all those originating from hole 2 act as if hole 1 is blocked. The disturbance suffered by a particle illuminated at one opening is enough to destroy its phase relationship with the other opening.

Look but don't touch? Impossible. In the Land of the Small, where even Planck's constant seems large, an observer cannot have it both ways. To probe a diffracting electron or photon too closely is to turn a wave into a particle. Knowledge of position means ignorance of momentum. One choice excludes another.

Out with the Old

Says the observer to the particle: *Where are you now, and where are you going next?* In other words, tell me your position and velocity along each of three axes:

Position			Velocity		
x	y	z	V_x	V_y	V_z
?	?	?	?	?	?

If the particle responds consistently with the same six numbers, then it is a classical particle. Our idealized observer, using the most delicate instruments imaginable, will then be able to make measurement after measurement without encountering any discrepancy. No matter in what order the measurements are carried out—whether position is registered before velocity or velocity is registered before position—the six numbers never change.

Why? Because the object is a classical particle, and a classical particle is a lump and nothing more. A localized lump of matter. A lump with no capacity to undergo interference. A lump with enough mass to exchange energy and momentum in what seem to be infinitesimally small doses.

A classical particle, all but exempt from the Heisenberg uncertainty principle, shrugs off whatever slight disturbance may accompany a measurement. The mechanical state of a classical particle is unambiguous. Its six initial values are certain and knowable, and once we know them, we know all there is to know. Position. Velocity. Momentum. Energy. Anything you like. Newton's equation of motion, powered by the known forces acting on a system, lays out an unwavering course. The classical observer of Chapter 4 is granted a complete mechanical history, an all-encompassing knowledge of future and past that comes from an exact knowledge of the present.

In a classical universe, a particular cause produces a particular effect, always the same, never a surprise. Everything is knowable, at least in principle, and there is no room for chance. To be "lucky" in a classical universe means simply to be well informed. Anyone who takes the trouble, for example, to measure the initial positions, velocities, and all the forces acting on the bouncing balls in the Sextillion Dollar Mega Bigga Lottery may well expect to win the jackpot. Never mind the enormous practical difficulty of making the required measurements. Never mind the hopelessly intricate interplay of forces. Never mind the horrendous complexity of the equation of motion. Nobody said that winning $1,000,000,000,000,000,000,000 was going to be easy, but in a classical universe it is certainly possible. Work hard enough, and it is even guaranteed.

But not in a quantum universe, not in a universe where the same cause will routinely produce a different effect. Not in a universe where observable events occur entirely by chance, the kind of chance that no amount of prior knowledge will suffice to tame. Not in a universe where a straightforward question (*Where are you now, and where are you going next?*) has no straightforward answer.

In a quantum universe, a particle cannot respond with only *one* position and *one* velocity, because a particle in a quantum universe is something more than just a lump. It is a lump that also exhibits the properties of a wave, and a wave is necessarily an ambiguous construction. A tension always exists between the spread in space (position) and the spread in wavelength or frequency (proportional to momentum).

Go back to our treatment of wave superposition at the end of Chapter 6 and see. The names change, but the mathematics does not. Quantum mechanical waves, electromagnetic waves, water waves—waves of any sort, even abstract curves with no explicit physical meaning—all follow the same blueprint. If the oscillation contains just one component with just one wavelength and frequency (a single, well-defined momentum), then the wave extends over all space. It is entirely delocalized. The position of the disturbance remains wholly indeterminate, infinitely uncertain:

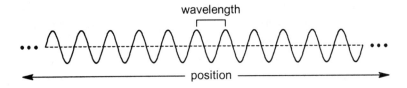

At the other extreme, if the oscillations are confined to a single point in space, then now it is the wavelength and frequency that become completely indeterminate. An infinite number of sine and cosine waves, each with its own momentum, contribute to produce an infinitesimally localized disturbance:

In between, we find precisely the same tradeoff between position and momentum demanded by the Heisenberg uncertainty principle: that the spread in position multiplied by the spread in momentum must equal or exceed some finite number. The more we know about one aspect of the wave, the less we know about the other. The uncertainty in position varies in inverse proportion to the uncertainty in momentum.

And so it comes to pass in a quantum universe that a particle has no simple answer to the observer's simple question. "My *position*?" says the particle-that-is-also-a-wave, "my position is different whenever you look at me. If you poke me right now, maybe I will turn up over here. But if you poke me again, in exactly the same way, maybe I will turn up over there:

Trial	Outcome
1	possible position 1
2	possible position 2
3	possible position 3
⋮	⋮

"You will have to be content with a statistical average of all my possible positions, because every measurement might yield a different random value."

For momentum, too. Contained within the particle's wavelike essence is a spread of possible wavelengths and frequencies (momenta), and any one of them is likely to register on an observer's measuring device. Since we have no way to predict which number will come up in a given trial, we rely again on statistics to establish an average value:

Trial	Outcome
1	possible momentum 1
2	possible momentum 2
3	possible momentum 3
⋮	⋮

That is how the game is played in the world of the quantum. We make the same measurement in the same way under the same conditions and yet, despite our ingrained notion that "things must happen for a reason," we find no rhyme or reason at all. Constrained only by the laws of probability, we get different numbers each time. We deal in averages, not fixed values.

What lies ahead now, in Chapter 8, is to make sense of the quantum world: to flesh out this embryonic notion of a "wave–particle duality"; to construct a suitable state and equation of motion for whatever system is at hand; to look for probabilistic averages rather than deterministic paths; to outline, at least, the legal framework under which atoms and electrons and nuclei and protons and neutrons manage to exist.

8

THE PATH NOT TAKEN

Say goodbye to certainty. Say goodbye to a clockwork macrocosm in which the Moon never wavers from its course, and say hello to a guesswork microcosm in which an electron follows no predetermined path at all. Say hello to a world in which the sharp values of position, velocity, momentum, energy, and all the other certitudes of classical mechanics blur into the statistical averages of quantum mechanics. Chalk it up to the indeterminacy intrinsic to light and matter.

Yet, despite the differences, think also of what Moon and electron have in common. Each, in its own way, acts as a moving part in a mechanical system, a machine of sorts. Over time, each machine passes through a sequence of recognizably different states; and each machine, progressing from state to state, operates in response to a controlling influence. For the Moon, it is the gravitational interaction. For the electron, it is the electromagnetic interaction.

The dimensions are different. The masses are different. The mechanical states are different. The interactions are different. But for all that, there remains a common thread. The changes undergone by both Moon and electron, although different in so many ways, still follow a set of rules consistent for each system in its own world. Classical or quantum, a mechanical system obeys an equation of motion. Classical or quantum, a mechanical system evolves from one state to the next in a fixed sequence. State 1 leads inevitably to state 2, as surely as one city follows another along the highway. Classical or quantum, a mechanical system goes where the road takes it, visiting each state in turn:

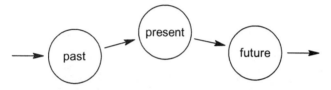

There is a difference, though, and no small one. A classical system, like a traveler who always follows the same routine in a given city, transacts the same mechanical business whenever it passes through a certain state. A quantum system, ever spontaneous, may do something different during each visit.

Tell us the name of a city, and we know to an absolute certainty how a classical traveler will spend a night there: the same hotel, the same restaurant, the same meals during every stay. Likewise, tell us the mechanical state of a classical system (the positions and velocities of its particles) and we know precisely what to expect: the same energy, the same momentum, the same mechanical properties attached to that one particular state.

For a quantum traveler, a visitor of a different sort, we learn to expect the unexpected. We know all the hotels and restaurants in town, as well our traveler's preferences, but we do not know which establishments will be chosen on any given day. Two visits out of ten, it will be hotel 1. Three visits out of ten, hotel 2. Five visits out of ten, hotel 3. And so it is, too, for a quantum mechanical system, where the same state, measured under the same conditions, may yield different mechanical values for each trial. Two measurements out of ten, energy 1. Three measurements out of ten, energy 2. Five measurements out of ten, energy 3.

Our program for dealing with a quantum mechanical universe will be twofold. First, within each city–state we need to establish all possible options and their associated probabilities: the statistical likelihood that hotel 1 will be selected, or energy 3, or restaurant 6, or momentum 2.

Second, we need a road map to determine the sequence of cities en route. The latter part of the plan requires construction of a quantum mechanical equation of motion, a state-to-state conveyor belt to take the place of Newton's second law. The former requires specification of a quantum mechanical state, and it will be in the fashioning of that state—above all, in the renunciation of sovereign position and momentum—that the predictable unpredictability of the microworld begins to emerge.

States and Measurement

The question is not so much what an electron *is*, but rather what an electron *does*; and what an electron does, at least in some of its actions, is behave like a wave. It diffracts. It interferes. It has a wavelength. It has a phase. Whatever a wave can do, an electron can do as well.

It is a particle with the properties of a wave. An electron trades purity of wavelength for purity of position, and it cannot be razor sharp in both. At one extreme, as we saw in Chapter 7, the particle purchases an exact wavelength at the expense of an infinitely uncertain position:

wavelength

[figure: continuous sinusoidal wave along a position axis]

position

At the other extreme, the choice is reversed. An infinitesimally sharp position carries with it an infinitely broad spread in wavelength (and consequently momentum). The roster of sines and cosines is endless:

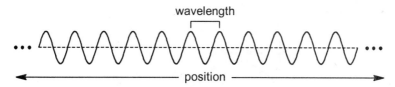

position = [wave] + [wave] + [wave] + ...

momentum 1   momentum 2   momentum 3

Whereas a classical particle manifests just *one* position and *one* momentum at any instant, an electron simultaneously allows for the possibility of many. The quantum mechanical state of an electron accommodates a multiplicity of options, packaged together as if the particle were a wave.

As if the particle were a wave. Among other things, recall, a wave can travel ahead or stand in one place, and in analogous fashion so can the wavelike state of an electron. Allowed to roam free, an electron has its pick of a continuous range of wavelength, each value differing infinitesimally from the next. The possibilities are unlimited:

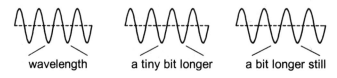

wavelength   a tiny bit longer   a bit longer still

But not always. True freedom—freedom from the influence of all other particles and fields—is ever hard to find, and an electron typically becomes enmeshed in a web of interactions, its movements constrained. Trapped between walls of potential energy (in an atom, say, or a molecule), the particle exists only in certain stable configurations. The associated waves are restricted, *quantized* in ways that call to mind the standing waves we met earlier in Chapter 6. Picture the disturbances schematically as bounded, confined to small regions, with maxima and minima and nodes in between:

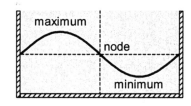

It is a fact of life in the microworld, no less than in the macroworld. Confinement brings quantization to an electron, just as confinement brings quantization to a piano string, or to a column of air in an organ pipe, or to a laser beam trapped between mirrors, or to any other kind of wave in the repertoire. And a wave, by any other name, will still obey the same generic equations common to all such disturbances.

If we want to understand the quantum mechanical electron, then we do well to invoke the properties of a classical wave. After all: "A wave *is* what a wave does," and the mathematics of an electron is the mathematics of a wave.

Classical Quantization

Take a piano string, for example. Pinned down at each end, it can vibrate only as one or more standing waves, with frequencies and wavelengths fixed by the length of the string:

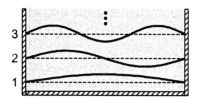

Nothing else is allowed, because the oscillations must fit into place. They must fill the allotted space exactly, with the disturbance forced down to zero at both endpoints.

Lowest in energy, at the bottom of the diagram, we have the so-called fundamental mode or "first harmonic": *one* half-cycle, spanning one lobe of a sine curve. Next, oscillating at twice the frequency, comes the second harmonic: *two* half-cycles, a full sine wave with its two lobes fitted into the same space. After that, the third harmonic: *three* half-cycles, with three lobes…and after that, four half-cycles, and five, and six, and the rest, theoretically up to infinity.

They are all possible, and they are all uniquely independent. The fundamental mode cannot be created from any combination of harmonics, nor can the second harmonic, nor can the third harmonic, nor can any of them be built up from the others—no more than a step to the north can alter a hiker's course by even a millimeter to the east or west. Each basic mode of oscillation, if excited, makes its singular contribution to the overall vibration. Standing in place, it moves only up and down:

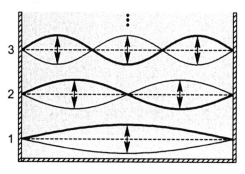

For a classical observer, the path is well marked. Every element of the string has a definite position. Every element of the string has a definite velocity. All one has to do is specify, first, the form of each mode (the shape and frequency of the standing wave) and then the extent to which that mode contributes to the total vibration (how much of the fundamental is present, how much of the second harmonic, how much of the third, and so forth):

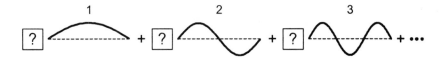

That done, our classical observer will have formulated the requisite mechanical state. Newton's equation of motion, given a description of the forces acting on the string, stands ready to do the rest.

Now an electron is no piano string, but an electron does obey the mathematics of a wave. Trap the particle within a well of potential energy, and the evidence of confinement and quantization is plain to see. There develops a hierarchy of permissible quantum waves, characterized by discrete energies and simple integers (like the *one* half-cycle of a string's fundamental frequency, or the *two* half-cycles of the second harmonic, or the *three* half-cycles of the third harmonic). Sometimes, too, under just the right conditions of confinement, the quantum waves of an electron become indistinguishable mathematically from the standing waves of a piano string and assorted other classical oscillators. The same sets of numbers and the same curves describe two wholly different phenomena.

What differs, of course, are the physical meanings we ascribe to the numbers—our understanding of what the waves represent—and for that, a quantum mechanical interpretation means all the difference in the world.

Quantum Waves and Probability

What shall we make of it, then, this wavelike disturbance associated with an electron? Is it a variation of force or potential, like the electric and magnetic fields of an electromagnetic wave? Or is it perhaps a material disturbance, plain and simple, like an ocean wave or a vibrating string? Might we be dealing literally with a wave of *matter*,

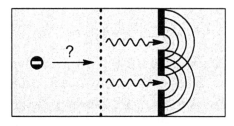

whereby we imagine a thinly spread paste of mass in which each little bit somehow displays both wavelength and phase?

No, banish the thought. The diffraction experiments of Chapter 7 prove that an electron or photon cannot possibly spread out like the continuous fabric of a water wave. Particles arrive at the detector one by one, intact, each landing as a lump at some random point. The full interference pattern appears only gradually, one grain at a time, in a way inconsistent with our naively imagined "matter wave." Not like this:

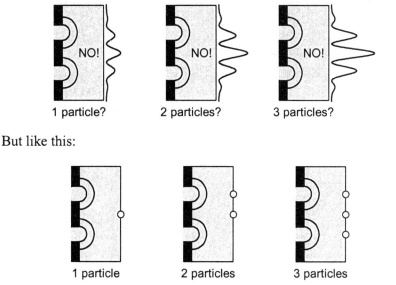

But like this:

Whatever it is that does the waving—and thus creates the interference—it is not simply a disturbance of matter and mass.

A quantum mechanical wave is instead something strangely intangible, something that cannot be seen, touched, or tasted. It is not force. It is not mass. It does not register directly on the dial of any instrument. A quantum mechanical wave, rather, is a wave of information. It tells us something. In the diffraction experiment, it tells us the statistical probability of finding the particle at every possible point. Out of an infinity of choices, a quantum mechanical wave tells us the odds of stumbling upon any particular one of them when we make a measurement.

We are given a set of numbers, one number corresponding to each point in space, and every number is pregnant with possibility. For there is indeed a genuine physical meaning to be extracted from a quantum mechanical "wave function," a meaning that shows up not so much in the numbers themselves but more properly in the values *squared*. It is the square of a wave, remember, that gives rise to interference; and just as the square of an ordinary wave corresponds to energy, the square of a quantum wave corresponds to probability.

Suppose, for instance, that some arbitrary wave function looks just like the second harmonic of a vibrating string, and suppose further that it has the value 2 at point A and −1 at point B. The squared values are then 4 at point A and 1 at point B, making the electron four times more likely to be found at A:

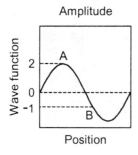
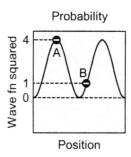

We do not know *which* four times out of five the electron will show up at B rather than A, but we do know the long-term odds. And under the rules of quantum mechanics, the wave function and its attendant probability distribution represent all there is to know.

It is the best a quantum observer can do. We make measurement after measurement, carefully restoring the system to its pristine original state after each one. We carry out the measurements in exactly the same way on exactly the same system, and nevertheless each result is different. In one trial, the electron—all of it, too, the whole particle-like lump—turns up at point A. In another trial, point B. In another trial, point C. Every trial, another point. Every trial, another lump. Not until we tally all the results (counting up how many times the electron appears at each point) do we realize that we are dealing with a wave. The distribution of detected electrons traces out the square of the wave function:

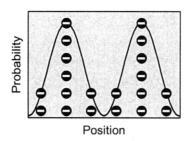

Position

What appears in retrospect is an album of unconnected snapshots, not a movie. The probability distribution offers us a tableau of where the electron might be, but not how fast it will be moving nor in what direction it will be heading. To look is necessarily to touch, according to Heisenberg, and with each of our measurements we alter the wave function in a way that cannot be undone:

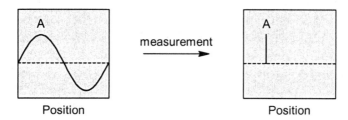

It's either one or the other, position or momentum. By forcing the electron into a definite position we collapse a delocalized wave function (with a narrow distribution of wavelengths and momenta) into a sharp spike (with a broad distribution of wavelengths and momenta). On the one hand, the measurement reduces an infinity of spatial possibilities to a final, make-up-your-mind-once-and-for-all realization of just one: a single position, randomly generated in conformance with a set of probabilities. On the other hand, the very same disturbance leaves the wavelength and momentum of the particle indeterminate. Heisenberg's uncertainty principle, law of the land, keeps an electron from telling us simultaneously where it is and what it is doing. The quantum mechanical particle follows no set path.

A State of One's Own

A classical observer would find it appalling: that a measuring device, exquisitely crafted and tuned to perfection, could fail to yield the same number time after time. "In *my* universe," says Newton, "the first law of nature is the law of common sense. When you do the same thing to the same system in the same way, you should at least expect to get the same result!"

A quantum observer, coming to terms with a different kind of reality, has no such expectation. Macroscopic intuition means little in a world where Planck's constant looms large (and where, quite the contrary, a measure of indeterminacy does not offend an observer's common sense). "And now that you mention it," responds our microworld observer, "you overstate the case. Not every last one of my measurements turns on a roll of the dice. I can show you infinitely many quantum states that always produce the same values."

They are called *eigenstates* (from the German *eigen*, meaning "own" or "particular"), and the fixed numbers they yield upon measurement are called *eigenvalues*. There are eigenstates for energy. There are eigenstates for momentum. There are eigenstates for position. There are eigenstates for every observable quantity, and complete sets of eigenstates—those that span the full range of possible values—become the bricks from which more general quantum states can be built.

Suppose we observe, for example, that a particle inhabiting quantum state 1 always shows 10 units of energy. If so, then quantum state 1 serves as an eigenstate of energy. With the particle occupying a portion of space that conforms to the prescribed wave function,

Position

we measure the same 10 units of energy every time. For *this* specific state, a measurement of energy occasions no randomness, no indeterminacy, no violent disruption. As if sheltered on an island of certainty, the system offers but one choice in response to an observer's probe: a probability of 100% for registering an energy of 10 units. Measurement after measurement, the eigenvalue never varies; and measurement after measurement, the eigenstate suffers no perceptible change:

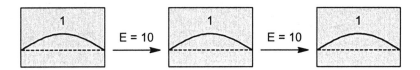

We make our measurement. We get our expected number. We get back our eigenstate unscathed. State 1, impervious to the measurement of its "own" particular property, is an eigenstate of energy—although *not* an eigenstate of position and *not* an eigenstate of momentum, because the wave function does allow for more than one possible position and more than one possible momentum. State 1 is instead an eigenstate of energy, no more and no less, and it stands firm with its eigenvalue of 10 units in the face of indeterminacy everywhere else.

If the system happens to be in a different eigenstate of energy (let's say state 2, with an eigenvalue of 20 units), then we get a different number, but it still comes with a 100% guarantee. Measurement of the energy of eigenstate 2 yields the same 20 units every time:

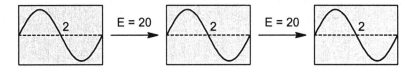

For eigenstate 3, another fixed number. Eigenstate 4, still another. For every allowed value of energy, expect to find a corresponding eigenstate and eigenvalue. We have to take them all into account.

Now if a quantum system always existed in a single, unique eigenstate that embraced all possible mechanical variables (position, momentum, angular momentum, energy...everything), then there would be no uncertainty principle. There would be no element of chance. There would be no quantum mechanics. An infinitely careful observer would be assured a fixed value for every measurement, and the state of a system would remain untouched before and after.

But a quantum system rarely exists as a unique eigenstate, a condition that would erase all doubts concerning its measurable properties. Instead, an observer typically deals with an ambiguous, neither-here-nor-there *superposition* of eigenstates, a smorgasbord of possibilities offered up at the same time:

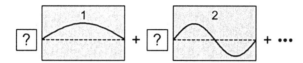

There is a little bit of this and a little bit of that, and so the question for a quantum mechanical observer now becomes: What should I expect? Will my measurement yield energy 1 today and energy 2 tomorrow, or will it be energy 3 or energy 4 or energy 5? Or might all the possible values blend together simultaneously, the way a vibrating string harmonizes its standing waves into the rich timbre of a piano?

For an answer, we look to the principle of superposition.

Superposition

It comes up again and again, one of the great recurring themes in nature's design: the idea of using irreducibly simple building blocks, each one independent of the rest, to put together an arbitrarily complex phenomenon. However different the events may appear—the sounds of a symphony, the transmission of a television signal, the flow of heat—the method of superposition brings them all under a common mathematical umbrella. Construction of a quantum mechanical superposition becomes but one more example.

Nowhere is the idea illustrated more transparently than in the partitioning of space into perpendicular dimensions. Beginning in Chapter 3 and echoed ever since, this joint notion of *independence* and *completeness* has already informed our understanding of relativity, classical mechanics, electromagnetism, and waves of all kinds. In a word: East is east and north is north, and neither one of them can alter the course of the other. Three steps east and four steps north bring a marcher five steps to the northeast,

but no combination of steps north–south is able to augment or diminish the progress east–west. So long as we refer positions consistently to the same fixed grid, then three steps east and four steps north remain just that—a displacement wholly to the east (with no northerly contribution) superposed onto an independent displacement wholly to the north (with no easterly contribution):

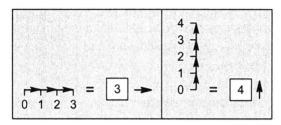

Given a complete set of unit steps, one in each perpendicular direction, an observer is then sure to locate any point on the grid. In one dimension, we need only one step along the one unique axis to make the set complete. In two dimensions, two steps. In three dimensions, three.

So, putting one foot in front of another, we march three steps east and four steps north. We can do so by zigzagging east–north or north–east one step at a time, or by moving entirely east and then entirely north, or by moving partly north and partly east in any sequence whatsoever. The outcome does not change. Three steps east and four steps north add together not as scalars but as vectors (see again, Chapter 5), and in partnership they point five units in the direction 53 degrees north of east.

Working independently, maintaining fixed lengths along their own axes, the perpendicular components combine according to the Pythagorean theorem. The geometry is not the geometry of a straight line ($3 + 4 = 7$), but rather the geometry of a right triangle ($3^2 + 4^2 = 5^2$). The *squares* add together:

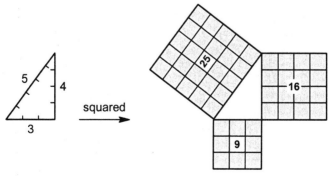

The scalar distances, moreover, prove invariant to any overall rotation of the grid, as we established first in Chapter 3. Our definitions of east and north may change, but a distance of five units remains five units whether it points northeast in one coordinate system or east in another:

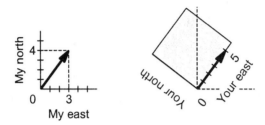

According to my axes, a point lies three steps east and four steps north. According to yours, the same point lies five steps east. It makes no difference. Since neither of us can claim any absolute east–north basis for navigation, then your choice of coordinates is as valid as mine. I know how to express my principal directions in terms of yours, and you know how to express yours in terms of mine. Provided that each of us chooses two *perpendicular* directions for "east" and "north," we will subsequently agree on everything of physical significance.

With that, we acquire a tool of immense power: a nearly unlimited capacity to build up vectors from independent components, whatever those components may happen to be called. For we need only change some of the names—from "point" to "quantum state," say, and from "east and north" to "eigenstates 1 and 2"—and we have a ready-made template for manipulating a quantum superposition. The mathematics is admittedly more complex for quantum mechanics than for triangles, but there are valid analogies to be made nonetheless. One by one, the connections fall into place.

First, the components are independent. Each of the standing waves of a vibrating string, for instance, stands alone. Just as east is east and north is north, the fundamental mode is equal to the fundamental mode and nothing else. Neither the second harmonic nor any other makes a contribution:

The second harmonic, likewise, is the second harmonic and nothing else,

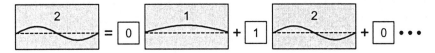

and so it goes for the third harmonic and the fourth and the fifth and all the rest. Each independent mode behaves as if belonging to a dimension entirely its own, perpendicular to every other.

Second, a complete set of such modes (*one* half-cycle, *two* half-cycles, *three* half-cycles, and so on, up to infinity) provides enough flexibility to build up a curve of any shape, as complicated as you like. Combine a sufficient number of harmonics in the correct proportions, and you can trace out the coastline of Australia, just as you can locate any point in the country itself by combining unit steps east–west, north–south, and up–down. It is a property that mathematicians guarantee beyond a doubt, and it is a property that embraces more than just sine and cosine waves. The twin virtues of independence and completeness are shared by sundry other special curves as well, many of which figure prominently in quantum mechanical states.

Third, the independent modes obey a generalized Pythagorean theorem. They combine as *squares*—and conveniently, too, because the square of a wave function tells us the likelihood of measuring each of the possible values implicit within. So now, for example, if we chance upon a superposition containing *three* parts component 1 and *four* parts component 2,

we know immediately what to expect of the squared wave function and hence the eigenvalues: the same 9–16–25 relationship as for a triangle in which *three* steps east are superposed onto *four* steps north. For every 25 measurements, there will be (on average) 9 trials that yield eigenvalue 1 compared with 16 trials that yield eigenvalue 2. Each measurement results in one eigenvalue or another, but repeated trials yield an average somewhere between the two extremes.

Fourth, since our independent quantum modes behave just like perpendicular components, why not dispense with the laborious drawing of curves? For once we realize that "component 1" merely means such-and-such a curve and "component 2" means another-and-such kind of curve, then we can track them concisely as two perpendicular contributions to a vector—either symbolically and metaphorically on a graph,

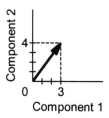

or in a corresponding table:

Component	Amplitude	Probability
1	3	$\frac{9}{25}$
2	4	$\frac{16}{25}$

To account for additional components, we simply add lines to the table.

Fifth, and finally, the choice of independent components for a quantum state is no less arbitrary than it is for a point in a plane. Just as I can rotate my spatial axes to coincide with yours (and all the while change nothing of physical significance), I can analogously rotate my quantum axes as well:

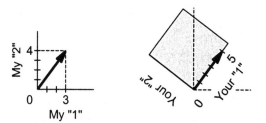

And since our observations must ultimately be immune to a meaningless rotation of axes, then the laws of quantum mechanics must also withstand any overall turning of the reference frame.

Count on it. The quantum mechanical equations of motion, in whatever form they take, will have to permit flexible recombination of the building blocks we call 1 and 2. Your east is as good as mine.

A New Mechanics

Where do we stand? We confront a world seemingly divorced from macroscopic reality, a world where particles diffract through crystals, where particles interfere like waves, where particles appear to be in two places at once. Welcome to a world where particles, clustered into nuclei and atoms and molecules, behave less like billiard balls and more like the restricted harmonics of a vibrating string. A world where particles have wavelength, where particles have frequency, where particles have phase. A world where the watchwords are *quantization*, not continuity; *indeterminacy*, not certainty; *superposition*, not uniqueness; *probability*, not determinism. Such are the realities of the quantum mechanical microworld, and such are the realities addressed by the quantum mechanical vector–state described just above.

Like a pair of dice waiting to be thrown, a quantum mechanical state offers a promise not of what *will* be but rather a prediction of what *might* be: a 2 or 12 once every 36 tries (but who knows when?), together with a 3 or 11 twice every 36 tries, together with a 4 or 10 three times, together with a 5 or 9 four times, together with a 6 or 8 five times, together with a 7 six times. The entire array comes packaged as a superposition of independent components, one for every possible outcome, all of them present simultaneously.

A quantum mechanical state, so conceived, is tailor-made to deal with the way things are in the microworld. It provides for interference. It provides for quantization. It provides for the random disturbance of a measurement.

For a microworld observer, it represents all there is to know. The quantum mechanical vector–state, taking the place of the initial positions and velocities of Newton's world, becomes the first step on the road to a new mechanics.

Changing States

For some systems, there may be just two options—the "up" and "down" magnetic fields of a single electron or proton, for example (about which, more to come). For other systems, there might be three options, or four, or twenty-four, or even infinity. For all quantum mechanical systems, though, an observer deals with some countable list of possible values, any one of which might pop up randomly as the result of a measurement. *Energy 1, energy 2, energy 3. Angular momentum 1, angular momentum 2. Position 1, position 2, position 3, ..., position ∞.* However many independent eigenvalues there are, whether two or two zillion, our mechanics must recognize that the system has access to them all.

To each, accordingly, we assign a perpendicular axis in an abstract space, a space large enough to accommodate the entire spectrum of possibilities. An iconic two-dimensional coordinate system is sufficiently general to make the point:

And the point, remember, is that a quantum mechanical system has the capacity to exist simultaneously in more than one eigenstate. When the vector falls between the axes, the state acts as if it were a superposition of components 1 and 2. The length of a component along its own axis gives us the amplitude of that component (the number of steps "east" or "north"), while the square of the amplitude gives us its associated probability of occurrence. The total probability (the likelihood that one or the other component will appear) is the sum of contributions 1 and 2:

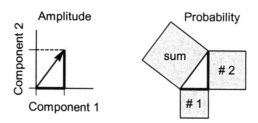

Keep in mind further that a superposed state, reminiscent of an electron emerging unobserved from two openings at once, has an amplitude to be in component 1 *and* component 2 at the same time. Not until later, not until an observer intrudes with a measurement, must the system choose either component 1 *or* component 2:

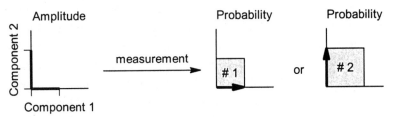

Now look at what measurement has wrought. There has been a change of state, marked by a new vector with new components. The system has gone from one composition of eigenstates to another, and our mechanical model must be able to follow along.

It is just one example of many. To cope with quantum mechanical change in general, we need an assortment of tools—sets of mathematical instructions called "operators"—to alter the makeup of a vector. We need, among others, operators that can make a vector rotate,

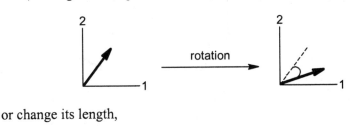

or change its length,

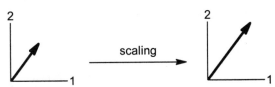

or pick out each of its perpendicular components (the eigenstates) one by one:

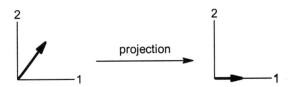

We need operators for position, operators for linear momentum, operators for angular momentum, operators for energy. We need operators for every measurable property, and we need operators that reflect the vagaries of measurement as well: operators that yield not only the eigenstates and eigenvalues of a quantized system, but operators that also respect the Heisenberg uncertainty principle. For that, especially, we need operators that render indeterminate the simultaneous measurement of complementary quantities such as energy–time, linear momentum–position, and angular momentum–angle, as anticipated in Chapter 7.

Angular momentum, for instance, is associated with rotation, and thus a measurement of two components of angular momentum (along two perpendicular axes) entails two consecutive rotations. See what happens, though, when we rotate a vector unthinkingly about two axes in succession, say 90 degrees counterclockwise about x followed by 90 degrees counterclockwise about y. A vector that starts along y ends up along x, flipping first from y to z and then from z to x:

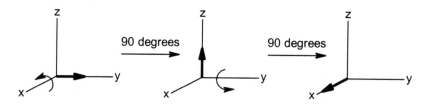

But if now we reverse the order, rotating first about y and then about x, the outcome turns out to be entirely different. A vector that starts out along y, unaffected by a rotation about its own axis, ends up along z:

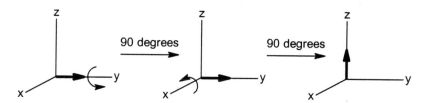

Since the effect of two rotations depends on the order in which they are performed, any measurement of angular momentum in one direction inevitably compromises the accuracy in another. The operations do not "commute," and once again an observer is faced with a choice: to determine one quantity or the other to arbitrary precision, but not both at the same time. Recognize in that dilemma a manifestation of the uncertainty principle, rooted in the unpredictable disturbance brought about by a measurement.

Bridging Two Worlds

The quantum observer may be forgiven a slight feeling of superiority vis-à-vis the classical observer. "My reality is more comprehensive than your reality," says Heisenberg to Newton, "because your world is put together entirely with building blocks taken from mine. Your billiard balls and planets are nothing more than clusters of microscopic particles run amok, and therefore your laws of mechanics must ultimately derive from mine."

Nobody can deny it. Quantum mechanics is a more general model than classical mechanics, in the same way that Einsteinian relativity is a more general model than Galilean relativity. One picture subsumes the other. Quantum mechanics, pushed to the limit of the large, goes over smoothly into classical mechanics, whereas classical mechanics remains resolutely classical even when pushed to the limit of the small.

Consider it all the more remarkable, then, that quantum mechanics (the more general theory) had to be formulated by working backward from classical mechanics. Galileo, Newton, and their successors developed the equations of mechanics by experimentation and theoretical refinement. So, too, did Heisenberg, Schrödinger, Dirac, and the other early quantum mechanicians, but they needed to do something still more. They needed to peek at the classical equations in order to set the quantum equations on the right track. They needed a hint, a point of departure, an idea of where the broader theory must eventually lead.

They needed to know, for example, how momentum and energy were expressed classically before they could postulate operators to take their place quantum mechanically. Given just the barest of clues, however, they could subsequently develop consistent rules for constructing all the necessary forms.

> *Whenever you see 'momentum' in classical mechanics, substitute such-and-such an operator. Whenever you see 'position,' substitute this-and-that kind of operator. Whenever you see 'angular momentum,' substitute* this *particular operator. Whenever you see 'energy,' substitute* that *particular operator.*

Thus the classical momentum (mass × velocity) becomes one kind of quantum mechanical operator, and the classical angular momentum (mass×velocity×radius), becomes another kind of operator, and the classical kinetic energy ($\frac{1}{2}$×[momentum]^2÷mass) becomes yet another kind. Properly constructed, the various operators exhibit the commutation properties demanded by the uncertainty principle, and they also honor the special links between conjugate quantities such as momentum and position. What was true in classical mechanics proves to be powerfully and universally true in quantum mechanics as well. A linear momentum operator generates a linear displacement. An angular momentum operator generates an angular displacement. An energy operator generates a temporal displacement.

Borrowed and adapted from the classical world, the operators cross over into the quantum mechanical realm by taking on the sign of the quantum mechanical beast: Planck's constant, that tiny-tiny number h we encountered back in Chapter 7, with its suggestive dimensions of (momentum)×(distance) or, same thing, (energy)×(time). Operators for linear momentum, angular momentum, and energy all contain Planck's constant, a fundamental constant of nature found nowhere in classical mechanics or electromagnetism.

The rules for operators apply even to properties lacking direct classical counterparts, notably the "spin" angular momentum of a particle. Coming not from any explicit motion about a point (as exemplified by the *orbital* angular momentum of a planet revolving around the Sun),

orbital angular momentum

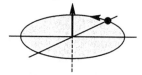

the spin angular momentum, so called, is inherent to the particle itself. An electron standing stock still, orbiting nothing, has an angular momentum already built in, an endowment as fundamental as charge and mass. The electron does nothing to earn it.

One might suppose that the particle literally spins like a top, but the easy comparison fails. Because to spin like a top a body must have an internal structure, and to the best of our knowledge an electron does not. An electron seems instead to be a truly fundamental particle, a structureless speck of mass, charge, and intrinsic angular momentum. And as if to emphasize the nonclassical character of the electronic spin even further, nature doles out the associated angular momentum not in steps of $h/2\pi$ (the value expected for a spinning top), but rather in quanta half the size: $\frac{1}{2} \times (h/2\pi)$ for any one perpendicular component.

This combination of electric charge and a "spin-$\frac{1}{2}$" angular momentum gives the electron (the proton and neutron, too) a magnetic dipole moment, suggestive perhaps of a current loop. But unlike the classical current loop of Chapter 6, the magnetic moment of an electron is not free to point any which way in an external magnetic field. The quantized angular momentum is restricted to only two possible orientations, "up" and "down":

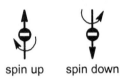

spin up   spin down

Nevertheless, even though the spin angular momentum has no exact classical antecedent, we can still model it with operators analogous to those representing orbital angular momentum.

By their fruits we shall know them. Regardless of origin, angular momentum operators share the same commutation properties and the same characteristic rotations. The genetic relationship between classical and quantum mechanics remains ever strong, honored even in the breach.

Niels Bohr, one of the founders of quantum mechanics, called it the "correspondence principle": the requirement that the statistical averages of quantum mechanics become identical to the deterministic pathways of classical mechanics when small turns into large. Under conditions where Planck's constant seems effectively to shrink to zero (as it does when particle masses and system dimensions are sufficiently large), the transition is complete. The average motion of a system in the classical limit conforms to Newtonian expectations, and purely quantum mechanical phenomena like spin disappear entirely. The fuzzy indeterminacy of the microworld passes over into the crystalline certainty of the macroworld.

Toward an Equation of Motion

Given the state of a system today, what will it be tomorrow? The great question of mechanics, *What comes next?*, remains forever the same.

For a quantum mechanical state, one overarching certainty rises above all else: the certainty that whatever the options, whatever the probabilities, whatever randomness there may be, the system must provide for all eventualities. Just as a worker must account for every hour and every minute of every day, a quantum mechanical state must also account for every possible outcome. The probabilities have to total 100%.

Suppose we determine that our system has only two admissible values of energy, each with a definite probability. So be it, but the state now has to guarantee that either one or the other value is always certain to be picked. If the probability of observing energy 1 is 36%, then the probability of observing energy 2 must be 64%. If the probability of energy 2 is 100%, then the probability of energy 1 must be 0. If the probability of energy 1 is 50%, then the probability of energy 2 must also be 50%. *Not* 50% energy 1 and 40% energy 2. *Not* 50% energy 1 and 60% energy 2. *Not* 50% energy 1, 40% energy 2, and 10% "other." Unlike the statistics of an opinion poll, there are no undecideds in quantum mechanics. There are no overcounts, and there are no undercounts.

To stay within the realm of possibility, a changing state has but a single course to follow while evolving in time. Progressing smoothly from one superposition to the next, not yet disturbed by any measurement, the vector can only *rotate* within the allotted space. Its length remains fixed:

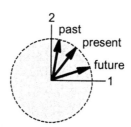

The mixture of components changes, but the sum of the squares (the total probability) does not. All other routes would make the total probability sometimes greater than 100% and sometimes less. They are not allowed:

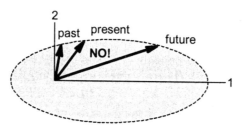

No, the vector–state submits to a length-preserving rotation operator that derives its influence from (what else could it be?) the fundamental relationship between energy and time. It is energy that drives the classical machine, and it is energy—acting now under a new set of rules—that also drives the quantum mechanical machine. Look for the total energy of the system, kinetic and potential, to work its way into the operator that causes the state to change with time. Look for energy, incorporated into a suitable equation of motion (for atoms, the Schrödinger equation), to give a quantum mechanical system its marching orders. Look for energy, always energy, to bring about time and change.

It brings about *deterministic* change, too, for there is nothing random in the way a quantum mechanical system evolves from state 1 to state 2. There is no throwing of dice as the vector–state travels the Schrödinger highway, moving from one city to the next with never an unforeseen destination. State to state, the Schrödinger equation decrees a mechanical history for an electron no less determined than the Newtonian path of a planet. State to state, one quantum mechanical superposition follows another as reliably as the Sun dawns in the eastern sky. State to state, the quantum mechanical equation of motion charts an unwaveringly deterministic course. Give Schrödinger the amplitude of each possible outcome today, and he will tell you the amplitude (and hence the probability) of each possible outcome tomorrow.

What Schrödinger cannot say, however, is exactly which of those outcomes will occur if you interrupt the smooth evolution of the state by making a measurement. The probability of each outcome, yes. The average value of a series of measurements, yes. A guaranteed result for each trial, *no*. To look is to touch, and to touch is to terminate with extreme prejudice the systematic progression of a vector–state from one superposition to the next. A world of possibilities collapses into a single reality for the intrusive observer.

If you like your universe orderly, with all its manifold possibilities intact, then keep your hands off. A quantum superposition is not to be trifled with.

9

SYMMETRY PERFORCE

Turn away for a moment from looking at a blank sheet of paper (make it a geometrically perfect square, with an unblemished white surface), and then look again. Do you still have the very same object in the very same position?

Are you sure? *Can you prove it?* In the twinkling of an eye, perhaps, a leprechaun might have turned the paper by 90 degrees, or by 180 degrees, or by 270 degrees, or by 360 degrees:

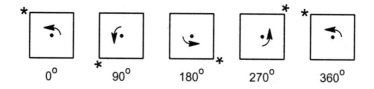

If so, would you even know that the object has been touched? *No, the leprechaun leaves no fingerprints.* Absent any distinguishing marks (like the asterisks), would you be able to determine the degree of rotation?

No, the rotated squares all end up in equivalent positions. Would you be able to tell the difference between the original sheet and a duplicate the leprechaun might have put in its place? *No, if you've seen one sheet of unmarked white paper, you've seen them all.*

There is no difference to tell, of course, and so an open-minded observer can only say: "It's all the same to me. Perform whatever 'symmetry operation' you like behind my back, so long as you restore the system to an equivalent configuration. It makes no difference."

But indeed symmetry does make a difference (it must), because symmetry dictates design. Symmetry introduces constraints. If a building is to look the same when rotated by 120 degrees, then the architect has only a limited set of blueprints upon which to draw. A world of unrestricted possibilities shrinks to a world circumscribed by equilateral triangles, hexagons, circles, and other such figures that withstand threefold rotations:

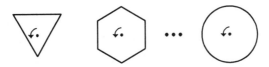

So, too, with the laws of nature. For every symmetry, there comes a constraint. If physics is to look the same when the origin of time is shifted by a fixed amount—so that, with eyes closed, we could not tell whether a leprechaun has added one second to all our clocks—then a menu of unlimited mechanical possibilities must be cut down to just a select few. Only those processes that conserve energy are allowed. *A constraint.* If physical law is to be immune to the arbitrary displacement of our spatial axes, then nature requires the conservation of linear momentum. *A constraint.* If the laws are to be unaffected by the arbitrary rotation of a coordinate system, then angular momentum must be conserved. *A constraint.* If the laws are to be the same for all inertial observers, then the space-time interval must be invariant. *A constraint.*

And now another constraint, a facet of nature so beautiful as to make one's jaw drop in wonder: that both the interactions between particles and the fields that mediate them are the inevitable consequences of symmetry...that atoms exist and nuclei exist and electromagnetic fields exist precisely because of the special symmetries nature imposes on quantum mechanical states... that, in short, *symmetry creates force*.

We begin with the indistinguishability of identical particles and the restrictions that fall upon their associated wave functions.

By Halves: Identical Particles

If they were balls on a pool table, we could give each one a number and track it individually:

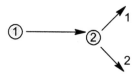

We know where every ball is going. A classical particle follows a path.

A quantum mechanical particle does not. We cannot paint a number on an electron (or a proton, or a neutron, or a quark, or a photon, or any other particle of the microworld) and follow its comings and goings amid a host of fellow particles that look exactly alike. "Is that electron 1 over there," asks the uncertain observer, "or is it electron 2? I can't determine which is which." An indeterminate path renders all particles of the same class utterly indistinguishable, and any attempt to label them as individuals becomes an arbitrary exercise.

You say that particle 1 is on the left, and I say that particle 1 is on the right. Each of us is entitled to our opinion,

and our difference of opinion seems at first to make no difference. Nothing of significance can depend on which of two identical particles an observer chooses to label as "1" or "2."

Yet make a difference it must, for we know that symmetry always has repercussions. Wherever there is arbitrariness and apparent indifference, there is an underlying constraint as well, and here the constraint shows up in the form of the quantum mechanical state itself. Only certain kinds of wave functions, with a symmetry all their own, can support the arbitrary interchange of indistinguishable particles.

The cause is subtle, but the effect is dramatic—nothing less than a division of the world into two camps, as we are about to see.

In a Spin: Fermions, Bosons, and the Pauli Principle

We prepare to observe the quantum mechanical doings of two identical particles, looking for a wave function (Chapter 8) that will give us the probability of finding particle 2 at some position *r* relative to particle 1. Three relationships are possible:

1. When *r* is positive, particle 1 lies closer to the origin than particle 2:

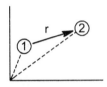

2. When *r* is negative (although otherwise the same number), the particles trade places. The numerical labels are swapped, with particle 2 now lying closer to the origin than particle 1, but the underlying distance *r* remains unchanged:

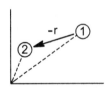

3. When *r* is zero, the two particles occupy the same point in space. They fall one atop the other:

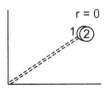

With these three configurations in hand (there are no others), we now begin to lay out the wave function point by point. And aside from the particulars of *how* to do it, which will depend on the specific particles and interactions, the goal in general is clear enough. For every possible value of r, we seek a corresponding number with a special property: a probability amplitude ψ (Greek "psi"), chosen so that the *square* of ψ is proportional to the likelihood of observing the given configuration. All we have to do is establish the proper amplitude ψ for each separation r,

r	ψ
⋮	⋮
-1	?
0	?
1	?
⋮	⋮

and our job will be done. We will have mapped out the joint wave function for the two particles.

But beware the leprechaun, who, unbeknownst to us, might switch the identities of particles 1 and 2, putting 2 in the place of 1 and 1 in the place of 2. Unable to tell the difference, we would remain blissfully unaware of any physical change. We would observe the same average values for position, the same average values for momentum, the same average values for energy; and so, to justify our ignorance, the square of the wave function (ψ^2) would have to appear the same to us both before and after the leprechaunic mischief. If not, we would know that something has been done to the system. The probabilities would be different.

There are only two sure ways to preserve ψ^2. Either the values of ψ can all stay exactly the same when the particles are interchanged (here are some made-up numbers to demonstrate the arithmetic),

r	ψ	ψ^2
⋮	⋮	⋮
-1	2	4
0	3	9
1	2	4
⋮	⋮	⋮

r	ψ	ψ^2
⋮	⋮	⋮
-1	2	4
0	3	9
1	2	4
⋮	⋮	⋮

or every value of ψ can be multiplied by -1 (and for that, here are some more made-up numbers):

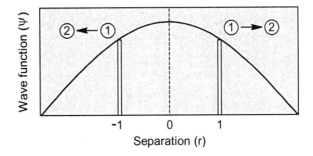

r	ψ	ψ²
⋮	⋮	⋮
-1	2	4
0	0	0
1	-2	4
⋮	⋮	⋮

r	-ψ	ψ²
⋮	⋮	⋮
-1	-2	4
0	0	0
1	2	4
⋮	⋮	⋮

To realize the first option (a state that is *symmetric* to particle interchange), each number ψ must be the same for both r and $-r$. The two halves of such a wave always appear as if reflected left–right through a mirror, maybe something like this:

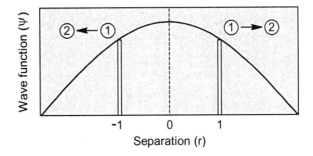

To realize the second option (an *antisymmetric* state), the two halves must appear as if reflected both up and down and left to right—for example, like this:

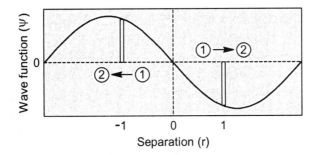

Nothing else is allowed.

"*Either … or*," declares the governing "Pauli principle." Either the state of a system is symmetric with respect to interchange of identical particles, or it is antisymmetric. Either the wave function remains as it is when all the particles are switched, or it shifts by 180 degrees. Either the curve looks the same left–right (so as to be unaffected by particle interchange), or it appears inverted (so as to undergo multiplication by -1). *Either ... or.* There is no in-between.

Particles with wave functions that are symmetric with respect to interchange are called "bosons," in honor of the physicist S. N. Bose:

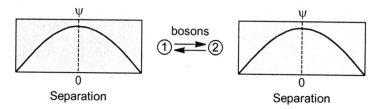

They include all particles carrying spin angular momentum in integral multiples of $h/2\pi$, as well as zero. The photon, with its spin of 1 unit, heads the family.

Particles with wave functions that are antisymmetric with respect to interchange are called "fermions," in honor of Enrico Fermi:

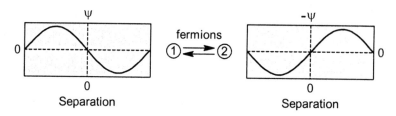

They carry *half*-integral spins, quantized in units of 1/2, 3/2, and so forth. Electrons are fermions. Quarks are fermions. Protons are fermions. Neutrons are fermions.

Fermions, by virtue of their peculiar symmetry, have zero amplitude at the point $r = 0$ (where the total wave function goes from positive to negative). With a probability of exactly zero to occupy the same position at the same time, fermions thus tend to keep their distance. They "exclude" one another, as if pushed away by a force. No two of them can simultaneously occupy the same quantum state, unless they possess opposite spins.

If, for instance, two discrete levels of energy are available to four electrons, then the particles—as fermions—have just one way to distribute themselves. They fill the levels two at a time, one electron with spin up (↑) and the other with spin down (↓):

energy 2

energy 1

There is no alternative. The Pauli principle forbids all other configurations, not only those considered (and rejected) below but everything else as well:

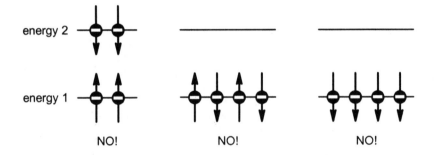

energy 2

energy 1

NO!   NO!   NO!

Any one of them would violate the fundamental symmetry of a fermionic state. They are not permitted.

Bosons suffer no such restrictions. Nothing prevents a bosonic wave function from having a nonzero amplitude at $r = 0$, and consequently nothing prevents two bosons or three bosons or even three trillion bosons from piling up in the same quantum level. There is always room for one more. Bosons are gregarious particles, able to perform feats of great cooperation and coordination (superconductivity, for one; laser beams, for another), and they can do so simply because their interchange symmetry is positive rather than negative.

Positive or negative, though, the symmetry of identical particles forces matter to take a stand: to become one or the other, to be enrolled as either fermion or boson. Bosons, typified by the photon, carry the fundamental forces that cause fermions to attract and repel. Fermions, led by electrons and quarks, become the constituents of ordinary matter. They produce a material world of nuclei and atoms founded on principles of hierarchy and exclusion.

It takes both kinds. Woven together, fermion and boson constitute the warp and weft of the quantum mechanical universe. A minus sign makes all the difference.

Material World

To design a city is one thing; to populate it is another. Not the buildings, not the streets, not the neighborhoods—what matters most is the *life* of the city, the ways in which the people make use of what they have. A city that finds everyone clustered tightly around the core, far from the outskirts, differs greatly from one in which the population is spread out more evenly. The first remains isolated, cut off from surrounding communities,

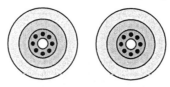

whereas the second is likely to form social and business ties with the world outside:

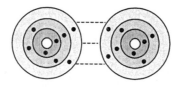

The one herds all its inhabitants into close quarters, growing ever more dense as the population increases. The other expands outward to absorb new inhabitants. The first kind of city behaves like a system of *bosons*, while the second (analogous to the electrons in atoms and molecules) behaves like a system of *fermions*. And if the rules were any different— if electrons, say, behaved like bosons rather than fermions—then we and everything else in the chemical and electromagnetic world simply could not exist. The Pauli principle, even more so than the Schrödinger equation, lends shape and structure to atoms and molecules.

The Schrödinger equation acts as city planner and architect, designing streets and buildings without dictating specifically how the installations are to be used. Given the number of particles, their charges, their masses, their kinetic energy, their potential energy, their attractions and repulsions, the equation merely establishes a hierarchy of quantized energy levels and associated wave functions:

Schrödinger equation

It says nothing about who goes where.

The Pauli principle, playing the role of zoning board, directs the filling of the energy levels. Electrons enter from the ground up, with spins alternately up and down, restricted to a maximum of two particles per quantum state:

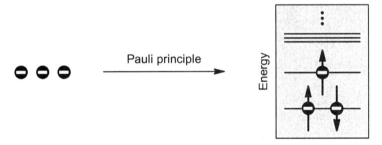

Pauli principle

The atom thus grows in layers, expanding outward from the center. Electrons near the nucleus, closer to a naked positive charge, are attracted more strongly than electrons near the edge. The outer electrons, shielded from the full force of the nucleus by a screen of negative charge, are bound more loosely. They are better able to enter into combinations with the outer electrons of other atoms and thereby form molecules. They are better able to come and go, better able to undergo chemical reactions, better able to make one kind of atom different from another.

Different atoms, different properties. Some atoms (such as hydrogen and carbon) react readily; others (such as helium and xenon) react hardly at all. *It depends on how the electrons are configured.* Some atoms (like sodium and calcium) are givers of negative charge; others (such as oxygen and fluorine) are takers. *It depends on how the electrons are configured.*

Some atoms (such as hydrogen and iron) are magnetic; others (such as helium) are not. *It depends on how the electrons are configured.*

Consider magnetism, for example. In systems where all the energy levels are filled, there is an "up" magnetic moment for every "down" magnetic moment. One spin cancels the other, and the total magnetism falls to zero:

But in configurations that leave one or more electrons unpaired,

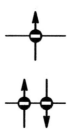

each atom retains a net magnetic moment of its own. Like a quantum mechanical compass needle, the internal atomic magnet will then assume a restricted number of orientations when placed in an external magnetic field. Indeed for some systems (iron, especially), the individual magnetic moment of each atom is strong enough to command the orientation of its neighbor. Such a material becomes a permanent magnet, exerting and responding to magnetic forces even without the influence of an external field.

How remarkable, too, that the various atoms (every one of them built from the same indistinguishable particles) prove ultimately to be so very different. Different elements differ in their magnetic properties, their electrical properties, their optical properties, their thermal properties, their sizes, their melting points, their boiling points, their reactivities. They differ in every way, and they owe their differences largely to the symmetry demanded by the Pauli principle: the proposition that electrons are thoroughly and hopelessly indistinguishable, that they can be interchanged at will, that "nobody can tell the difference."

Compelling Force:
Local Symmetry and Global Interaction

Imagine waking up one morning to find that the price of everything has increased by 10%. Every good. Every service. Every wage. Every tax. Yesterday, bus fare was $1.00; today, $1.10. Last month, a pair of shoes cost $100; this month, $110. Last year, you earned $100,000; this year, $110,000. Is the world suddenly a different and more expensive place?

Not at all. A pair of shoes still costs 100 times as much as a ride on the bus. With your old salary of $100,000, you could buy 999 pairs of shoes and have enough money left over to take the bus 100 times. With your new salary of $110,000, you can buy the same 999 pairs of shoes (at $110 per pair, for a total cost of $109,890) and still have just enough change for 100 bus fares at $1.10 each. Whatever you could do with $1000 or $10,000 or $100,000, you can do the same with $1100 or $11,000 or $110,000.

Neither you nor anybody else will be the wiser. None of us will be able to tell the difference, for our entire economy will have undergone an indiscernible symmetry transformation: a uniform addition of ten cents to every dollar in every transaction, an adjustment applied consistently over the entire range of goods and services. Nothing of consequence will have changed, and (except for the total amount of money in circulation, an arbitrary number) the financial world will be the same today as it was yesterday:

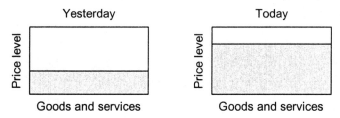

Since we all shift our accounts by the same fixed amount, not a single buyer or seller gains or loses.

Now suppose that different regions of the world have the power to act independently, to raise or lower their prices without coordinating that action with anyone else. Suppose that prices rise uniformly by 10% in North America and 30% in South America, fall uniformly by 20% in Asia, rise uniformly by 50% in Europe, and fall uniformly by 5% in Africa.

Suppose, in short, that symmetric shifts in prices are made *locally* rather than globally,

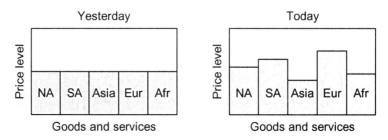

and ask: Will an across-the-board shift that goes unnoticed in the United States be registered nevertheless as a perceptible difference in Germany? Can each of us, waking up one fine morning in our respective countries, determine that the balance of trade has changed in some way?

Of course we can, because that is how an interconnected capitalist world works. Shifts in prices are never coordinated globally, and thus local changes are sure to be recognized for what they are. The only way we could possibly overlook them would be if some mediating agency (an invisible army of gnomes, let's say) were to smooth out the ups and downs from region to region. The mediators would inject money here and extract money there, subsidizing price increases in one area and confiscating windfalls in another. Exercising a kind of intercontinental economic "force," an exchange of influence, the gnomes would ensure that any locally symmetric transformations would also be undetectable globally.

Is that the way things really work out? In the economic world, no. There is no perfectly efficient mediating agency to smooth away the differences, and so (in anything short of a World Socialist Paradise) local and global economic symmetry are incompatible.

In the physical world, yes. In the physical world, nature decrees not only that local symmetry shall be permitted, but that a transformation in one place shall not impair the symmetry of physical law everywhere else. A leprechaun in Alabama, acting alone, can arbitrarily shift the local baseline of potential energy without asking a leprechaun in New Hampshire to make exactly the same change. Observers in both places always describe phenomena in the same way with the same equations, and neither of them can tell whether a leprechaun might have surreptitiously and independently shifted the local plateau. Whatever mischief may have occurred (and wherever the arbitrary baseline may now lie), each observer agrees, for example, that a ball resting undisturbed on a flat surface stays put:

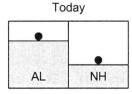

Yesterday   Today

The universe could well have been ordered differently, but apparently it is not. Like the American Congressman who once asserted that "all politics is local," nature seems to require something of the same sort for physical law: that all *symmetry* must be local, that legitimate transformations must be permitted at all points in space-time, that an observer in one place must have the right to make arbitrary shifts without worrying about observers in other places.

To enforce those rights, there must be a mediator. There must be an agency, a go-between, an intermediary to smooth away the ups and downs and thereby make the universe safe for local symmetry. For if there is to be local symmetry, then there must be a means by which to exchange influence; there must be a web of interconnected *force*.

Gravity. Electromagnetism. The strong force. The weak force. Each fundamental interaction is called into being by the requirements of a particular local symmetry. To appreciate now, in broad terms, the ways in which such connections may arise, we turn first to Einstein's macroscopic conception of gravity.

Leveling the Playing Field

To an observer locked inside, it makes no difference whether a sealed compartment is accelerating through empty space or sitting at rest in a gravitational field. Remember the lesson of general relativity (from Chapter 5): Whatever gravity can do, an accelerated reference frame can do just as well; and so, with nothing to differentiate one environment from the other, the unbiased observer has no choice but to remain agnostic.

> OBSERVATION: An object released near the ceiling falls to the floor, picking up speed at every instant. Is the fall occasioned by gravity

(the influence of mass), or is it just an illusion created by an external acceleration? Nobody knows. The same apparent motion can be produced either way.

OBSERVATION: An object already on the floor, left to its own devices, stays where it is. It moves neither up nor down, as if pinned in place by a force. Does the presumed attraction come from a gravitational interaction, or does it come from an acceleration of the reference frame? Nobody can say for sure. Either way, the effect would be the same.

Were it not for gravity, we would be able to tell the difference. We would be able to distinguish (and distinguish absolutely) between a reference frame accelerating at one rate and a reference frame accelerating at another rate. Observing the behavior of falling bodies, we would be able to determine that our compartment is increasing its velocity by, say, 10 meters per second from one second to the next, or by 20 meters per second, or by 30 meters per second. Why? Because in a world without gravity, our only explanation would be to attribute the accelerated motion of a freely falling body to the overall acceleration of the reference frame. There would be no doubt about it.

But if we are going to have a world in which acceleration is *not* absolute—a world in which the laws of physics are the same to all observers, even to those in accelerated reference frames—then it must perforce be a world that includes gravity. Nature, by providing for a gravitational field, renders all reference frames equivalent and equally valid for observation. The mere possibility of a field erases any distinction between accelerations produced gravitationally (by the influence of mass) and those produced artificially (by such contrivances as rocket engines and rotating wheels). The connection is inescapable. If there is going to be a symmetry of accelerated reference frames, then there must be gravity.

It is a local symmetry, too, a right accorded independently to observers anywhere in space-time: the freedom to apply whatever specific adjustment is necessary to mimic the effects of gravity at one's particular location. For an observer over here, the appropriate external acceleration will differ from the value needed by an observer over there. So it must be. Nature allows observers 1 and 2 the flexibility to transform their local coordinate systems without mutual consultation. They use the same method but not the same numbers.

What we have, then, is a restatement of Einstein's principle of equivalence (Chapter 5), the basis of his general theory of relativity and a justification of gravity's macroscopic place in the universe. Because if local symmetry, enforcing the principle of equivalence, is called upon to ensure a level playing field—so that all accelerated reference frames are created equal—then a gravitational influence must be called into existence as well. It is no coincidence.

And not just a macroscopic and classical field, like Einstein's gravity, but the microscopic and quantum mechanical fields (the electromagnetic, the weak, the strong) also seem to be conditioned on the demands of local symmetry: a fearfully significant, new kind of symmetry; a symmetry rooted in the abstract phases of quantum mechanical states; a symmetry that motivates the interactions between particles; a quantum mechanical symmetry that, in one form or another, creates the glue that holds together neutrons and protons and nuclei and atoms. Here, to finish the chapter, is a hint of how local symmetry brings quantum mechanical force into the microworld.

Going through a Phase

Start with the simplest system of all, a free particle. If the wave function below, with a phase of zero, is a solution to the Schrödinger equation,

phase = 0° (beginning of cycle)  • • •

then so is this:

phase = 90° (shift of ¼ cycle)  • • •

And this:

phase = 180° (shift of ½ cycle)  • • •

And this:

phase = 270° (shift of ¾ cycle)  • • •

And so is every other waveform that has the same shape and wavelength. If one of them happens to be an eigenstate, then so are all the others. All of them yield the same eigenvalue; and all of them, when squared, carry the same probabilistic information. They differ only in absolute phase, the arbitrary point at which a cycle is deemed to begin.

One starting point is as good as another. With eyes closed, an observer has no way to tell whether the overall phase of a wave function has been shifted. It is a fundamental symmetry enjoyed by all quantum mechanical states, whether represented as a single independent component or as a superposition of many.

Two observers, separated by time and space, construct their superpositions in the way described in Chapter 8. They do so, and their freedom as individuals to impose a uniform phase shift remains unabridged. To shift the phase is merely to redefine and recombine what one calls "state 1" and "state 2,"

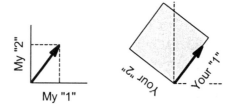

and we are long accustomed to tolerate the rotation of an arbitrary coordinate system. The length of the state vector holds constant during any rotation, and so do all measurable properties of a system.

Assume that observer 1, located in New York's Grand Central Station, chooses to take advantage of this peculiar quantum mechanical symmetry. For whatever reason (perhaps simply for convenience), observer 1 decides to twiddle the overall phase of a quantum mechanical state, advancing or retarding it by some fixed amount. Nothing in observer 1's world seems to change as a result. The system obeys the same laws, the same equations of motion, the same conservation principles as before. Its properties are invariant.

Yet what about the universe as perceived somewhere else? Does the right of observer 1 to apply a phase shift in Grand Central Station infringe on the rights of an observer 2 on the planet Melmac? Must observer 1, in order not to alter the laws of nature, ask observers everywhere else in the universe to impose exactly the same shift at exactly the same time?

If it were true, then quantum mechanical phase shifts would obey a global rather than a local symmetry, and the laws of physics would remain invariant only if such changes were applied uniformly and simultaneously all throughout space-time. Does observer 1 want to rotate a set of axes counterclockwise by 45 degrees? Fine, observer 1 is free to do so, except all the rest of us would then have to follow suit—if the symmetry were not local.

But the symmetry *is* local. In the quantum mechanical universe, each of us has the freedom to alter the phase of a system independently, seemingly without consultation or cooperation with observers in other locations. Hansel can shift by 90 degrees, and Gretel can shift by 76 degrees, and somebody else can shift by 150 degrees, and in the end, despite the asynchronous phases, everyone understands the same laws in the same way:

The various environments cannot be distinguished by observation, and any mathematical description of them (if it is to be valid and complete) must allow for a similar ambiguity. In whatever form they ultimately may take, the quantum mechanical equations must remain untouched globally by the imposition of a local phase shift.

Physicists call the operation a "gauge" transformation, and it is precisely this local gauge symmetry, coupled with special relativity, that gives birth to the quantum mechanical forces of nature: the electromagnetic field, the weak field, the strong field, and maybe (although a proper quantum theory has yet to be formulated) the gravitational field as well.

They appear as if by enchantment. Once we insist on local symmetry and relativistic compliance, a suitable mediating field forces itself onto a quantum mechanical equation of motion. Without such fields, quantum mechanics would not be invariant to local shifts in phase.

They are the middlemen, the physical realization of our economic gnomes in the fantasy a few pages back. The quantum fields, with their "messenger particles" (see below), convey information about local phase conventions to all points beyond. They are the emissaries that enable every particle to communicate with every other particle in the universe—and as a consequence of that global intervention, wonderfully, they establish local autonomy for all.

There is something aesthetically pleasing about the chicken-and-egg relationship. Global interaction is the guarantor of local symmetry, and local symmetry is the motivating principle for global interaction. Each serves as raison d'être for the other; each is simultaneously cause and effect.

The Medium and the Message

In the macroscopic world, we summon up the idea of "electric charge" and the field that it engenders without a hint of embarrassment. And why not? Classical electromagnetic theory is an engineer's paradise. Armed with Maxwell's equations, supremely confident, we make prediction after prediction and never fail. Our mathematical description of the macroscopic electromagnetic field is exact. Our numbers are never wrong. Let's even go as far as to say (because it is true) that within the boundaries of Maxwell's world, we know all there is to know.

We know that some particles are negative and some particles are positive. We know that opposites attract and likes repel. We know that stationary charges produce electric fields and moving charges produce magnetic fields. We know that a fluctuating magnetic field induces an electric field and, vice versa, a fluctuating electric field induces a magnetic field. We know how to compute the forces between charges and currents. We know how electromagnetic waves behave in a vacuum, and we know how they behave in matter.

Yet for all that (and "all that" is nothing short of miraculous, an intellectual triumph of the first order), we cannot give meaningful answers to some of the most basic questions. "Mr. Maxwell, please," asks a precocious child, "what, exactly, is an electric charge, and how does it go about creating an electromagnetic field?"

"That is not for us to know," replies Maxwell, "not in the Land of the Large, not in the Land of the Many Many Quanta. Be content to think of an electric charge in terms of what it does, not what it 'is' in some deeper sense. Think of field and charge in relation to the pushes and pulls you register on your instruments. Be realistic. Be practical."

Sorry to disappoint, but our macroscopic understanding is more empirical than fundamental, poorly suited to exposing the ultimate origins and causes of things. Viewed through a macroscopic lens, our view of the world is just too coarse. We can do little else than describe electric charge as an "endowment" of matter that makes a particle receptive to an electromagnetic force.

Awakened to the microscopic world, however, we understand the electromagnetic field in a new, more substantive light: as something grainy, not smooth; something tangible, not ethereal; something particulate, not continuous. The quantized electromagnetic field takes hold as an assembly of individual photons, each with its measured dose of energy, linear momentum, and spin angular momentum as noted in Chapter 7. They are massless *bosons*, to be specific, equipped with one unit of spin angular momentum per particle, and with zero rest mass a photon travels unceasingly and untiringly at the speed of light. Photons come into existence, moreover, not by accident and not without reason, but as the inevitable consequence of local symmetry and special relativity. They have a job to do. They serve as messengers of the electromagnetic field.

Picture this: An electrically charged particle, seen up close, acts as an inexhaustible source and sink of photons, emitting and absorbing them without letup. Back and forth, pitch and catch, one particle transmits a photon and another receives it. Like two sweethearts exchanging letters, the charged particles interact. They attract or repel. Just as a certain kind of correspondence will decrease the emotional distance between sweethearts, an analogous form of attraction results when oppositely charged particles exchange photons. The net effect is to bring the particles of matter closer together:

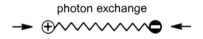

And just as another kind of letter, with a different wording, has the potential to drive sweethearts apart, so it is with photons as well. Photons exchanged between like-charged particles bring about a repulsion, a tendency to move farther away:

In either case, there is a connection between particles. A boson (a photon, a quantum of the electromagnetic field) links together two fermions. A particle of matter interacts with a particle of the field, and a particle of the field interacts with a particle of matter. The exchange of a photon makes the association possible.

It is, oddly enough, a surreptitious exchange that borders on the illegal, since the emission of a photon—out of the blue, apparently—increases the global energy account by one quantum. Such action would never be tolerated in a classical universe, but in a quantum mechanical universe there is a loophole in the law: Heisenberg's uncertainty principle, which allows a temporary violation of energy conservation.

In Chapter 7, recall, we stated the uncertainty principle in terms of an indeterminacy in momentum (call it Δp) and an indeterminacy in position (call it Δx). When the indeterminacy in momentum is large, the indeterminacy in position is small. When the indeterminacy in momentum is small, the indeterminacy in position is large. The law requires that the product $\Delta x \, \Delta p$ be greater than Planck's constant, h, or roughly so.

An equivalent principle holds for energy and time. Nature permits an amount of energy (ΔE) to be "borrowed" for a time Δt commensurate with the intrinsic indeterminacy of these two complementary quantities. So long as the product $\Delta E \, \Delta t$ remains less than approximately h, the value of E is undetermined and a borrower can redeposit the energy in question without penalty. Nobody misses it. If the amount taken is small, then the time to repay is long. If the amount taken is large, then the time to repay is short. Outside the allotted time, nature enforces energy conservation without exception. Within the allotted time, anything goes.

Gone but not forgotten. A fermion over here spits out a photon, and a fermion over there swallows it up. A quantum of energy rises out of the void and disappears before the loan comes due. "No harm, no foul," as basketball players like to say, and thus the exchange of a messenger boson takes place as if under the cover of an indeterminate quantum fog. The transitory photons are called "virtual" photons, as opposed to the enduring "real" photons of electromagnetic radiation. They (the virtual photons) create the force between two fermions.

Toward Unity

Fermion...boson...fermion. With this basic pattern as prototype, we discover a formula that extends beyond the electromagnetic interaction to quantum fields in general. See it as the start of something small:

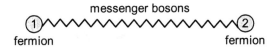

A particle of matter, a fermion, communicates with another particle of matter (another fermion) by the agency of a stealthy messenger particle: a massless boson, an emissary that goes forth to preserve local symmetry everywhere throughout space-time, staying all the while within the energy-time constraint of the uncertainty principle. For every mission, there arises a messenger or group of messengers appropriate to the particular endowment of matter and the particular local symmetry. The electromagnetic interaction, with its exchange of photons between electrically charged particles, proves to be just one cousin in an extended family of forces.

We consider each of them in turn, starting with a capsule summary of quantum electrodynamics, which serves as a model for the rest.

THE ELECTROMAGNETIC INTERACTION: The controlling endowment is the electric charge, and the messenger particle it dispatches is the photon. Viewed quantum mechanically, an electric charge gives a fermion the license to emit and absorb photons. Candidates for electromagnetic interaction include electrons and quarks, together with all charged particles built from quarks (such as protons). The local symmetry invoked is a relatively simple phase rotation involving only one angle, similar to what we imagine for our archetypal two-state system.

THE WEAK INTERACTION: From Chapter 2, we already have one example of beta decay. A down quark in a neutron changes into an up quark, and the result is to convert the neutron into a proton, electron, and antineutrino (an electrically neutral fermion with little or no rest mass):

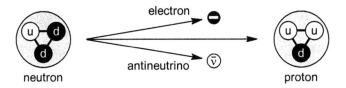

Another example of a weak interaction is the collision of a neutron with a neutrino, which can either leave the particles intact (like billiard balls),

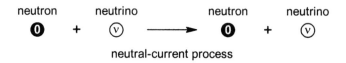

or transform them into a proton and electron:

Now, whereas the electromagnetic interaction derives from electric charge, the weak interaction derives from an appropriate "weak interaction charge." Particles so endowed are able to communicate via three messenger bosons (labeled W^+, W^-, and Z^0), which do for the weak interaction what photons do for the electromagnetic interaction. They link together the fermions and allow the various transformations to take place. For instance: A down quark emits a W particle and turns into an up quark, thus converting a neutron into a proton. And: A neutrino emits a Z particle and *stays* a neutrino, as in the neutral-current process above. Or: A neutrino and neutron exchange a W particle and turn into an electron and proton, respectively, as in the related charged-current process:

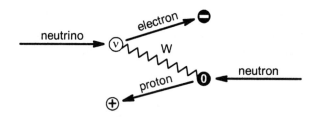

And so on. The local symmetry that begets the W and Z bosons traces back to the arbitrary resolution of a state into, for example, an "electron"

component and a "neutrino" component:

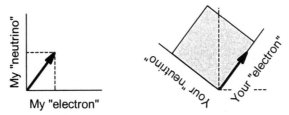

What you call an electron, I describe equally well as a superposition of electron and neutrino. What our correspondent in London calls an up quark, our correspondent in Rome calls a down quark. What John says is a 50:50 blend of this particle and that, Mary says is a 75:25 blend. We are all correct. Everybody has the right to an opinion, and nature gives us the W and Z bosons to guarantee our local freedom of choice. The force is tailored to a symmetry that takes in the entire group of possible transformations.

If it were a perfect symmetry, then the messenger bosons would be massless, as are photons, and consequently be able to travel great distances before running afoul of energy conservation. The symmetry is not perfect, though, and the imperfection is manifested in the masses of the weak bosons. The W and Z bosons are indeed heavy particles, more like bowling balls than Ping-Pong balls, and as messengers they can travel only absurdly short distances. Borrowing large amounts of energy to acquire their considerable masses (as prescribed by $E = mc^2$), they have only a brief time to hop from fermion to fermion before the loan comes due. As a result, the weak interaction operates at lengths at least a hundredfold shorter (maybe more) than the already absurdly short millionth of a billionth of a meter staked out by the strong interaction.

But consider this: The W and Z particles, notwithstanding their short range, do have some things in common with photons. Like photons, the weak bosons all carry a single unit of spin angular momentum; and, like photons, at least one of them (the neutral Z particle) does not alter the identity of a fermion when absorbed or emitted. An electron that emits a photon stays an electron. A neutrino that absorbs a Z particle stays a neutrino. Might that be a sign of kinship, evidence of a common ancestry?

Perhaps, but think again of all the differences. Photons are massless; the W and Z particles are not. Photons can travel from one end of the universe to the other; the W and Z particles struggle just to get out the door. Photons mediate the electromagnetic interaction and are plain as day. The W and Z particles mediate mostly beta decay, obscure radioactive processes that we hardly notice at all. The energy involved, another clue, is also far less for the weak interactions than for electromagnetic phenomena.

Different? Clearly so; and yet, deep down, the electromagnetic and weak interactions are really as close to each other as electricity is to magnetism. They are variant expressions of a single force arising from a broader, more inclusive, more perfect symmetry—but it is a symmetry that we do not normally see. It is a symmetry apparent only at very high energies, a primeval symmetry believed to have existed briefly during the Big Bang but long since broken. To recreate it, experimenters must orchestrate collisions of exceptional violence in high-energy particle accelerators.

Symmetries are made to be broken. It happens all the time in a less than perfect universe, an unavoidable and prosaic cost of doing business. Think of how gravity and mass create a misleading sense of "up" and "down" in a space otherwise lacking direction. Think of how a magnetic field introduces an accidental asymmetry into the world of a compass needle. Think of all the circumstances under which a deeper symmetry can be masked, and then be prepared to accept a similar fate for the electromagnetic and weak interactions. Their apparent divergence arises not from any fundamental difference, but merely from broken symmetry. It is an accident.

For when the true, underlying symmetry of the electromagnetic and weak fields is realized, then the photon and the trio of weak bosons merge into a quartet of massless messenger particles, all emissaries of a single, unified force: the "electroweak" interaction. The weak interaction, no longer in a class by itself, becomes simply another face of the electromagnetic interaction, just as the magnetic interaction is an equivalent face of the electric interaction. Rather than treat electricity, magnetism, and beta decay as three separate forces, we can now roll all of them into one. It is a step toward unity.

THE STRONG INTERACTION: Quarks have it; electrons do not: the fancifully named attribute of "color," source of the strong interaction.

It makes a quark a quark. Endowed with color, a quark participates directly in the strong force, something no other elementary fermion can do. Lacking color, particles such as electrons and neutrinos are immune.

We touched on the strong interaction at the end of Chapter 2, and we return to it now in the context of quantum fields and their messengers. Let the quantized electromagnetic field serve as a standard for comparison:

1. In electromagnetic interactions, charge comes in one basic variety (positive) and one corresponding "anti" variety (negative). *In strong interactions, charge comes in three basic varieties (the nonliteral colors red, green, blue) and three corresponding anti varieties (the anticolors negative red, negative green, negative blue).*

2. In electromagnetic interactions, opposite charges attract and like charges repel. *In strong interactions, different colors attract and like colors repel.*

3. In electromagnetic interactions, negative cancels positive to leave a system uncharged and thus electrically neutral. *In strong interactions, anticolor cancels color to leave a system colorless: chromatically neutral. Positive blue plus negative blue makes zero.*

4. In electromagnetic interactions, equal weights of red, green, and blue wavelengths produce chromatically neutral white light. *In strong interactions, equal weights of red, green, and blue quarks produce chromatically neutral (colorless) protons and neutrons.*

5. In electromagnetic interactions, absolute phase has no meaning. A cycle may be advanced or delayed by any amount without prejudice to an observer. *In strong interactions, absolute color has no meaning. Quark colors may be shifted uniformly from red-green-blue to cyan-magenta-yellow or to any other mixture without prejudice to an observer.*

6. In electromagnetic interactions, a one-dimensional phase rotation calls forth a field of photons: massless bosons, carriers of the force, each with one unit of spin angular momentum. *In strong interactions, a symmetric rotation through all three colors calls forth a field of "gluons": massless bosons, carriers of the force, each with one unit of spin angular momentum.*

7. In electromagnetic interactions, possession of an electric charge enables a quark or electron to traffic in photons. *In strong*

interactions, possession of a color charge enables a quark to traffic in gluons.

8. In electromagnetic interactions, photons bind oppositely charged nuclei and electrons into electrically neutral atoms and molecules. *In strong interactions, gluons bind quarks of different colors into chromatically neutral protons and neutrons.*

9. In electromagnetic interactions, photons exist in only one form and are electrically uncharged. They broker the electromagnetic interaction but do not experience it themselves. *In strong interactions, gluons exist in fully eight forms and carry flickering color charges as well. They transcend the role of mere broker and take part directly in the interactions.*

Quantum electrodynamics, our long-standing theory of the quantized electromagnetic field, paints a picture of photons passed between electrically charged quarks and electrons. Quantum "chromodynamics," our evolving theory of the strong field, paints a picture of gluons passed between color-charged quarks. It is a richer, more complex quantum field born of a more complex symmetry.

The color-charged messenger particles, the gluons, enforce the symmetry. Quarks, remember, come in six "flavors" distributed over three families (up–down, charm–strange, and top–bottom, as we asserted back in Chapter 2), and it becomes the job of the gluons to change the colors while preserving the flavors. *W* particles, by contrast, preserve the color of a quark but change its flavor—say from down to up, as when a neutron undergoes beta decay to become a proton. And that, in a nutshell, is the difference between the weak force and the strong force: a manipulation of flavor versus a manipulation of color.

Of the six flavors offered on nature's menu, only the up–down quarks contribute to the protons and neutrons of ordinary matter; yet still, even with this restriction, the opportunities for mischief are many. Think of just some of the possible ways a leprechaun might play symmetry games with the color. A *red* up quark emits or absorbs a colored gluon and becomes a *blue* up quark or a *green* up quark. A *blue* down quark emits or absorbs a colored gluon and becomes a *red* down quark or a *green* down quark. A *green* up quark emits or absorbs a colored gluon and becomes a *red* up quark or a *blue* up quark. Or, if you prefer, a *cyan* quark becomes a *magenta* quark, or a *magenta* quark becomes a *yellow* quark, or a *yellow* quark becomes a *cyan* quark (or, *ad infinitum*, a quark of whatever arbitrary color you contrive to mix is put through its chromatic paces).

The choices are all equally acceptable. Observer 1's definition of primary colors makes little difference to observer 2, provided that a field of eight gluons arises to guarantee the same law for both parties. Switching continually from red to green to blue themselves, the gluons keep a proton or neutron chromatically neutral even as they rotate the three component quarks through all their colors. The strong force between quark and quark shines through.

There is a twist, too, because the attribute of color also gives rise to gluon–gluon interactions, and these additional exchanges lend a peculiar profile to the strong force. Unlike electromagnetic attractions and repulsions, quark–quark interactions become stronger with distance. Quarks that are relatively far apart are attracted more powerfully than quarks that are near. Like prisoners on a chain gang, the particles move about freely when they stay close together (and the chains are loose), but they face severe constraints if they try to move farther away:

   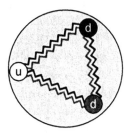

freedom   confinement

Past a certain point, the chains tighten and the quarks are unable to escape. They exist only as inseparable bundles of two and three—as *three* quarks of different colors in protons and neutrons,

proton   neutron

and as *quark–antiquark pairs* in a class of particles called "mesons":

Mesons, which come in assorted varieties, happen to be bosons with nonzero masses, and they act sometimes as short-range messengers. Passed between protons and neutrons, for example, certain kinds of mesons supply the glue that holds together a nucleus. Proton interacts with proton. Neutron interacts with proton. Neutron interacts with neutron:

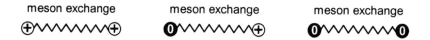

Each trio of quarks is a chain gang unto itself, and together—tossing mesons back and forth—the neutrons and protons in a nucleus play a familiar game of not-too-near and not-too-far. Close together, they repel; far away, they attract; and somewhere in the middle, at some comfortable distance in between, the interacting chain gangs find stability:

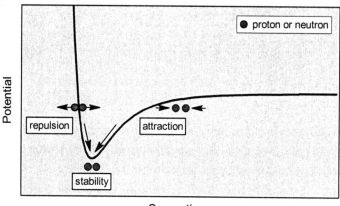

But don't think we have some kind of new interaction here, because the meson plays only a secondary role in implementing the strong nuclear force. The real action occurs at the level of quarks and gluons, not protons and neutrons. Meson exchange, a deceptively simple game of pitch and catch, traces back ultimately to the underlying color interactions. It is a high-level *effect* of the strong force, not a primary cause.

Finally, there is the question of what physicists call "grand unification" of the quantum fields. Does the strong force (the color force) stand alone, or is it part of a larger family? Might the strong force and the newly unified electroweak force be different aspects of a more general, more inclusive field? Are there circumstances under which the photon, the *W* and *Z* bosons, and the gluons all become indistinguishable members of the same team, just as the photon, *W*, and *Z* have already done?

The answer, consistent with our present understanding, is a firm *maybe* (many would even say, *probably*), but theoreticians and experimenters have yet to elucidate the full symmetry of a grand unified field. Despite considerable progress, serious questions remain. Time will tell.

THE GRAVITATIONAL INTERACTION: Gravity, for now, remains on the outside looking in: as a field we understand macroscopically (and beautifully so), but nevertheless a field for which there is still no satisfactory quantum theory. Einstein's general relativity, however, does allow for the existence of oscillatory "gravitational waves" analogous to Maxwell's electromagnetic waves.

Like its electromagnetic counterpart, a gravitational wave would emanate from a disturbance of its source (mass); and like an electromagnetic wave, it would propagate through space at the speed of light. If true, we should then expect every clump of mass to send ripples through the fabric of space-time whenever it moves, like a stone dropped into a pond.

Yes? No? *Maybe.* Gravitational ripples may indeed be all around us (and many experts firmly believe they are), but they are so extremely weak as to elude direct detection...at least with our present technology. Indirect evidence tends to confirm their existence, though, and someday, possibly soon, we may know for sure.

Meanwhile, as with the electromagnetic field, we expect the gravitational field to be composed microscopically of particles that play the role of photons, albeit particles with at least one intriguingly different property. Dubbed the "graviton," the messenger particle of a quantized gravitational field would be a massless boson with a spin of *two* units rather than the usual one. A subtle sign, perhaps, of apartness.

Again, it may well be. Minutely small energies make gravitons extraordinarily difficult to detect. Our inability to find them does not necessarily damn them to nonexistence.

But in the end, the answer may lie elsewhere, in something other than a quantum mechanical field as conventionally formulated. Severe theoretical impediments stand in the way, first, of constructing a quantum theory of gravity and, second, of unifying a quantized gravitational field with the electroweak and strong fields. A new approach is needed, and some of the world's physicists are working seriously to find one.

Will salvation come in a form of "string theory" or one of its derivatives? Will our traditional notion of pointlike particles be replaced by a picture of ultra-ultra-ultra-tiny lengths of space-time vibrating in a world embracing more than four dimensions? Will all the fields, including gravity, then fall into place—and then, even if they do, will some new fundamental interaction, yet to be discovered, show up to crash the party?

For the moment, nobody knows. Time will tell.

10

ENDS AND ODDS

Splat! A drop of ink falls into a glass of water, and immediately the color starts to spread. Dark at first, confined to a small spot, the stain keeps growing until eventually—some time *later*—it fills the entire liquid with a uniform tint:

now   later

And after that, nothing more seems to happen. Whenever we choose to look, we always see the same unvarying tinge:

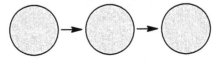

What we do not see (not in a million billion years), is the movie run in reverse. Once the ink has spread, we never see the color go spontaneously from uniformly pale to dark in the center. If someone were to show us a film alleging to portray such a process,

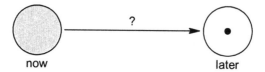

we would know right away that something is wrong. It would be like seeing Humpty-Dumpty put himself back together again. It would violate the natural order of things. It would be like swimming against the tide of time.

Time, as the saying goes, marches on. Time flows macroscopically in only one direction, from *now* (when the ink is concentrated in one place) to *later* (when the ink is spread evenly through the water)... from *now* (when the air bunches up at the mouth of the balloon) to *later* (when the air fills the balloon uniformly)... from *now* (when a lump of sugar floats in the coffee) to *later* (when the lump is gone and the coffee is sweet throughout). The world moves ahead incxorably, and in the macroscopic universe there is no turning back.

But how can it be? How can a macroscopic process be irreversible when the microscopic laws of motion—which, after all, should ultimately govern the macroscopic world as well—make no temporal distinction between forward and backward? Follow just one of the molecules in a large system,

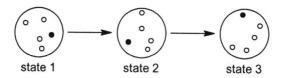

and try to determine whether the scene is unfolding in "real" time (presumably as above) or in reversed time, as below:

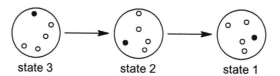

Is there anything particularly unusual, say, about the positions of the black dot in the sequence 3–2–1 as compared with the sequence 1–2–3? Is one series more "natural" than the other, in the way that a uniformly pale mixture of ink and water is a more natural final state than any other?

No. For each of the microscopic particles in a macroscopic system, time is a two-way street. Neither Newton's second law of motion (classical mechanics) nor Schrödinger's equation (quantum mechanics) distinguishes between time flowing forward and time flowing backward. If the equations of motion call for a particle to leave Chicago on Monday and arrive in Memphis on Tuesday, then those same equations will also allow it to leave Memphis on Tuesday and arrive in Chicago on Monday.

How can it be? Macroscopic experience notwithstanding, the microscopic laws of nature assign no significance to the direction of time; and yet somehow, although not by design, an unmistakable sense of past, present, and future is the macroscopic consequence. To understand why—to reconcile macroscopic and microscopic—would be to take a giant step toward understanding a universe ruled by blind chance.

As a start, we restrict ourselves to an important but special class of events: to processes, like the dissolution of an ink blot, that come to a predictable and stable end. Whatever their ups and downs, these stable systems settle into a unique and final state of equilibrium in which outward signs of change are no longer manifest. Such systems are isolated and self-sustaining, able to maintain their equilibria without importing or exporting energy and matter. They make do with what they have.

Systems in stable equilibrium are by no means the only ones of interest (to be alive, for instance, is to be decidedly *out* of equilibrium), and we shall address the issues of chaos and complexity in the next chapter. For now, however, we focus on the Big, the Many, and the Simple: complicated systems that, in the end, behave simply.

Strategies for the Big, the Many, and the Simple

Think big: a macroscopic system, a piece of matter big enough to suffer gross changes when we manipulate it, a system that by outward appearances betrays no evidence of any microscopic substructure. No molecules. No atoms. No electrons. No protons. Picture the material simply as a continuous fabric with no perceptible grain,

and prepare to deal with it strictly on a macroscopic basis, as something
we can see and touch.

Let it be, for example, a quantity of gas confined to a tube, a setup that
we can easily control. We provide valves to allow material in and out. We
install a weighted piston to impose a pressure, and we use a surrounding
heat bath (it can be a big tub of water) to impose a temperature:

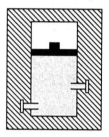

What can we do to our system, and what might we find?

By moving the piston, we can squeeze or expand the gas; and by
varying the weight from above, we can alter the pressure. We can heat or
cool the gas by changing the temperature of the surrounding bath. We can
add gas; we can take away gas; we can introduce other substances as well.
These gross macroscopic variables—volume, pressure, temperature,
amount of material, chemical composition—are all at our disposal. They
give us a way, with just a few numbers, to describe the bulk system with
mathematical precision and to characterize the changes it undergoes. We
measure the variables as if atoms and molecules do not exist, and we use
the numbers to answer "big-picture" questions such as the following:

QUESTION: How much space does the gas take up?
ANSWER: Measure the volume (length × width × height).

QUESTION: How hard does the gas push against the piston?
ANSWER: Measure the pressure, the ratio of net force to area.

QUESTION: How hot is the gas?
ANSWER: Measure the temperature, as shown on a thermometer.

QUESTION: How much stuff is there?
ANSWER: Measure the mass.

Before measuring anything, though, we wait for the variables to assume constant values. They always do, eventually, for a system of this sort always settles into a state of equilibrium (the word means "equal balance") when its opposing forces battle to a draw. Thereafter, everything stays the same. The pressure is constant. The temperature is constant. The distribution of matter is constant. There is a macroscopic tug of war fought over each variable, and there is no winner.

First, the pressure. Able to interact through a movable wall (the piston), the gas and its surroundings push one against the other until neither side can prevail:

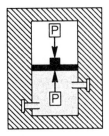

The equilibrium pressure exerted by the gas subsequently remains equal to the pressure exerted by the weighted piston. The weight presses down and the gas pushes up, but still the piston stays in place. There develops a standoff between system and surroundings, an ongoing stalemate that leaves the gas with unchanging pressure and volume.

Second, the temperature. Able to exchange heat through a conducting wall (the tube), the gas and its surroundings come finally to a common temperature. Heat passes from regions at high temperature to regions at low temperature, flowing back and forth in different amounts until everywhere the temperature is the same:

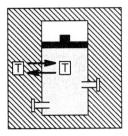

Another continuing standoff sets in, this time a persistent equality of temperature brought about by the free exchange of heat.

Third, the distribution of matter. Able to flow freely, portions of the gas move randomly throughout the container until all differences in concentration are erased. The random flows never cease, yet still the uniform distribution persists in a state of equilibrium. Once again, freedom leads to equality: freedom of position, equality of distribution.

From freedom and time, evidently, comes a state of equilibrium, a seemingly natural condition of rest for a system. The freedom to push against a movable wall, the freedom to exchange heat, the freedom to move internally—the freedom, in general, to *interact* and the requisite time to do so are the guarantors of an eventual equilibrium.

We start to develop a strategy. With our equilibrated gas as a first model, we imagine taking full control of the variables. If we increase the volume while holding the temperature and mass constant, what happens to the pressure? How does the temperature change when the pressure is varied slightly? How does the amount of material affect the volume? Under what combinations of pressure and temperature will the gas condense into a liquid? Are the relationships among the variables the same for all gases, or is each system a world unto itself? Are there perhaps certain conditions under which all gases seem to behave alike?

Guided by experiment, we set out to map the macroscopic changes occasioned by every such manipulation: to condense, as best we can, vast quantities of data into a manageable set of equations. The mathematical picture that develops (called "equilibrium thermodynamics") is extraordinarily robust and powerful, all the more so for its lack of microscopic justification. Reliably and accurately, the equations describe what *is* and predict what will be, and yet they represent bulk matter as structureless and continuous, making no mention of electrons, atoms, molecules, and their interactions. The model stands on its own, mathematically self-consistent, grounded firmly in observation, hard to deny. For matter of a certain size and scale, the thermodynamic view tells us all there is to know about a system in equilibrium.

We shall demand much more, of course, because no macroscopic model can tell us fundamentally how or why matter does what it what does. To discover, for instance, why a material contracts under pressure, we need to expose an internal structure unseen by an observer at the macroscopic level. We need to remember that bulk matter, smooth on its face, is grainy and particulate when viewed close up. Gas, liquid, or solid is built from molecules and atoms, each a small particle; and each atom is built from even smaller particles; and some of those smaller particles, too, are built from even smaller particles. Only from far away, like sand on a beach, do the tiny grains seem to blend together into a smooth continuum. The particles of matter are small but many, and every attribute of the large system comes ultimately from the small.

We return to our gas. Understood microscopically, its pressure arises from the impacts of individual molecules against the walls of a container. Each single impact is small, yet there is strength in numbers. The collisions come rapidly and in tremendous quantity, averaging together to produce a steady macroscopic pressure at equilibrium: a statistical average, emerging clear and sharp from the microscopic confusion. Microscopic randomness gives way to macroscopic reliability.

Temperature, also a statistic, derives from the average speed of the particles—again, clear and sharp. Can you tell me the speed of a single particle? There is a range of probabilities. Some particles travel slowly; some travel quickly; others (most of them) travel at speeds in between:

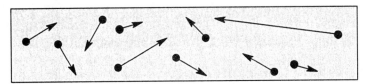

Can you tell me instead the average speed of many particles in equilibrium? Ah, now *that* becomes a statistical certainty, guaranteed by the tremendous number of individual players in the system. Each collision (each interaction) brings about a small exchange of energy, and eventually a large-scale state of equilibrium ensues. Microscopic randomness gives way to macroscopic reliability.

Distribution, too. Molecule by molecule, matter rearranges itself internally; everywhere there is motion, never rest. One molecule moves in and another molecule moves out, and collectively, statistically, an enormously large assembly of particles eventually fills the entire space uniformly. Microscopic randomness gives way to macroscopic reliability.

Right before our eyes, every day, we see the power of small particles to reshape the face of matter in bulk. Squeezed together, the molecules of a gas move closer as the pressure is increased. Cooled to a lower temperature, they move slower as well. Closer and closer, slower and slower, they interact with growing effectiveness. The position of one particle influences the position of another, and the molecules attract and repel:

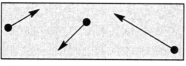

farther, faster   closer, slower

Pressed sufficiently hard, the particles behave collectively. They begin to cohere. The gas condenses into a liquid, and the liquid freezes into a solid. From the microscopic comes the macroscopic.

Always in motion, microscopic particles exchange energy and influence as they slip into a state of equilibrium. They move. They collide. They interact one by one; they interact in small clusters; they interact in large assemblies. We need to understand particles as crowds, and we need to understand them as individuals and small groups. For the many, we need statistics. For the few, we have the classical and quantum mechanical laws of motion: equations that determine the energy and structure of a microscopic system, equations that predict its mechanical past and future, equations that provide a microscopic foundation for a macroscopic statistics.

Balancing Act

We begin not at the beginning of the journey, but at the end: in a state of equilibrium, where systems go to die...and perhaps later to be reborn.

The End of History

To be in equilibrium is to lose track of time, to disappear into the gray sameness of an unchanging macroscopic state. The same pressure. The same temperature. The same volume. The same distribution. The same composition. Everything the same. With nothing to differentiate one instant from the next, past and future blur into an unvarying present, an eternal *now* unrelieved by any sense of earlier or later. Without change, time disappears. The clock ceases to tick.

It was not always so. Once upon a time, our simple system was probably undergoing all manner of change on the road to an eventual equilibrium. There were presumably imbalances in pressure, or temperature, or distribution, or any number of things. There may have been a surfeit of material here and a shortage there, a hot spot in this place and a cold spot in that place, a whirling, swirling flow of matter and energy all throughout. There would have been, moment to moment, a macroscopic "potential" for change, a driving force to impel the system from one state to another.

It might have been a tendency for heat to flow (compelled by an imbalance in temperature) or a tendency for matter to flow (compelled by an imbalance in distribution) or a tendency for one substance to change into another substance (compelled by an imbalance in chemical composition). Whatever the source, though, and whatever road a system may take, its thermodynamic driving force is strong when the difference in potential is large and weak when the difference is small. The steeper the slope, the greater the force:

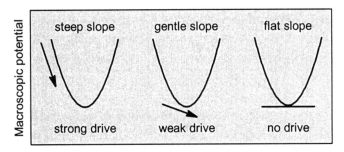

Far from equilibrium (above left), where the variation in macroscopic potential is large, the drive is strong. Close to equilibrium, where the variation is small, the drive is weak. In equilibrium itself (above right), where the potential is flat, the drive is zero. It is a balancing point, a position that offers no incentive for further change, and there the system sits. Its drive is spent. Whether nestled snugly in a valley or poised precariously on a hilltop, the equilibrated system comes to rest:

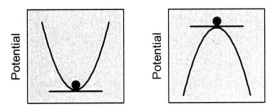

Game over. The system has done all it can do. It has no future, and it has no past.

For once a system attains equilibrium, all memory of the past is gone. Looking only at the present, nobody can say *when* the system got there, *how* it got there, *why* it got there. Many different routes might lead to the same final state, and an observer-come-lately has no way to tell which specific path was followed. The scant macroscopic information to be gleaned (constant pressure, constant volume, constant temperature, constant this, constant that) provides no clue to what came before. It is, at least for the present, the end of history.

Renaissance

Settled in a valley, trapped by walls of rising potential, a state of equilibrium maintains its balance in the face of small fluctuations. Like a ball given a gentle push up a hillside, the system relaxes back to equilibrium when disturbed only slightly:

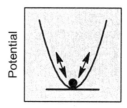

It enjoys a "stable" equilibrium, to be contrasted with the unstable kind of equilibrium established on a hilltop, where the potential is at a peak rather than a trough. Up at the summit, with no corrective force to nudge a fluctuating system back to equilibrium,

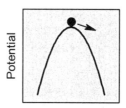

just the slightest touch is enough to get the ball rolling. An unstable equilibrium persists only until something comes along to disturb it. There is no place to go but down.

But even a stable equilibrium need not last forever, because stability is always a matter of where one sits in relation to some other possible state. A ball resting in a valley does find a measure of stability, yes, but with a sufficiently swift kick it will clear the neighboring hilltop and land somewhere else. And not just a ball in a valley, but all kinds of equilibria endure only as long as they receive no better offer. They are like inconstant lovers. Provide a system with the appropriate means, motive, and opportunity, and it will abandon one equilibrium condition for another.

See for yourself. Add enough heat to a pot of water, and the liquid will bubble up and start to boil. Evaporate enough water from a sugary solution, and granules of sugar will crystallize out of the mixture. Light a match. Burst a balloon. Recharge a battery. Hit any equilibrated system hard enough,

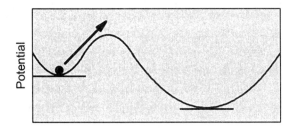

and it will awaken as if from a slumber. Setting off in search of a new, deeper valley of potential, the rejuvenated system will break the bounds of its former equilibrium and reveal a latent capacity for change.

Evidently there can be a second act after equilibrium, and more than that, too: there must also be activity *during* equilibrium. How else could a quiescent system, lost in the macroscopic timelessness of equilibrium, be able to accept a better offer and embark on a new history? To do so, it must tap a power that comes from within. It must draw upon a microscopic power belied by an overall macroscopic calm.

To see that power at work, we now need to look beneath the surface. We need to peer into the tumultuous microscopic depths of equilibrium.

Fight to a Draw

To Mack, our macroscopic observer, equilibrium is a static affair: a tableau, timeless and unchanging, a still photograph rather than a movie. Except for the occasional fluctuation, which flickers briefly and then disappears, there is nothing to report but a single set of macroscopic values. Mack submits his observations on just one sheet of paper, summarizing all there is to know about this uniquely persistent, one and only macroscopic state ("macrostate," for short).

To Mike, a microscopic observer, equilibrium offers a restless, dynamic picture of infinite variety. Atoms and molecules move this way and that. They bang into the walls. They bang into each other. Some speed up, and some slow down. They smash together and come away with new structures. They change partners. They form new associations. Over here, a gang of water molecules swarms around a sugar molecule; over there, a sugar molecule slips away and crystallizes out of the liquid. Things happen. Things change. Microscopic equilibrium is a movie with a cast of zillions, and every frame is different. Microscopic equilibrium unfolds as an ever-changing, kaleidoscopic sequence of different "microstates," each with its particular distribution of particles and energy.

But even as Mike's fine-grained movie plays on, with one inexhaustibly rich image giving way to another, Mack still sees the same scene frozen in time. For him, nothing changes. Each one of the microstates yields the same single set of macroscopic values. Each one of the equilibrium microstates, so different in its details, is consistent with just *one* macroscopic pressure, *one* macroscopic temperature, *one* macroscopic distribution; in sum, with *one* macroscopic state:

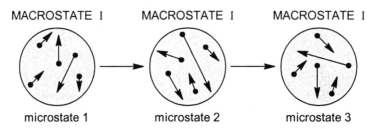

Run the tape backward, forward, in random sequence, in whatever way you like—it makes no difference. The macroscopic variables remain unmoved.

Meanwhile, the microscopic actors work furiously only to have the system stay in place. Atoms and molecules, colliding unceasingly, exchange energy and put it to work in innumerable small ways. They move through space; they vibrate; they rotate. They interact with fields. They reconfigure their electrons. They break into bits. They react chemically.

The give and take of energy never stops, but at equilibrium only the one macrostate endures. The microstates partition the total energy among different particles in different ways, yet still the equilibrium macrostate remains the same. Once established, the state of equilibrium is maintained actively from within.

Call it a draw. The microscopic frenzy of equilibrium succeeds only in preserving the status quo. For every molecule that pings sharply against a wall, another one rebounds all the more gently. *Averaged over time, the macroscopic pressure holds steady.* For every molecule that gains energy in a collision and begins to move faster, another one loses energy and slows down. *Averaged over time, the macroscopic temperature (a measure of particle speed) holds steady.* For every molecule that moves into a particular region, another one moves out. *Averaged over time, the macroscopic distribution of material holds steady.* For every molecule A that turns into molecule B, there is a molecule B that turns into molecule A. *Averaged over time, the macroscopic chemical composition holds steady.*

Tell that to Mack, and he is rightly amazed. He asks Microscopic Mike to describe the special force that guides a system unerringly (almost eerily) to equilibrium and subsequently defends the stable state so stubbornly against small fluctuations. "Mike," he says, "you can see what I cannot. Explain this great mystery to me."

"What are you talking about?" says Mike. "What mystery? What special force? I see nothing but a lot of little molecules obeying the ordinary laws of mechanics exactly as they should. Believe me, there is nothing unusual going on here."

No, nothing unusual at all, except maybe this: *dumb luck,* one of the most powerful, irresistible, compelling influences in nature. Call it "the law of averages" or "the law of large numbers" or (as we shall do later in this chapter) "the second law of thermodynamics," but recognize it as the law of the land in the Land of the Big, the Many, and the Simple.

It is the Law of Happenstance: the realization that things arrange themselves the way they do purely by accident, not through the intervention of any Master Strategist. A solitaire player, for example, never expects to deal 52 cards from a shuffled deck in strict numerical order, from ace through king in each of the four suits. Is it because the solitaire gods won't tolerate such perfection, or is it because there are more than

80,000,000,000,000,000,000,000,000,000,000,000,000,000,000,000,000, 000,000,000,000,000,000 other ways to rearrange 52 distinct objects?

Flip a coin ten times, and the results might be all heads. Such an outcome, although a thousand-to-one shot, is not beyond the realm of possibility. Flip a coin 100 billion times, however, and you will surely come within a hairsbreadth of 50 billion heads and 50 billion tails, certainly never all heads. Is it because the coin-flipping gods believe in fair play, or is it because there is only one way to throw 100 billion heads and uncountably many more ways to realize a fifty-fifty distribution?

Do the trillion trillion molecules in a small box of air spread out evenly over the volume because some providential hand purposefully directs their motion, or do they spread out simply because there are so many more ways (our minds cannot readily grasp the number) to occupy the entire space than to occupy just a portion of it?

Ask question after question of this sort, and the answer is always the same. Numbers in the microworld are staggering, beyond visceral understanding, and the sheer size of those numbers is what impels the universe forward, creating the illusion of purpose where there is indeed none. For processes destined to end in a stable equilibrium, look for the razor's edge of statistics to sharpen the arrow of time.

The arrow points the universe in the direction of increasing disorder, prescribing a "later" in which a fixed amount of global energy (the total never changes) will be spread over a greater number of recipients. Past, present, and future, the story of matter is told in the redistribution of energy among particles and fields. By following the trail of energy, macroscopic and microscopic observers alike bear witness to a universe that stumbles by chance from one configuration to the next.

Energy, Entropy, and the Arrow of Time

Atoms and molecules come and go, but energy lasts forever. Whether stored in a field or carried by a body in motion, energy provides nature with the capacity to do work. Energy moves matter. It rearranges particles. It builds up and tears down.

If atoms and molecules are the material goods of the universe, then energy is the capital invested in them. Just as money flows into and out of different accounts for different purposes, so does energy. Energy goes into the flight of a helium atom, into the rotation of a hydrogen molecule, into the vibration of a water molecule, into the flow of electrons in metallic gold. Energy goes into the wind and the rain, into the orbit of Jupiter, into the cosmos-shattering explosion of a supernova. Energy, potential and kinetic together, ebbs and flows from structure to structure, passing from particle to field and from field to particle.

Where there is a change in matter, there is a flow of energy; and in the midst of never-ending global change, there is this one great constancy: the total amount of energy stays the same, strictly conserved, constrained neither to grow nor to shrink. No power and no machine can create new energy, and no power and no machine can destroy existing energy. We can only make do with what we have, conserving energy wittingly or not. Nature allows no other way.

The energetic pie is divided and redivided, but not a crumb ever goes missing. One particle's loss is another particle's gain, and thus it happens, in dribs and drabs, in fits and starts, that the universe lurches forward.

The issue, then, is not the total amount of energy the universe has at its disposal (the sum is fixed), but rather how the universe disposes of what it has. Where does all the energy go, we wonder, and to what ends? Why does thermodynamic time seem to flow in only one direction?

The First Law: Work and Heat

Macroscopic observers like to think big. They see big things moving in orderly fashion—a flywheel turning, a piston pushing, one weight falling, another weight rising—and they speak of "work" being done. They speak, in mathematical language, of the work done by a force when it moves an object from point A to point B:

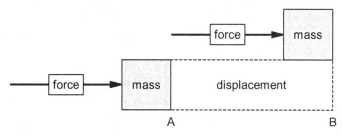

The greater the force and the greater the displacement, the greater the work. It costs a worker more effort to carry a fifty-pound sack than to carry a five-pound sack, just as it costs more effort to push an object fifty feet than to push it only five feet. The worker needs more energy to do so, because to do work is to expend energy. Energy is the currency used to purchase work.

Now if energy is conserved (and it is), then we should expect a dollar's worth of energy always to buy a dollar's worth of work. So when a quantity of energy is advertised as being able to move a mass from A to B, in a perfect world we hope to receive just that: the full displacement, every inch of it, an equitable exchange of energy for work. In the real world, however, the work delivered is inevitably less than promised. In the real world, where things rub and drag and scrape along as they move, the displacement invariably falls short of the mark:

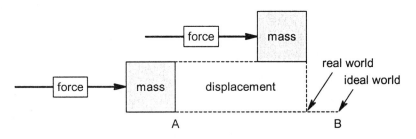

Maybe by a little and maybe by a lot, but the mass never makes it all the way to point B. There is always a shortfall.

Like money vanishing into the administrative expenses of a charity, some of the energy fails to show up in the work done to move the object. It appears, instead, in the form of heat, which Macroscopic Mack detects as a rise in temperature. He attributes this counterproductive diversion of energy to a phenomenon called "friction" (the rubbing of one object against another), and after long experience and many, many observations he detects a great regularity. He determines that the sum of all the work done and all the heat generated always adds up exactly to the total energy expended:

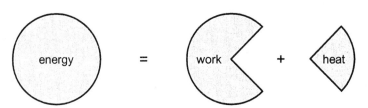

It is the "first law of thermodynamics," a restatement of the law of energy conservation, and nobody has ever reported a violation. Provided that we understand heat as a form of energy equivalent to mechanical work, then the totality of energy never changes. Energy that does not make its way into "useful" work goes into "wasted" heat. Nothing is lost, and nothing is gained. Energy withdrawn from one account shows up in another. The books always balance.

What do we get for our money? With work, the service purchased in exchange for energy is a movement of matter in bulk. We measure the resulting displacement with a ruler. With heat, the service purchased is a warming or cooling of a body—not any external displacement, but a change in temperature brought about by an internal flow of energy. We measure it with a thermometer.

So much for the macroscopic, thermodynamic view of heat and work. To a microscopic observer, who sees grains of sand where a macroscopic observer sees only beach, every particle in the universe acts as a buyer and seller of energy on a small scale. What Mack regards as a wholesale deployment of energy into work and heat, Mike understands as a retail deployment, bit by bit, into the motions of individual atoms and molecules.

Mike sees microscopic energy going into *translational* motion, the movement of a particle in a straight line:

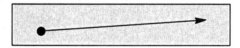

He sees it going into *rotational* motion, the turning of a molecule about an axis:

He sees it going into *vibrational* motion, the stretching and bending of a molecule:

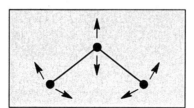

He sees it going into *electronic* motion, the promotion and demotion of electrons from one quantum level to another:

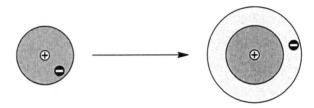

Into and out of these four basic modes, available individually to the myriad electrons, nuclei, atoms, and molecules of a large system, our microscopic observer follows the flow of energy.

"What you call heat," says Mike to Mack, "I attribute to the random, uncorrelated investment of energy into a trillion trillion separate accounts. You see a big object getting hotter while sitting in one place, whereas I see an unruly mob of small particles in wildly disorganized motion. As individuals they move every which way, and the motion of particle 1 bears no relation to the motion of particle 2:

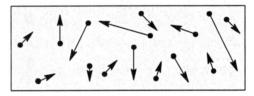

"What you call work, I see instead as a corporate effort by an enormous assembly of particles. The individual particles move together as a concerted unit, all in the same direction:

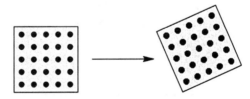

"To me," concludes Mike, "the difference between work and heat amounts to the difference between organized motion and disorganized motion."

Microscopic or macroscopic, statistical or thermodynamic, our two observers thus come to a practical understanding about energy, work, and heat. They agree to assess energy not just by its quantity, but also by its quality. They agree that a quantity of energy can either do qualitatively useful work (by moving matter in bulk) or dribble away as heat. They agree that motion can be organized or disorganized, coherent or random.

And they agree, finally, to keep observing. Recognizing a difference between the *quality* of work and the *quality* of heat, Mack and Mike realize that there must be a second law to complement the first: a law to regulate, in broad terms, the transfer of energy between nature's two biggest accounts, work and heat.

The Second Law: Running Down

Some things never happen. A warm house never becomes warmer when someone opens the door on a cold day. A pendulum at rest, unprovoked by any external force, never starts to swing. A randomly milling crowd never organizes itself spontaneously into a marching band.

Indeed not. A crowd of individuals, each one free to move in any direction, is unlikely—unlikely to the point of impossibility—to fall perfectly into lockstep. Why, after all, should we ever expect a mob to stumble by accident into just one particular organized configuration,

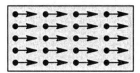

when so many other disorganized possibilities are available? Like these:

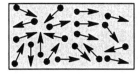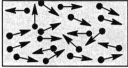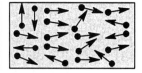

And these:

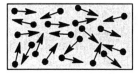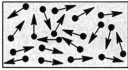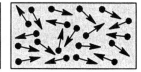

And a great many more. On the contrary, to organize a chaotic mob into a marching band requires the disciplined urging of a leader and the cooperation of all the followers. It can be done, but it takes sweat. To go the other way, however—to turn a marching band into a mob—requires no sweat at all, just patience. Watch a parade long enough, and the organized motion is sure to degenerate into disorganized motion. Accidents happen. A marcher or two might lose the beat. Another one might slip on a banana peel. Eventually all of them will succumb to hunger and fatigue. In the long run, order gives way to disorder. It is the way of the world.

A pendulum swings for a time and then comes to rest. Inevitably, sooner or later, the organized to and fro of macroscopic motion dissipates into the disorganized microscopic motions of atoms and molecules. Energy initially invested in work is dispersed into heat, and after that there is no turning back. An observer would have to wait a long time (an impossibly long time) for the particles in a crowd to find themselves serendipitously pushing the pendulum all in the same direction. The air around the pendulum would have to stumble out of a statistically likely *disorganized* configuration,

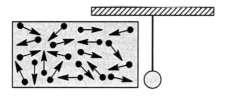

and stumble into a statistically unlikely *organized* configuration:

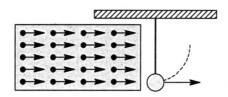

It does not happen. The natural course of a pendulum is gradually to stop swinging, not start. In the long run, order gives way to disorder. It is the way of the world.

A warm body transfers heat spontaneously to a cold body, not the other way around. To take away heat from an already cold body would be like robbing the poor to give to the rich. It would call for an ordered system in stable equilibrium (the cold body) to become even more ordered, an improbably steep climb up the statistical mountain. If it is to happen, we have to invest some capital in the process; we have to construct a heat pump and power it by an external source of energy. The natural course is for a warm house to give up its heat to the cold air outside, for order to give way to disorder. It is the way of the world.

If stable equilibrium is the goal, then it is clearly the way of the world for work to degrade into heat, for organized motion to degenerate into disorganized motion, for energy to dribble its way into more and more individual accounts. It is the way of the world for systems to run down and eventually stop, and the first law of thermodynamics has nothing to do with it. The first law demands only that energy, flowing in the form of work and heat, be conserved from start to finish. The first law makes no distinction between the conversion of work into heat and the conversion of heat into work. A complete transformation of heat into an equivalent amount of work conserves energy just as well as a complete transformation of work into heat. The second law, however, recognizes a world-altering difference between the two processes. The second law of thermodynamics, unlike the first, recognizes that work readily dissipates into heat, but heat does not readily turn back into work.

To transform work into heat is no trick. It can be done with 100% efficiency and without any special equipment. We pay a price, though, to go in the opposite direction, to extract work from heat and thereby transform disordered energy into ordered energy. Nature, speaking through the second law, says that no machine shall be permitted to take *all* the energy contained in a hot system and convert it fully into work:

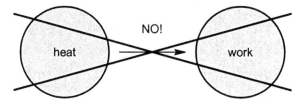

Instead, a portion of the heat must be dumped into the surroundings without ever being marshaled into the service of macroscopic work. The amount lost to friction and other dissipative channels may be large or small (it varies with the construction of the machine), but it cannot be zero. The conversion of heat into work can never be realized with 100% efficiency:

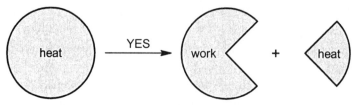

Squandering energy, according to the second law, is the natural thing to do.

What a pity, it seems, to let some of the heat go to waste, but such is the way of the world. For only by dumping heat can we commit an unnatural act against an equilibrated state—the creation of order from disorder—while ensuring that disorder grows to an even greater extent everywhere else. In general, nature demands of a system that it stir up the surroundings *outside* to compensate for any order created inside. If a piece of the universe is to become more ordered (as, for instance, when liquid water freezes into ice), then a certain quantity of heat is the price paid:

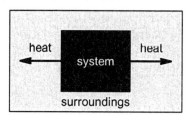

By the time the process runs its course, a larger slice of the global energy pie will have been allocated to disorganized motion. Less energy will be available to do organized work, just as a sponge becomes less absorbent as moisture seeps gradually into its pores.

The more heat, the greater is the disordering effect on the world outside. The lower the temperature, the greater is the disordering effect as well, because even a small amount of heat makes a relatively large splash in a cold system, not unlike a big fish in a small pond. Together, the combination of heat and temperature produces a change in a quantity called entropy (from the Greek, meaning "a turning within"), and it is in the direction of steadily increasing entropy that the universe turns.

Pessimists, rejoice! Governed by the second law of thermodynamics, the macroscopic world grinds ahead unrelentingly on a one-way course toward greater entropy. The clock runs down and stops, never to start up again on its own. The egg falls off the wall and shatters into pieces, never spontaneously to reassemble. Living things die and decay, never to live again. Time marches on, and nothing can turn the tide.

In the macroscopic world, every process that occurs of its own accord (not necessarily quickly, but eventually) is doomed by the second law to be irreversible, a journey from which there is no return. Done is done, and there is no going back. Even if all the king's horses and all the king's men manage to put Humpty-Dumpty together again, they do not restore the status quo ante. They do not turn back the clock. The world that they engineer does not reproduce the world that had existed previously. The horses and men must all expend energy to recreate an ordered Humpty-Dumpty from a disordered heap of broken eggshell; and in the process, huffing and puffing, they inevitably release a certain amount of heat into their surroundings. The waste heat stirs up the ambient air and brings about a net increase in global entropy, leaving Humpty-Dumpty to enjoy newfound order while the universe as a whole suffers an irreversible increase in disorder. Things are different. When equilibrium is reestablished, there is less usable energy on hand than before. It's later than you think.

Later, according to the second law of thermodynamics, means a world with greater entropy than now. *Later* means a world in which a fixed quantity of global energy has spread to a larger number of recipients. *Later* means a world in which useful energy (energy that can be harnessed to do work) has become just a little bit harder to find, a world that has yielded just a little bit more to the relentless pull of statistics.

Statistical Destiny

Here, in the end, is the question that haunts the grim determination of the second law of thermodynamics: When does a legitimate sequence of microscopic events, in full compliance with all mechanical laws, become so unlikely as to prove macroscopically impossible? What statistical compulsion turns a reversible microscopic process into a macroscopic one-way trip? If the laws of mechanics make no distinction between energy flowing from A to B and energy flowing from B to A...and if the laws of mechanics make no distinction between a particle moving from left to right and a particle moving from right to left...and if the laws of mechanics make no distinction between a second hand rotating clockwise and a second hand rotating counterclockwise—in short, if the laws of mechanics permit a single particle to move through space and time in any direction it likes, then why not a crowd?

To answer, we return to a system far less complex than broken eggs, the same simple system we routinely use as a proving ground for equilibrium thermodynamics: a gas confined to a container. This time, though, we acknowledge that the gas actually consists of particles in random motion (helium atoms, for example), and we allow the particles to go wherever the equations of motion take them.

Let the material be confined initially to the left-hand side of the container, separated from a vacuum on the right-hand side:

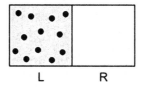

Then, after waiting for the gas to equilibrate, we suddenly remove the barrier and send the system on its way to a new equilibrium.

The gas expands. With a rush, it spreads out in all directions and fills the container. When equilibrium is reestablished, the temperature, pressure, and distribution of particles are uniform everywhere throughout the larger volume—and once that point is reached, there is no return. The gas never contracts spontaneously to its former state of equilibrium. It cannot go home again. It cannot revisit the past.

The particles, meanwhile, continue in constant motion, randomly shooting in all directions. Every atom has available to it the full container, and no position is specially favored. Alone or in a crowd, a single atom (particle 1) has an equal chance of being found somewhere on the left or somewhere on the right. There are two microstates:

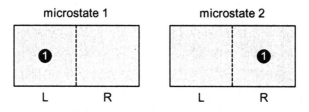

With either microstate equally likely to turn up, the odds are fifty-fifty: left or right, one chance in two. That is our assumption.

Add another atom now, for a total of two. Here, with each of four microscopic arrangements equally permissible (LL, LR, RL, RR), the likelihood of finding both particles on the left (LL) drops to one chance in four:

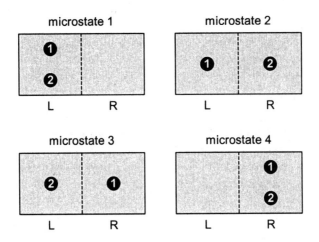

The other extreme, RR, shares the same low probability, whereas the most probable macrostate is again a fifty-fifty distribution: an equal combination of the microstates LR and RL, two chances in four to attain the same macroscopic outcome.

Now all we are doing is counting, but look what happens to the left-right distributions as the population grows. With three atoms and eight possible microstates (LLL, LLR, LRL, RLL, LRR, RLR, RRL, RRR), only one of them places all the particles on the left (LLL) and only one of them places all the particles on the right (RRR). The probability for either extreme is one chance in eight. With four atoms, the probability of either LLLL or RRRR drops to one chance in sixteen. With five atoms, the probability for LLLLL or RRRRR is halved yet again, to one chance in thirty-two. And thus the pattern continues, with the probability of an all-left or all-right configuration becoming progressively smaller as more particles enter the fray.

Indeed the probability of any lopsided distribution grows so small that before long the system has only one real option: to accept a macrostate in which half the particles fall on the left and half fall on the right. The sheer weight of large numbers—statistical probability, nothing else—pulls the system toward this most likely macrostate, sealing its fate long before the population approaches the trillion trillion particles that make up even a modest macroscopic assembly. Already for a hundred particles, a tiny system, the statistics allow only one infinitesimal chance in 1,000,000,000,000,000,000,000,000,000,000 of finding all the atoms on one side or the other. And when the population grows to thousands and millions and finally to a trillion trillion and beyond, the numbers boggle the mind and trap the system in its one overwhelmingly probable macrostate: half the particles on the right and half on the left, microscopically arranged and rearranged in countless equivalent ways. Nothing else comes close. The fifty-fifty distribution at equilibrium, supported by enormously more microstates than any other, dwarfs all the rest. Once it takes hold, it persists.

Understand that the rule here is strictly one-microstate-one-vote, provided that nature plays no favorites. The extreme, one-of-a-kind, everybody-on-the-left microstate

LLLLLLLLLLLLLLLLLLLLLLLLLLLLLLLLL … L

is therefore just as likely to occur as any particular one of the fifty-fifty microstates (the configuration below, say),

LLLLLLLLLLLLLLL … RRRRRRRRRRRRRRR …

but the all-left arrangement can be realized in just the single way shown. The fifty-fifty microstate, by contrast, has the statistical virtue of being macroscopically indistinguishable from

LRLRLRLRLRLRLRLRLRLRLRLRLRLRLR …
LLRRLLRRLLRRLLRRLLRRLLRRLLRRLL …
LLLRRRLLLRRRLLLRRRLLLRRRLLLRRR …

and all the other fifty-fifty microstates. Each one of them is equally likely to appear as the particles randomly bounce about.

So there, in the statistics of large numbers, we have a plausible explanation of why the gas expands spontaneously and never turns back. Staggering blindly from one configuration to the next, the system goes from the least likely arrangement (particles all on the left) to the most likely arrangement (particles evenly distributed). Nearly every lurch leaves the particles in an increasingly likely macrostate, allowing them access to an ever larger number of equivalent microstates.

Step by step, microstate by microstate, the distribution approaches fifty-fifty; and when it finally gets there, the particles can no longer improve on the statistics. The system falls into the probabilistic nirvana of equilibrium, and there it stays. The gas flickers randomly through an unending series of microscopic arrangements, each one presenting the same macroscopic face: equal distribution everywhere, all the particles randomly shuffled, no region with either too many or too few. Any other arrangement would mean less microscopic freedom, not more; any other macrostate would be backed by fewer microstates, making it a less robust alternative. The system holds a lottery, and the macrostate with the largest number of tickets comes out the winner. The odds are unbeatable.

With that prosaic yet startling realization, we have at last our statistical interpretation of the second law of thermodynamics. Nature's tendency at equilibrium is to spread out, to relax, to disperse energy and material into the largest possible number of microstates. The statistical imperative is to increase the microscopic disorder present in the universe and, as a result, to maximize the global entropy (which rises and falls along with the number of accessible microstates). Not by design but by chance, a fixed amount of energy works its way through particles and fields until an optimum distribution is attained.

In gamblers' parlance, it is a sure thing. A system that is so complex as to appear *simple* has a statistical destiny to fulfill.

11

SURPRISE ENDINGS

Some systems are obligingly simple. Their parts are few; they travel along well-marked roads; they are not particularly fussy about where their journeys begin. They permit precise measurements of their initial state, and they end up (like clockwork) in a stable, readily predictable final state.

We can trust them. These are simple systems, after all, incapable of surprise. They lend themselves to neat mathematical treatment, even if idealized sometimes to the point of abstraction. They are paragons of order and dependability. They adorn the pages of textbooks.

The Moon locked in orbit around the Earth. An electron confined around a nucleus. A pendulum gently swinging. Both classical and quantum mechanical, these textbook systems and others like them all have something in common. They take their marching orders from a deterministic equation of motion, and they evolve predictably in a world where knowledge of the present holds the key to knowledge of both future and past. Given *what* you are now (your initial state) and the influences that control you, the equation of motion determines what you will be the very next instant. It maps out the precise road to be followed, a two-way highway connecting one destination to the next:

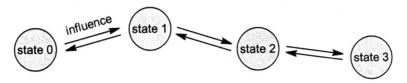

For a classical system, the initial specification will be a set of particle locations and velocities, whereas for a quantum mechanical system it will be the values of a probabilistic wave function, but make no mistake: the end result, whether determined by Newton's laws (Chapter 4) or the Schrödinger equation (Chapter 8), comes as no surprise. One state follows another as inevitably as night follows day. These are simple, deterministic systems that go where the road takes them, and they do not get lost along the way.

Sometimes, too, even the most complicated systems—structures so overwhelmingly complex as to defy the orderly progression of an equation of motion—prove to be no less obliging, as we have observed in Chapter 10. The trillion trillion atoms in a cubic foot of helium gas, for example, cannot forever postpone their rendezvous with statistical destiny. Colliding willy-nilly, they fall predictably into an equilibrium from which there is no return. The large system finds its way into the most likely macroscopic state, driven forward by little more than blind chance. Staggering, overpowering, unbeatable odds favor the final outcome, a one-way trip to a one-and-only destination. As individuals, the particles demonstrate a formidable microscopic complexity; as a mob, they behave with macroscopic simplicity.

Yet if in the end "complex" proves to be "simple" (under certain conditions, at least), then oftentimes simple proves to be complex. Indeed all throughout nature, we find systems simple in every way that nevertheless behave chaotically and unpredictably: *small* systems, systems with few interactions, systems that obey deterministic equations of motion...*simple* systems, by any measure, but systems exquisitely sensitive to the slightest change in initial state. A tiny shift in where they start makes a big difference in where they finish.

A casual observer would describe them as random and lawless, but they are not. Chaotic systems obey rules of the road as rigorously as

Earth and Moon, and yet their every action comes as a surprise. With their extreme sensitivity to initial conditions, where one small step is enough to change the world, chaotic systems effectively deny us knowledge of the present, and without knowledge of the present there can be no knowledge of the future.

Where will the chaotic road take you? Nobody knows. And don't expect a map to be of any help.

Lost with a Map

We begin with two stories, just to set the tone: first the hikers, then the bankers.

Strange Terrain

Three hikers set out independently on the same hike. It is going to be a long trek, with many twists and turns, but each hiker carries a complete set of instructions (the same for all three of them) to map out every step of the way. "Walk so many feet in this direction," they are told, "and then so many feet in that direction, and thus-and-such many feet in another direction," and so forth, one foot in front of the other, from the first step to the last. Nothing is left to chance.

Side by side at the outset, the hikers execute their instructions to the letter (and remember: except for the different starting points, the three plans are identical). Step by step, acting as unconnected individuals, each of them subsequently follows the same path:

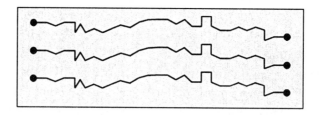

Viewing the trails from above, nobody is surprised. The three hikers start out together, and they stay together. From beginning to end, they go through the same series of steps and make the same series of turns. If

they are spaced five feet apart at the beginning, then at the end—
wherever that may happen to be—they remain five feet apart as well.

Imagine our surprise, then, if we were to see these same three hikers,
following the same three sets of instructions, trace out three wildly
divergent paths. Like the ones below, perhaps:

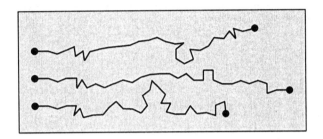

Inconceivable? Not necessarily, for if the individuals started out sufficiently far apart (maybe five miles rather than five feet), then we might
reasonably attribute the results to differences in terrain. Hiker 1 may have
had to deal with hills and ravines, while hiker 2 may have traveled entirely
on flat ground, while hiker 3 may have walked into a tree or fallen into a
hole. If so, we would have scant reason to expect the members of our trio
to remain on parallel tracks.

But suppose that, no, the hikers did not start out five miles apart.
Suppose that they really did begin with only five yards or five feet or
five inches between them. Suppose that no matter how close they were at
the outset, they always wound up (apparently at random) either farther
apart or closer together at the end of a long, intricately mapped route.
For if they did, we would know that they were negotiating no ordinary
ground; we would know that they were hiking over a chaotic landscape,
a terrain where one confronts the real world in all its complexity and
where one learns, right from the start, to expect the unexpected.

Lawful Anarchy?

"Money in the bank," we say, a metaphor for steadiness and dependability,
for certainty without surprise. Put away a thousand dollars at 10% annually, sit back, and watch the money grow: $1100 after one year, $1210
after two years, $1331 after three years, $1464 after four years, $1611

after five. Year after year, the same rule applies: take the amount on hand and increase it by 10%. There are no wild highs and no unexpected lows, just a steady growth from one year to the next. Whatever the plan lacks in excitement, it makes up for in regularity:

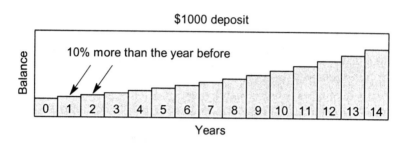

$1000 deposit

Start out with a thousand dollars or ten cents, a million dollars or a billion—the initial principal has no effect on the subsequent pattern of increase. The total still grows by 10% per annum, building each year on the amount already present. And even if taxes, bank fees, or other expenses should swallow up half the interest, the linear relationship (the direct proportionality) between the balance today and the balance tomorrow remains unaltered. The amounts after expenses are less because the rate of growth falls to 5%, but still the money compounds in the same steady way: $1050 after one year, $1103 after two years, $1158 after three years, $1216 after four years, $1276 after five; every year, 5% more than the one before. We can count on it.

Now conjure up a different kind of account, the kind that might be offered by First Chaotic National Bank, where an initial deposit of $1000 might follow one course over a number of years,

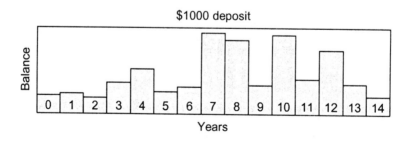

$1000 deposit

and an initial deposit of $999 (barely different from $1000) might follow quite another:

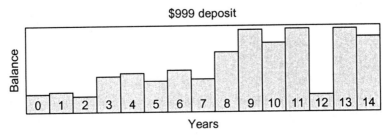

The amounts go up; the amounts go down. A small difference at the beginning makes a big difference at the end, and yet both accounts are governed by exactly the same terms. "What kind of bank is this," the depositors complain, "do you throw dice to decide how much money to credit us year after year? How can we plan our financial future if we don't know what rate of return to expect?"

To which the bank president replies, somewhat disjointedly: "Well, no, you cannot make any plans. Well, yes, everything is perfectly in order. Dice? No dice. We follow strict rules here at First Chaotic National, proud to say. Your accounts are behaving entirely as expected, just as they should, and you are getting exactly the returns promised. We apply the same formula and the same rates to calculate the balance in both accounts."

Strange as it sounds, everything the bank president says is strictly correct. First Chaotic National is a legitimate, honest institution which writes its deposit agreements in simple and straightforward terms, and neither depositor 1 nor depositor 2 has any reason to rejoice or complain. The disparate fates of their two accounts come about not as the result of any swindle, but rather as a consequence of "nonlinear feedback" in the method used to generate compound interest.

Compare: In the Bank of Linearity, a conventional institution, balances are adjusted annually in direct (linear) proportion to the amount on hand. Twice as much money in the fifth year generates twice as much interest and expenses in the sixth. Three times as much money generates three times as much interest and expenses. Four times as much money generates four times as much interest and expenses. Nothing out of the ordinary will ever happen under such an arrangement, provided that the feedback (the reinvestment of the system's output into its input) remains linear.

In First Chaotic National, however, the adjustment is calculated in proportion to a higher power. Twice as much money in one year might generate four times the interest and three times the expenses in the next, or five times, or six times, or nine times. And whenever the feedback in a system becomes nonlinear—as it does so often in the real world, where friction and turbulence give the lie to textbook examples—then all bets are off. A small change in an initial condition, like a single dollar in our fictitious bank account, makes an enormous difference down the road. More than that, a nonlinearity can destroy any notion of predictable cause and effect, even when an unambiguous rule determines the evolution of state 1 into state 2 along every step of the way.

Impossible, you say? You question how a deterministic equation of motion can ever lead to an apparently random outcome? Then see what happens now when we put a typical nonlinear process through its paces. It is a telling example, even though just one of many, and it is something more as well. This "logistic difference equation," explored below, becomes the paradigm for a new way of looking at a simple world made complex.

A Numbers Game

Pick two numbers, x_0 and A, and apply the following simple rules:

1. Start with x_0 anywhere between 0 and 1.
   For instance, $x_0 = 0.5$.

2. Subtract x_0 from 1 to form the number $1 - x_0$.
   Like this: $1 - 0.5 = 0.5$.

3. Multiply $(1 - x_0)$ by x_0 to form the number $x_0(1 - x_0)$.
   Like this: $0.5 \times 0.5 = 0.25$.

4. Multiply $x_0(1 - x_0)$ by A to form the number $Ax_0(1 - x_0)$. Call it x_1.
   With $A = 1$, for instance: $x_1 = 1 \times 0.25 = 0.25$.

5. Using x_1 in place of x_0, reapply rules 2 through 4 to obtain x_2, the next number in the sequence: $x_2 = Ax_1(1 - x_1)$.
   Like this: $x_2 = 1 \times 0.25 \times (1 - 0.25) = 0.1875$.

6. Continue the game for as long as you like, using x_2 to obtain x_3, and then x_3 to obtain x_4, and then x_4 to obtain x_5...and, well, you get the picture.

All it takes is a calculator and considerable patience.

The insight gained more than makes up for any computational tedium, because the game we are playing provides a simple but enlightening model of nonlinear feedback: *feedback*, first of all, because the output of one calculation becomes the input for the next (rule 5); and *nonlinear* feedback because part of the response is proportional to x *times itself* (rule 3), not just to the current value of x alone.

What a difference it makes. If there were no nonlinearity—if rule 3 were simply to multiply x by 1, rather than to multiply x by $(1 - x)$—then the effect would be merely to generate compound interest. We would multiply x_0 by A to obtain x_1, and then multiply x_1 by A to obtain x_2, and then multiply x_2 by A to obtain x_3, and so on. For any value of A (the equivalent of an interest rate), the value of x (the equivalent of a current balance) would grow directly and without restraint, ever increasing. By incorporating a nonlinear response, however, we build in a mechanism for x to rise and fall in a rich variety of ways, sometimes delightfully unexpected. The details depend on both the initial condition (x_0) and the strength (A) of the nonlinearity driving the system.

Try it. First, note that under certain conditions the nonlinear feedback eventually drives x down to zero, where it remains forever trapped. The graph below, for example, shows what happens when A is equal to 1 and x_0 is equal to 0.5:

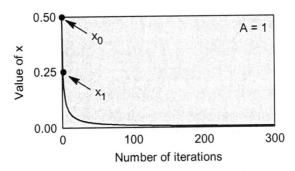

If x represents a bank account, then the long-term outcome corresponds to financial ruin. If x represents a population of wolves, then the long-term outcome corresponds to extinction. Whatever value x takes in the beginning, it falls to zero after a sufficient number of iterations.

Another possibility, realized over a different range of A, is for x to settle into a constant but nonzero value, taking either a "monotonic" or "oscillatory" approach to equilibrium. During a monotonic approach, the numbers increase or decrease in only one direction until at some point they flatten out:

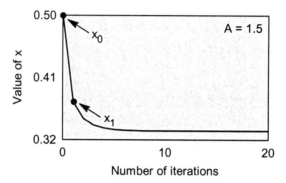

During an oscillatory approach, the sequence swings rhythmically about its equilibrium value before finally settling down. Now high, now low, the oscillations gradually become smaller and ultimately disappear:

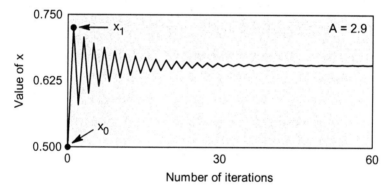

Either way, though, whether the course is oscillatory or monotonic, the game ends with a single fixed value for x—so long as A does not rise above a certain point.

And what happens after that? What happens when A grows larger? Up to now, admittedly, we have yet to observe any particularly unusual behavior, certainly nothing that would challenge our faith in a deterministic equation of motion. But suddenly, like the proverbial straw that breaks the camel's back, the growing nonlinearity starts to push the numbers into an entirely new realm. When the parameter A surpasses 3, the system becomes abruptly unable to make up its mind. Instead of converging to a single value at equilibrium, the sequence flip-flops between *two* final points without ever settling down. It repeats, endlessly, the same two-point cycle, like Persephone alternating between Earth and the Underworld:

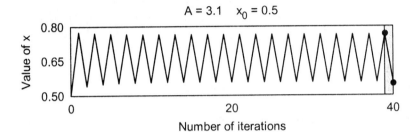

Think of the implications. If someone were unfortunate enough (or dumb enough) to open a bank account under such terms, then the balance might flip between, say, $1264.32 in the odd years and $881.97 in the even years, caught for all time between two fixed values with nothing in between. Or a forest ranger might count 654 wolves in 1930, 1932, 1934, 1936, and 1938, but only 376 in 1931, 1933, 1935, 1937, and 1939. Or a meteorologist might record wind velocities of 10 knots on the first, third, fifth, seventh, and ninth of the month, and 15 knots on the second, fourth, sixth, eighth, and tenth. An observer would think that two separate equilibria were fighting for control of the same system, each one taking its turn with clockwork regularity.

The funny business is just beginning. With the system driven a little bit harder, the length of the repeating cycle soon doubles from two to four,

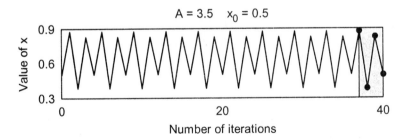

and then from four to eight, and from eight to sixteen, and from sixteen to thirty-two, more and more, until finally something snaps. Beyond a certain threshold value of *A*, the nonlinear system turns "chaotic"; its cycle becomes infinitely long and never repeats, as in the arbitrary example shown below:

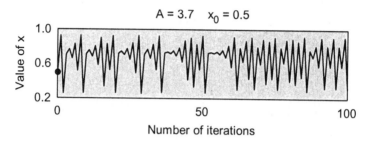

The numbers seem to be random, but of course they are not. Each value is determined precisely by the one that comes before. If the sequence has fallen into chaos, then it is evidently a *deterministic* kind of chaos, *a chaos that follows rules*, not the kind of chaos where chance reigns supreme.

For although an uninstructed observer might find no regularity in the list of numbers (0.5000, 0.9250, 0.2567, 0.7060, 0.7681, 0.6591, 0.8313, 0.5189, ...), those of us who are in on the secret know better. We know that looks are deceiving. We know that our simple rules generate precisely the same sequence under precisely the same initial conditions, albeit a sequence so haphazard that the numbers could have come equally well from a lottery. We know, despite appearances, that there is order in chaos.

Nevertheless, we can make little practical use of the deterministic certainty programmed into the equation, because the end of the game (even if preordained) depends with hair-trigger sensitivity on the initial point x_0. Make the slightest change in x_0 (say from 0.500000 to 0.499999),

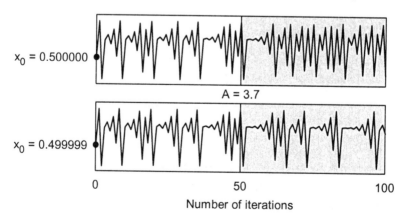

and the chaotic outcome varies disproportionately. Here, where the initial point is shifted by just one part in five hundred thousand, the two sequences start to diverge irreconcilably after only fifty iterations—to the point where they soon have nothing in common. So unless we know to an absolute certainty that the sequence begins with x_0 equal to 0.500000 (not 0.500001, not 0.499999, not 0.499998, not 0.500002, but *0.500000 exactly*), then we have no idea what to expect later on.

A higher initial precision will only postpone the difficulty, not remove it, because somewhere along the line we are always bound to go astray. After some finite number of iterations, large or small, the chaotic sequence begins to vary widely even if the starting point is jiggled by the most trifling amount. Shall we try one part in five million, rather than one part in five hundred thousand? Not good enough. Past a certain number of steps, the sequence generated when x_0 is 0.5000000 bears no relation to the one generated when x_0 is 0.4999999 (a difference of one unit in the seventh decimal place). One part in five billion? Not good enough. One part in five *trillion*? Five quadrillion? Five quintillion? No. None of it is good enough. However tiny the initial shift may be, it plants a seed of instability that grows inexorably into a chaotic tangle; and in a chaotic world, "almost" does not count. Either hit the number exactly, or give up any claim to forecasting a deterministic outcome beyond a limited number of steps. You would do just as well throwing dice.

Renounce all hope, those who pass through the Gates of Chaos, because the promise of mechanical determinism is broken from the start. No matter how finely we craft our instruments, no matter now carefully we make our measurements, no matter how minutely we control the environment—no matter what pains we take, ultimately we cannot overcome the extreme sensitivity of a chaotic process to its initial conditions. Driven for a sufficiently long time, the system gets lost along the way. A pinball player launches the ball with slightly less force, and at the end of a long game, with many, many rebounds, the final score changes from 1,000 to 10,000,000. The temperature of a turbulent fluid fluctuates unexpectedly by a fraction of a degree, and the flow is altered irrevocably. A butterfly flaps its wings in Tokyo, and two weeks later rain falls in Madrid.

It is called, in the not-entirely-whimsical language of chaos theory, the "butterfly effect": the cascade of ignorance that enables a chaotic system to mock the deterministic equations that govern it. We learn the rules of a process only to discover that the rules do not apply, or, rather, that the rules merely legitimize a systematic kind of anarchy.

Accept it and embrace it. Chaos is as much a part of the mechanical landscape as the conservation of energy and momentum, a capricious feature of the law that puts limits on our capacity to know. There are goings-on in this world so delicate that an observer can only guess what comes next, unable to say for sure.

Just ask a meteorologist.

Chaos Unleashed

"If only I hadn't called Gladys that morning! Then she wouldn't have missed the bus, and then she wouldn't have gotten to work an hour late, and then she wouldn't have skipped her lunch appointment with Sally, and then Sally wouldn't have gone window shopping on Main Street, and then Sally wouldn't have stopped to buy a newspaper, and then Sally, oh, poor Sally, she wouldn't have been reading that newspaper when she stepped into the street at half past twelve, and she wouldn't have gotten flattened by the speeding vegetable truck. *It's all my fault.*"

We blame ourselves, wrongly, for all kinds of happenings in everyday life, when really the fault lies in the chaotic unpredictability of the butterfly effect. Small consolation, perhaps, to Sally and those who mourn her, but nonlinear dynamics and the chaos it engenders are woven inextricably into the fabric of the world. Seek and ye shall find chaos

> ... in long-range weather patterns that defy prediction,
> ... in dripping faucets,
> ... in pinball machines,
> ... in animal populations,
> ... in river rapids,
> ... in irregular heartbeats,
> ... in swirls of smoke,
> ... in brain waves,
> ... in oil pipelines,
> ... in epidemics,

...in lasers,

...in gearboxes,

...in electric circuits,

...in chemical reactions,

...in baseball games,

...in flapping flags,

...in the rings of Saturn,

...in the orbit of Pluto,

...in the Great Red Spot of Jupiter,

and virtually everywhere ye look. The list grows and grows. Nearly everything in the universe, from colliding molecules to clustering galaxies, is susceptible to chaotic behavior under the appropriate conditions. We have only to open our eyes.

Simple is beautiful. When a simple system remains within its linear bounds, it behaves with mathematical grace and elegance. We can write down the equations with an economy of symbols. We can solve the equations exactly, once and for all. We can use the same equations, remarkably, to describe phenomena of diverse kinds, from water waves to electromagnetic waves to heat waves to quantum waves. Linear systems are similar in their simplicity, and once we understand them, we fool ourselves happily into thinking that we know all there is know. We fool ourselves into thinking that a linear simplicity is the rule rather than the exception.

But there is a world of complexity, too, and complexity has a beauty all its own, especially when it is a complexity born of simplicity. Complex systems, *nonlinear* systems, *chaotic* systems—wherever we find them, we deal with them not as textbook examples but as varied individuals. Each has its own set of equations, its own range of parameters, its own idiosyncrasies. In contrast to linear systems, one size does not fit all. Like Tolstoy's unhappy families, each chaotic system is chaotic in its own way.

Between Now and Forever

"In the long run," remarked the economist John Maynard Keynes, "we are all dead." In the long run, systems exhaust their capacity for change. They come to equilibrium. Order gives way to disorder, and the entropy of the universe increases. The second law of thermodynamics has its way.

The end may be a long time coming, but come it must: the odds cannot be beaten. Sooner or later, the timelessness of equilibrium sets in. In the gambling casino of nature, the house always wins. It is a question only of when.

In the meantime, though, between beginning and end, anything is possible. There are fluctuations. There are surprises. States that are statistically improbable flicker into and out of existence. Ordered structures arise from disordered structures and sometimes persist. To be alive, don't forget, is to exist defiantly in a nonequilibrium state, holding the second law of thermodynamics at bay, at least for a while.

Nobel laureate Ilya Prigogine, who devoted decades to the understanding of nonequilibrium processes, speaks of the difference between *being* and *becoming*: the difference between equilibrium and evolution, the difference between the destination and the journey, the difference between the end of an affair and the middle of one. It is an insight that coincides with a nascent appreciation of complexity in many diverse fields, a recognition that systems may be more than just the sum of their constituent parts. Complicated assemblies can evolve cooperatively into structures capable of extraordinary doings, all the while bucking a natural tendency to slip into disorder.

A system needs resources to swim against the current. To maintain a steady state away from equilibrium—to preserve order where there would otherwise be disorder—a system needs to import energy and matter from the outside. It needs to work actively rather than succumb passively to the second law. It needs to dispose of waste material. It needs to replenish spent fuel. It needs to make a living. If it can do so, then such a system may persist indefinitely in a nonequilibrium state.

And when the system happens to be nonlinear (and configured in just the right fashion), then occasionally it will surprise us in wholly unexpected ways. A nonlinear system, drawing on energy and matter from the surroundings, will sometimes organize into a more ordered, more complex structure with unforeseen new properties—as, for instance, when a lumpy, structured universe of stars and galaxies emerges from the inchoate seeds of the Big Bang, or when life on Earth first emerges from a collection of inanimate molecules. To accomplish these feats, however, a nonlinear system needs something more than just an external supply of energy and matter. It also needs a little bit of luck. It needs a random fluctuation to trigger a cascade of events that turns an intrinsic instability into a provisional stability.

Chance thus becomes an essential part of natural law, an unavoidable roll of the dice needed to select just one particular complex outcome from a set of many. Exactly which one, we can never be sure; we can only rely on statistical probability and prepare to be surprised at what results. When the roulette wheel stops turning, we might see a system emerge with new and unsuspected characteristics, a system that must be treated as a whole rather than as a collection of independent parts.

With that, we have the makings of a new science: the science of complexity, of chaos, of "emergent" phenomena; in a word, the science of surprise. It is a science barely out of its infancy. Expect the unexpected.

12

LOOSE ENDS

Nobel physicist I. I. Rabi, reacting to the discovery in 1937 of a new and unanticipated particle (the muon), asked famously, Who ordered *that*? Half jokingly, half seriously, the question reminds us that our understanding of physical law is never more than provisional: that the theory must fit the facts, not the other way around. Intellectual honesty requires that a model be challenged, revised, and broadened continually in the face of newly won knowledge.

We know things now, in the early years of the twenty-first century, that nobody—not even the most farsighted of scientists—could have suspected forty years ago, twenty years ago, five years ago, one year ago, the week before last. Day after day, we probe matter at lower temperatures, higher pressures, smaller distances. Month after month, our instruments look deeper into space and farther back in time. Year after year, we interrogate nature at higher energies. The more raw facts we discover, the more we want to understand.

We want to understand nothing less than the origin and fate of the universe itself, wondering even whether our one-and-only universe (or

284

so we have always thought) may indeed be only one of infinitely many. If so, must the "universal" laws of physics be demoted to the status of a local code, applicable within one particular jurisdiction and nowhere else? Still more, we ask, are the laws changing with time? Do physical constants such as the speed of light have the same value today that they did in ages past?

We want to understand the nature and role of *dark matter* in the universe, stuff that we know is out there (it leaves a gravitational fingerprint) but nevertheless cannot be detected electromagnetically. Is dark matter made up of particles already familiar in other contexts, or is it something entirely new? The curious mind wants to know. Dark matter, by far the greatest source of mass in the cosmos, can hardly be ignored.

We want to understand, similarly, the *dark energy* that permeates the universe, evidenced by a repulsive effect (apparently acting against gravity) that becomes increasingly strong over cosmological distances. Does the phenomenon arise from one of the four fundamental interactions already known, or does dark energy herald a new force entirely? Is our cosmic quartet really a quintet?

We want to understand, most profoundly, the ultimate structure and unity of nature on the smallest of scales. Is the wall that divides bosons from fermions an impenetrable barrier, or might "force" particles and "matter" particles be able to interconvert under a higher symmetry (a *supersymmetry*) yet to be recognized? Are quarks, electrons, and neutrinos built from constituents more nuanced and more versatile than hitherto imagined? Is there a limit, at long last, to the subdivision of matter? Are we wrong to partition space into infinitesimally tiny points with no extension? Are we wrong to assume that the space-time we readily perceive, with its three dimensions of space and one of time, is the only space-time that actually exists? Will the various fundamental forces ever find conceptual unity in a single, all-encompassing interaction?

To be able even to ask such questions is cause for excitement, and some of the answers (not to mention many new questions) may be forthcoming faster than anyone might have dreamed. And even if the quest has no end, we can at least take a provisional, tentative peek over the horizon. Let that be the goal of this twelfth and final chapter, by no means the final word.

Reconcilable Differences?

The lessons of history are clear: If you think you know everything, think again. Whenever physicists proclaim that the end is near—that our understanding of the fundamental laws is almost complete—then usually a revolution is due to begin. Some radical new discovery lies just around the corner.

Think back to the late seventeenth century. Newton's laws of mechanics, in their time, were tantamount to a Theory of Everything, a user's guide to the entire known universe. Here was a stunningly concise explanation of what made everything tick, applicable equally to the heavens above and the Earth below. For observers ignorant of forces other than gravity (and ignorant, also, of gravitational forces stronger than those in our solar system), it was a model that left little to be desired. It took into account all there was to know.

Along came the discovery of electricity and magnetism, and by the end of the nineteenth century the Classical Kingdom had split into a duarchy: a mechanical realm governed by Newton's laws of motion, and an electromagnetic realm governed by Maxwell's equations. But still, as far as anyone knew, the combination of classical mechanics and electromagnetism took in all of the universe known at the time. Together, the two branches of physics constituted a new Theory of Everything, second edition, revised.

Add to that the drastic modifications Einstein imposed with his theory of relativity, and by 1920 physics had an even more refined, more inclusive Theory of Everything—everything, that is, in a classical universe ruled by only two fundamental forces. The special theory of relativity, first, threw out absolute time and rewrote Newtonian law at high velocities, making mass equivalent to energy. The general theory of relativity, second, replaced Newton's gravitational force with a mass-induced curvature of space-time. It was the most complete description of the universe yet, the last word in a world without atoms and the uncertainty principle. A classical observer could ask for nothing more.

Then came quantum mechanics, which introduced a pair of new interactions, the weak and the strong, to share the stage with gravity and electromagnetism. It has been a phenomenal success, too, validated for systems as large as molecules and as small as quarks. The quantum mechanical view of the microworld, with its principles of superposition

and indeterminacy, now includes a theory of the electromagnetic interaction (atoms and molecules, photons, the electromagnetic field), the
unified electroweak interaction (beta decay, photons, the W and Z bosons),
and the strong interaction (protons and neutrons, quarks and gluons).
Quantum mechanics thus accounts for all of the fundamental interactions
except gravity, which remains unamenable to the kind of gauge field theory developed for the three other forces.

Now three out of four ain't bad, but three out of four is also not as
satisfying as four out of four; and the absence of a quantum mechanical
theory of gravity gives us pause. For if big things ultimately come from
small things (they do), and if small things obey the laws of quantum mechanics (they do), then why shouldn't Einstein's macroscopic description of gravity derive ultimately from a microscopic, quantum theory as
well?

That is the great challenge physicists now face, how to bring together
a kingdom of forces not quite united: the electromagnetic, weak, and
strong interactions on the one hand, gravity on the other. Microscopic
versus macroscopic. Quantum mechanics versus general relativity.

Searching for a link, one looks for circumstances under which the
gravitational interaction (normally so very weak) can compete with the
other forces. Where, in particular, might we find a suitable proving
ground? *In an atom?* No. The electromagnetic interaction between
electrons and protons in an atom outweighs gravity by some
1,000,000,000,000,000,000,000,000,000,000,000,000,000 times. *In a
nucleus?* No. Even at distances a hundred thousand times shorter than
those between electrons and nuclei, the gravitational force between
protons in a nucleus is still vanishingly small compared with the
electromagnetic and strong forces.

In a particle accelerator? Maybe, provided that we can supply an
unimaginably large amount of energy to squeeze the particles into
unimaginably close quarters, perhaps a hundred billion billion times
closer than in a nucleus. Is it feasible? For the present and foreseeable
future, no. All the engineers and all the money in the world will not
suffice to build such a machine.

In a black hole? Presumably yes, since here the particles already
come packed at extraordinarily high densities, no artificial contrivances
required. When distances become small enough, then gravity has at least

a chance. A black hole is an extreme case, admittedly, but nothing less than an extreme case will do.

In a small point containing all the energy and substance of the universe? Yes! And it is precisely this most extreme of extreme cases— the Big Bang, the instant at which cosmologists believe our particular universe was born—that would have provided nature with the best opportunity to deploy a complete set of unified quantum forces, gravity included. Gazing out into the unknown, then, we find ourselves wondering first about a possible beginning of time and space.

Once Upon a Time

The battle lines were already drawn in ancient times, with the Greeks and Hebrews arrayed on opposite sides. Greek philosophy maintained that the universe has always been here and will always continue to be here, that matter is eternal and unchanging, preexistent. The Hebrews believed instead that the universe arose *ex nihilo*, from nothing, in a creative instant (a Big Bang, we might say) that marked the beginning of space and time.

Neither the Hebrews nor the Greeks, though, had any objective evidence to support their beliefs, unlike cosmologists today who place growing confidence in the model of the Big Bang. Ancient philosophers did not know about the expansion of the universe, or about the microwave background that permeates all space, or about the telltale distribution of the elements hydrogen and helium. But today we do know, and for us—knowing all these things and still asking for more— the history of the universe is written in the stars.

Expanding Horizons

We hear it in the wail of a siren as an ambulance passes by: a higher pitch as the vehicle approaches, a lower pitch as the vehicle pulls away. It is the "Doppler effect," the variation in frequency registered by an observer relative to a moving wave source. It tells us, prosaically, that a train or ambulance is coming, and it tells us something grander as well. The Doppler effect, a simple property of waves, announces to us the expansion of the universe.

Imagine a traveler (his name is Hubble) who regularly mails one postcard each week to his friend Doppler back home. Near or far, Hubble always adheres to the same strict schedule of one card per week, never deviating in the slightest. Doppler, however, receives the mail with varying frequency, sometimes more than once a week and sometimes less. The cards that Hubble sends while moving away from Doppler travel successively longer distances and take longer to arrive. Those that Hubble sends when he is moving toward Doppler travel shorter distances and consequently arrive more frequently.

Waves do the same thing. Crests from a source moving toward an observer arrive with greater frequency than those from a source moving away from an observer:

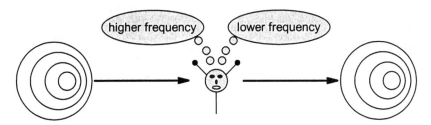

source moving toward observer   source moving away from observer

Water waves do it. Sound waves do it. Light waves do it. The electromagnetic radiation we receive from a distant galaxy appears shifted to lower frequency and longer wavelength, suggesting that the source is receding from view. Rather than observing, say, blue light from a stationary star (each chemical element has its own characteristic wavelengths), we see the radiation skewed toward the red end of the spectrum. It has a longer wavelength. The star is moving away from us, and we are moving away from the star:

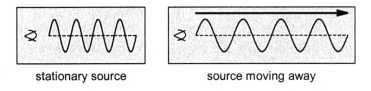

stationary source   source moving away

The farther away the galaxy, the greater is the shift in wavelength. All the galaxies, our Milky Way included, are moving away from each other at speeds proportional to their current distances. The farther, the faster. The phenomenon, called the intergalactic redshift, was discovered by astronomer Edwin Hubble in 1929.

It comes as no small surprise, this signature of a dynamic, evolving cosmos. The universe is evidently not static and enduring, as long believed, but rather space-time itself is a continual work in progress. The galaxies, understand, are not simply expanding into previously *empty* space, like squatters moving into unoccupied land. No, the universe is far more subtle than that—the galaxies are being dragged along, like it or not, by an expanding tide of freshly manufactured space. Think of what happens not as the filling of a fixed vessel (space) with energy and mass,

but instead as the expansion of the vessel and its contents together:

The overall fabric of the universe is expanding, creating new space every instant, opening up and occupying territory heretofore nonexistent.

General relativity demands it. The interplay of mass-energy and space-time makes for a restless universe, a playing field that is forever being shaped and reshaped. Mass and energy warp the structure of space and time, which in turn affects the motion of every object nearby. Those same bodies, gliding inertially in the clefts and channels of a warped space-time, generate more warping still, which in turn affects their motion yet again, and again, and again...and so it goes, back and forth in unending reciprocity.

The result, according to Einstein, is an activist space-time that cannot stand still. Space can grow and space can shrink, but in the presence of mass and energy it must do something. It cannot remain as is.

Given that requirement, consider the following proposition. If space-time is greater in extent today than it was yesterday, and if it was greater in extent yesterday than the day before, then what was it like way back when, back when the universe might have been very small indeed? Might there have actually been a time zero, when all the energy of the universe was compressed into a space infinitesimally small? Might the universe have erupted all at once, in a Big Bang of creation, spreading out its birthright energy and creating space-time along the way?

The theory of general relativity makes the notion plausible, and the observational fact of the redshift supports it. But there is further testimony as well, two additional souvenirs of cosmogenesis that we take up briefly below: the cosmic background radiation and the distribution of the chemical elements. Listen to what they have to say.

Witness to Creation

An accidental but astounding discovery, it earned a Nobel Prize for astrophysicists Arno Penzias and Robert Wilson in 1978: the observation that our expanding universe is filled everywhere with a uniform background of low-level microwave radiation, faint electromagnetic ripples emanating from a primeval disturbance of immense proportions. Like ripples in a pond, the oscillations persist far away and long after the initial shock that creates them:

Bang! A stone falls into the water. *Bang!* The door of a furnace blasts open. *Bang!* A burst of energy appears out of nowhere. The disturbance at the beginning soon fades into history, but the rippling heat waves remain. Farther and farther, weaker and weaker, they are a continual remembrance of things past.

The cosmic microwave background radiation bears witness to a universe that was once far smaller and far hotter, a tiny pressure cooker heated to a billion degrees, a trillion degrees, a trillion trillion degrees—the farther back in time, the hotter and smaller the cosmos must have been, all the way back to that unique moment when the pressure cooker blew its lid. The heat that escaped has been cooling ever since. It is now down to only 2.7 degrees above absolute zero (–455 degrees Fahrenheit), spread evenly throughout the universe with a variation of only 0.001%.

Look this way, look that way. Look over here, look over there. The temperature of the cosmic microwave oven differs by no more than a few hundred thousandths of a degree. Measuring it, we measure the relic heat of creation, the afterglow of the Big Bang.

And here is another reminder of the way things began: the present-day abundance of the lightest elements, principally hydrogen and helium. Don't search for them on the Earth, which is a tiny and atypical speck in the cosmos, but look deep into space and inventory what you see. Where there is ordinary matter to be found, most of the mass takes the form of hydrogen (one proton) and helium (two protons), with approximately three times as much hydrogen as helium. It is an ancient legacy, a chemical endowment funded at birth, because all of the hydrogen and most of the helium nuclei in the universe were created in the beginning. They remain with us to this day, the raw material from which nature manufactures the chemical elements.

It happens in the stars. Light nuclei, brought together under extraordinarily high temperature and pressure, fuse into heavier nuclei, building up first from hydrogen to deuterium ("heavy hydrogen," containing one proton and one neutron),

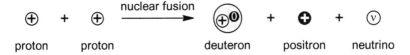

which is then consumed to produce various isotopes of helium, beryllium, carbon, and higher. The reactions are complex, throwing off by-products such as neutrinos, antineutrinos, electrons, and positrons (positive electrons, or "antielectrons"). Some of the nuclei formed are stable; others are not. Some decay radioactively into lighter elements, whereas others are transformed into heavier ones. Many nuclei are born in the cataclysmic explosion of a "supernova," which occurs when a burning star finally collapses under its own weight and generates even heavier elements in the process. The newly minted atoms fly like sparks from the exploding star, raining down fresh chemical possibilities throughout the universe.

However it happens, though—by whatever route the chemical nuclei come into being—the process of "nucleosynthesis" traces back to the primeval hydrogen formed at the outset. To look into a telescope and see a cloud of hydrogen is thus to see some of nature's original building material, still unused, like bricks piled up at a construction site.

The Big Bang in a Nutshell

From little acorns do great oaks grow, and the Primordial Acorn of the Big Bang theory was of a littleness and greatness that can scarcely be comprehended. The entire universe, the sum of all substance—its energy account fully funded, its four fundamental interactions switched on, its capacity to create space-time ready to burst forth—began as an impossibly small, impossibly dense, impossibly hot speck in the midst of nothing. Like a fertilized egg, this self-contained germ of a universe carried within itself everything that was yet to come, either actual or latent. It was a cosmos in microcosm, infinitesimal and infinite at the same time.

Where was it? Nowhere...and everywhere. There were no dimensions and no space in the beginning, nowhere to lay out a coordinate grid and have X mark the spot.

When was it? Approximately 13.5 billion years ago, according to the evidence of the redshift.

What came *before* that? Don't ask. The chronology of the Big Bang begins not at the very beginning, not at "time zero," but rather the tiniest fraction of a second afterward. At the extrapolated zero of time, with the universe presumably squished into an infinitesimally small and infinitely hot point (a mathematical "singularity"), the familiar laws of physics could not have been in force. We know nothing for certain about what existed at the exact instant of the Bang, let alone what may have come before.

Nevertheless, we do know—from the redshift, from the cosmic background radiation, from the distribution of the light elements, from the success of general relativity—that the universe must have started out as a very small, very compressed, and very hot concentration of energy, something like the hot gas in a freshly pumped-up tire. And from those small but potent beginnings, it erupted and grew. It expanded. It cooled. Some of the energy congealed into mass, and the first elementary particles began to appear. There were photons. There were quarks and gluons. There were particles of matter (like electrons) and evil-twin particles of "antimatter" (like positrons), both kinds materializing simultaneously out of an intangible store of energy. At that early stage, too, it was arguably a tenuous beginning for a material universe, because matter and antimatter do not easily coexist. Facing off with identical mass but diametrically opposed in all other attributes, a particle and its antiparticle annihilate on contact and return to the energy whence they came. Were it not for a slight excess of matter over antimatter in the beginning, the universe would have never evolved beyond a burst of pure energy.

But it did. As the dust settled and temperatures cooled, associations between the surviving particles of matter began to take hold. Quarks clumped into protons and neutrons. Hydrogen nuclei (single protons) fused into both stable deuterium nuclei (one proton and one neutron) and short-lived, radioactive tritium nuclei (one proton and two neutrons):

hydrogen (H-1)   deuterium (H-2)   tritium (H-3)

Other fusion reactions produced helium-3 and helium-4, along with small quantities of lithium-7,

helium-3   helium-4   lithium-7

and thus it happened, right at the beginning, that the universe was granted its chemical patrimony. Nearly all these nuclei, products of the Big Bang, continue to exist today in roughly the same proportions established not long after the beginning. Models other than the Big Bang cannot explain why the universe contains about 75% hydrogen and 23% helium, together with small amounts of lithium and deuterium.

We are still early in the throes of cosmogenesis, perhaps some three minutes after the Bang, and at that time the newly created electrons and nuclei had too much energy of their own to stick together. They would shoot past each other, unable to interact. Whatever electromagnetic attractions might have existed were easily broken, and the infant universe remained a dense paste (a "plasma") of charged particles, each one a free agent going about its own business:

No light shined through. Photons, mediating between electrically charged particles, found themselves trapped in the hot plasma. Bouncing this way and that, scattered from one particle to another, they were unable to run the gauntlet of positive and negative charges. Not until almost four hundred thousand years later did the soup cool sufficiently for neutral atoms to hold together, and only then could the imprisoned photons break free. There was light:

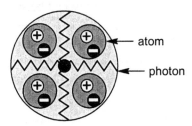

atom

photon

The universe became transparent to electromagnetic radiation, as if a cosmic switch had suddenly been thrown. Slipping through the cordon of electrically neutral atoms, photons shot forth in all directions (isotropically), favoring no particular orientation in a uniformly expanding space:

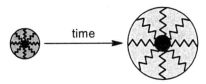

time

They are still traveling. Carried along by the expansion of space-time, these ancient electromagnetic escapees suffer a continual Doppler shift but are otherwise the same photons of old. We observe them today, their wavelengths stretched all the way to microwave frequencies, as the isotropic background radiation first detected by Penzias and Wilson. They are remnants of the primordial heat of the Big Bang, cooled now to faint embers.

Meanwhile, back in the world of matter, the temperature kept dropping and atoms began gradually to clump together into clouds of gas. A large-scale structure started to emerge in the universe as gravity asserted its long arm, leaving behind slight irregularities like lumps in the soup. Clouds of hydrogen (and some helium) clustered into stars, and stars clustered into galaxies, and galaxies clustered into clusters of galaxies. It took hundreds of millions of years, but the universe was finally on its way to becoming the cold, sparse place we know today.

And (from *our* point of view) what a wonderful world it is: a perfect world, the best of all possible, the only one from which conscious observers such as we could possibly have sprung. For if things had been just the slightest bit different—if a few numbers, a few parameters, a few physical constants had taken on other values—then the universe as we know it would never have been. If gravity had been a little stronger...or if the electromagnetic force had been a little weaker...or if the amount of mass had been a little greater...or if Planck's constant had been a little larger...or if one of the energy levels of a carbon nucleus had been a little lower...of if one of the energy levels in an oxygen nucleus had been a little higher...or if any one of a number of tiny details had been just a little different, then none of us would be here today trying to reconstruct the story. There would be no stars. There would be no planets. There would be no water. There would be no life.

The improbability of it all makes one think, and the Big Bang raises other questions as well. For example, how are we to explain the apparent homogeneity of the cosmic background radiation? A uniform temperature, remember, arises from a dynamic equilibrium maintained by a great many particles in constant communication (Chapter 10). They bang into each other, and they bang into the surrounding walls. They trade energies. They exchange information:

Relativity, however, limits the rate of any information transfer to no more than the speed of light, and in the early universe envisioned by the Big Bang there would not have been enough time for communication even at 186,000 miles per second (the so-called "horizon problem" of cosmology). Calculations show that the particles would have been close together, but not close enough.

How, then, absent some sort of collusion at the beginning, is the universe able to exist today at a common temperature? And if indeed there was such perfect homogeneity at the beginning, then what gave rise to all the little lumps and bumps in matter and energy—irregularities manifested now both as stars and galaxies and as the small (but not zero) variation in the cosmic background temperature?

The infinitely dense, infinitely hot singularity of the Big Bang, although called for by general relativity, does not explain why everything eventually worked out the way it did. For those questions we need something more, and cosmologists are increasingly convinced that they have it: the model of cosmological inflation, proposed in 1980 by Alan Guth and modified significantly since.

A number of different versions of the inflationary theory currently exist, but all of them share a common theme. They explain features of the observable universe in a way that traditional Big Bang cosmology cannot, and they describe early cosmic history in expressly quantum mechanical terms.

The quantum mechanics of symmetry breaking and inflation (taken up in the sections below), together with the general relativity of the Big Bang, link the big with the small at the beginning of time. To understand the vast world of the present, we must therefore go back to the minute world of the distant past, a world where symmetries were broken and inflation ran rampant.

Ancient History: Broken Symmetry and Runaway Inflation

To say that that Big Bang happened in the blink of an eye would be a cosmic understatement, because the die was cast over times so short as to defy any visceral grasp. The first milestone, called the Planck time, was reached after only a ten-millionth of a trillionth of a trillionth of a trillionth of a second after the Bang (a decimal point, 42 zeros, and then a 1). The second era, the time of grand unification, ended a trillionth of a trillionth of a trillionth of a second after that.

These are worlds beyond small, worlds in which the blink of an eye would have been an eternity. Since a signal traveling at the speed of light advances only about a billionth of a trillionth of a trillionth of a centimeter during the Planck time (call this distance the "Planck length"), the mind can hardly begin to fathom what such numbers really mean. The diameter of a proton, for instance, is already some hundred billion billion times greater than the Planck length. The diameter of a hydrogen atom is ten trillion trillion times greater. The width of a fingernail, a billion trillion trillion times greater.

Time and space were thus in their infancy, barely begun, when great changes started to occur. Momentous events took place with breathtaking speed. During the first period of time, the Planck epoch, all four forces— gravity, electromagnetism, the strong force, the weak force—are assumed to have been merged into a "superunified" quantum field of perfect symmetry, as yet unknown. Each of them would later distinguish itself from the others, but in this unimaginably brief moment after the Bang, with the universe at its smallest and hottest, the four forces of nature all looked to be of one piece. Just as the blades of a propeller blur together indistinguishably at high speed,

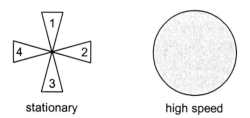

stationary   high speed

so also were the gravitational, electromagnetic, strong, and weak forces indistinguishable in the roiling ultra-ultra-ultra-miniature universe of the Planck era. Viewed from any perspective, they all appeared the same.

The era of a single force ended as quickly as it began. Cooling just enough to mar (but not entirely break) the original perfect symmetry, the universe abruptly began to distinguish two forces where before there was only one: a still unified strong–electroweak force, separate now from a gravitational force. The effect is akin to our propeller slowing down to a point where the individual blades become recognizable, even though each one continues to trace out a full circle. A measure of symmetry still remains:

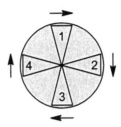

It was not to endure. This brief period of "grand unification," during which the strong and electroweak forces remained united, ended a scant trillionth of a trillionth of a trillionth of a second after the Planck era. The electromagnetic and weak forces broke apart some time later, perhaps a trillionth of a second after the Big Bang.

And then there were four. Observers thereafter would distinguish four fundamental interactions, each with its own source, strength, range, and messenger particles. The unity of ancient cosmic history, by now long forgotten, was thus wiped away in a rapid-fire succession of spontaneously broken symmetries.

Forgotten, but not gone—so the theory goes. An underlying symmetry remains hidden in the icy and desolate universe of today, ready to reemerge if ever the hot, dense conditions of the Big Bang could be recreated. And it was in those first few moments, too, on the heels of grand unification, that something truly big *and* quantum mechanical appears to have happened as well: an astonishingly rapid burst of expansion, all at once, during which the newborn universe underwent an exponential growth spurt of a kind unequaled in the almost fourteen billion years since. In regular intervals, each less than a trillionth of a trillionth of a trillionth of a second, the cosmic radius increased first by a factor of two, then four, then eight, then sixteen, and so on, completing roughly a hundred such doublings before the inflationary period was over. The initial blast itself is presumed to have come from a particular type of quantum field, often called the "inflaton," about which we shall have more to say later.

That, in brief, is the scenario put forth in the inflationary model, and if it is true, then the foundations of an enormous universe must have been laid in the very first instants. Expanding far faster than allowed under a standard Big Bang scenario (general relativity alone), space would have increased by some thirty or forty powers of ten (a factor of at least a million trillion trillion) during this dizzying but brief inflationary spiral. Rather than weaken and succumb gradually to the pull of gravity, like a tossed ball reversing course, the expansion would have accelerated explosively before running out of its inflationary fuel.

The initial impetus provided by inflation has been opposed inexorably by gravity ever since, but its effects still linger. Whatever else may have happened between the Big Bang and now, an early burst of inflation would have provided the universe with a tremendous head start, leaving us with a true cosmos far larger than meets the eye. Our telescopes, looking back no farther than the time when photons first broke free (almost four hundred thousand years after the beginning), would be able to see only a small fraction of it today. The great real-estate boom of the inflationary period would remain hidden behind an opaque curtain, its original boundaries swelled immeasurably by billions of years of continued expansion:

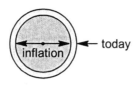

The cosmic background temperature, moreover, would be the same in all directions (as indeed it is, to within one part in a hundred thousand), since particles in a pre-inflationary universe would have been sufficiently close to establish thermal equilibrium. At the same time, any small fluctuations in the quantum mechanical soup would be magnified disproportionately by inflation and later grow into the lumpy galaxies and galactic clusters of today. With one stroke, then, inflation solves the vexing horizon problem of conventional Big Bang theory and also accounts for both the large-scale homogeneity and local *inhomogeneity* of the observable universe.

The inflationary model offers still another concrete prediction, one that can be tested against present and future observational data: that the visible universe should appear geometrically flat out to its edge, conforming to the laws of Euclidean geometry on a large scale. General relativity, by contrast, allows for both Euclidean and non-Euclidean geometries, indeed favoring such possibilities as spherical space (where lines initially parallel eventually draw near),

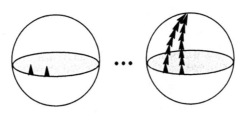

and saddle-shaped space (where lines initially parallel eventually draw apart). Inflation says, no: the space of the universe should be geometrically flat overall. Parallel light rays should remain parallel, never meeting. The three angles of a triangle should add up to 180 degrees. The ratio of the circumference of a circle to its diameter should be equal to π:

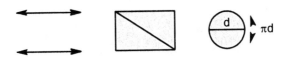

Why? Because the leaps-and-bounds expansion of an inflationary period would be like the sudden blowing-up of a giant balloon. No matter what wrinkles or puckers the universe may have had at the beginning, a small patch would later appear flat to an observer looking out into a vast, seemingly endless space. Without inflation, a flat geometry develops only if the initial density of mass happens to have one very specific value, *exactly*, with scarcely any room for error. Under an inflationary expansion, though, a broad range of initial conditions leads to the same flat outcome. Although both mechanisms are possible, the more forgiving of the two (inflation) appears far more likely—if, of course, the visible universe is really flat.

The final verdict is not yet in, but the first tests do support the inflationary prediction of flat space in the Great Beyond. Stay tuned.

Something in Nothing

Bang! A burst of energy appears out of nowhere. *Crack!* Symmetry is broken. *Whoosh!* A phenomenal rush of expansion gets the infant universe off to a flying start.

Three great events, cosmologists say, took place during those first infinitesimally brief moments, and each of them raises questions of the most profound sort. How did the initial energy of the Bang come into being? What mechanism caused the initial symmetry to break? What source fueled the inflationary expansion?

Nobody knows all the answers, not yet, but there are glimmers of understanding on the horizon. The first clues come from the strangely busy picture that quantum mechanics paints of a place otherwise thought to be literally nothing: the vacuum.

Half Empty, or Half Full?

Vacant. Vacuous. Vacuum. The words all derive from a Latin root meaning "empty," but the vacuum of quantum mechanics is anything but vacant. Bustling with activity, the deceptively empty space of a vacuum is filled with energy, with potential, with *fields*. It is a place full of surprises, never the same from one moment to the next. There are appearances and disappearances. There are fluctuations. There is a blurring of wave and particle. With no warning, a particle and its antiparticle (an electron and a positron, say) suddenly materialize out of "nothing"—the vacuum—and just as suddenly disappear:

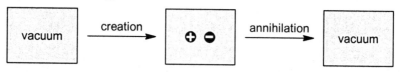

One particle, the positron, has a charge $+e$. The other, the electron, has a charge $-e$. Coming together, they annihilate. They come and go in a flash of $E = mc^2$, vanished before Heisenberg (Chapter 9) has time to establish the value of E. No conservation laws are broken. Energy is conserved, at least within the limits of the energy-time uncertainty principle. Charge is conserved, since the combined value $(e - e)$ is zero throughout. Momentum is conserved. $E = mc^2$ giveth, and $E = mc^2$ taketh away.

If the vacuum were truly an empty space, an absolute nothingness, then it would lack the wherewithal to create and annihilate pairs of particles and antiparticles. It would be unable to serve as a source and sink of photons. It would be incapable of change. But the quantum mechanical vacuum, far from being the archetype of emptiness, emerges as a physical state in itself, permeated by quantized fields of all kinds and subject to evolution under an equation of motion.

What we think of casually as a vacuum—mere emptiness—proves not to be a void at all, but rather a quantum mechanical system resting in its state of lowest possible energy. Energetically, there is no place to go but *up*. The excitation of a field, whether owing to direct interaction or to a random fluctuation, rouses the vacuum from its normally quiescent state and brings forth the bundle of energy and momentum (the quantum) that we associate with a particle. A photon, for example, comes about from the excitation of an electromagnetic field existing in a nominal vacuum:

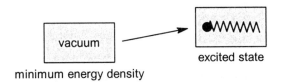

So it goes, too, for all the other particles. For every particle, there is a corresponding quantum mechanical field. A field for electrons. A field for protons. A field for quarks. A field for gluons. A field for whatever particle and whatever influence nature has to offer. Associated with each field is a set of energy levels and probabilities, and it is the material action of these fields that literally makes something out of nothing. Each particle in the repertoire comes about from the excitation of a quantized field from its rock-bottom "vacuum" state.

Now most fields are obliging enough not to burden the vacuum with additional energy when their source values are zero. Take, for instance, the ready example of electromagnetic energy (treated classically in Chapter 6), which varies macroscopically as the square of the electric and magnetic fields. The energy is large when the magnitude of the field is large, small when the magnitude is small, and zero when the magnitude is zero. As a result, any point where the electromagnetic field falls to zero (marked as dots in the diagrams below) contributes no energy to the vacuum:

It is the intuitively reasonable way that nature usually does business: no field, no energy.

Nevertheless, it is not the only way possible. Suppose, instead, that nature also fills the vacuum with fields of a different sort, fields that might be extraordinarily difficult to detect but still might have a critical role to play in shaping the balance of forces—fields, unlike the usual kind, that contribute a nonzero *energy* even when the field strength drops to zero. For if we postulate that such atypical fields exist (physicists call them "Higgs fields"), then we can imagine both a source for the inflationary expansion and a mechanism by which the primeval symmetry was broken. Everything becomes the fault of the Higgs fields, quantum mechanical serpents in the Garden of Symmetry.

Before going all the way back to time zero and the Higgs fields, however, we warm up first with a homey, more familiar example of symmetry and its loss, a garden-variety process that happens every day: the condensation of a gas and the freezing of a liquid.

Symmetry Lost and Symmetry Found

To a micro-sized observer, the space inside a gas looks both homogeneous and isotropic. Particles are distributed evenly and randomly throughout, the same at every point and in all directions. Nobody can tell the difference between up or down, left or right, near or far. No particular set of axes lays claim to any special preference. A gas is a paradise of spatial symmetry, a place where the concept of direction has no meaning.

Do something now to cause trouble in paradise. Cool the gas. Squeeze the gas. Siphon off part of the system's kinetic energy. Force the particles into closer quarters and make them move more slowly. Do these things with a sufficient amount of persuasion, and eventually something will happen. The spatially symmetric gas will undergo a "phase transition," a symmetry-breaking change of state. The every-which-way indistinguish-ability of particles in utterly random motion will degrade first into the partial order of a liquid and then into the more complete order of a solid:

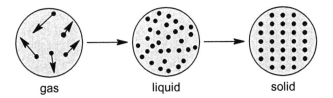

gas   liquid   solid

The rotational symmetry of the gas literally "freezes out" of the solid, where space may look markedly different along different axes. It is a new world, a world in which directions mean something. Each particle has its own address now, and it rarely wanders far from home.

Must we say, then, that the laws governing a gas differ fundamentally from those governing a solid? Are the pertinent equations symmetric for one system but not the other? Does an impassable barrier separate the two states, a difference in outlook so fundamental that the inhabitants of ordered Ice Land cannot even hope to understand the isotropic world of disordered Steam Land? No, of course not, says anyone who has ever boiled water or melted an ice cube. What turns a gas into a solid is merely an environmental and historical accident, a tale of Symmetry Misplaced rather than Symmetry Lost. Particles in a solid stay close to home not because they obey different laws, but simply because they have insufficient energy to do anything else. If we wish to buy back the original symmetry of the gas, we need only apply heat. The solid will melt, the liquid will boil, and any spatial distinctions will be erased.

Suppose, though, that we ask to know beforehand which particular directions will prove special as a gas undergoes its transition to a solid. We set out to predict whether a particular axis of the crystal will point north, east, southeast,

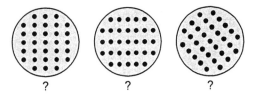

?   ?   ?

or in any one of an infinite number of possible directions. Can we do so?

No. We might just as well spin a roulette wheel with an infinite number of compartments and wait for the bouncing ball to settle down, because the overall orientation develops entirely by accident. A crystal that forms with a given axis pointing north is no more advantaged or disadvantaged than one that has the same axis pointing east. It may be a speck of dust, or a slight vibration, or a fluctuation in temperature, or some other wholly unpredictable event that causes the crystal to grow in one direction and not another. Who can tell? For whatever reason, a few particles begin lining up in a particular orientation and the neighbors play follow the leader. The innumerable macroscopic outcomes are all equivalent, and so we throw up our hands and leave the matter to chance. Let the issue be settled by a spin of the wheel.

Quantum mechanical symmetries, the kind of "gauge" properties we encountered in Chapter 9, go through analogous transitions, and here is where the proposed Higgs fields (mentioned above) may come into play: as a world-shaking, symmetry-breaking, leave-it-all-to-chance roulette wheel. Let's go on to imagine now how such a game might be played.

Spin of the Wheel: The Higgs Fields

As long as the ball is in play, anything is possible. Later on, a choice will be made—maybe a 2, maybe a 15, maybe a 32—but for now there is equal opportunity for all, a symmetry of possible outcomes. The bouncing ball samples every one of them:

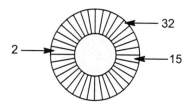

Eventually the wheel slows and the ball drops decisively into just one compartment, thus breaking the symmetry of potential outcomes. The slight convexity of the disk guarantees that there will be a winner, sooner or later, because the ball cannot come to rest either on the top of the wheel or on a slope. Where once there was a symmetry of possibility, in the end there is an asymmetry of actuality. Just ask any of the losers.

Different numbers have different consequences for different players. For Jane, who risked $100 on the number 15, the happy appearance of 15 means a gain of $3500. For Sally, who bet on 2, the same outcome means a loss of $100. For Dick, who bet $100 on each of 2, 15, and 32, the appearance of 15 means a net gain of $3300. To each his own.

Picture now a different kind of roulette wheel, one in which every compartment (and there can be an infinity of them) specifies the value of a particular set of Higgs fields. One slot may mean that Higgs field A has the value 12.2, Higgs field B has the value 17.4, and Higgs field C has the value −46.2. Another slot may mean Higgs field A has the value 15.1, Higgs field B has the value −0.7, and Higgs field C has the value 6.3. Another slot will mean something else, and another slot something else again, and so it shall be, for as many fields and combinations as needed, but only at *one* special point—the top of the wheel, a slightly rounded peak—will all the field values be equal to zero. The fields there will be zero, but the potential energy will be at a maximum. Just pretend:

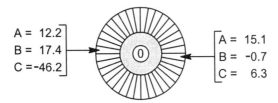

Now here is the object of the game: to acquire mass. For players of Higgs roulette, the field values determine not how many dollars the winners will collect, but rather how much mass a given particle is destined to have. Particles get their properties, recall, by interacting with fields, and so we look for our sundry particles to couple with Higgs fields as well.

Draw the analogy. An electron interacting with an electromagnetic field acquires a particular electromagnetic potential energy; an electron interacting with an appropriate Higgs field acquires a particular mass. For an electromagnetic field, the message is carried by a photon; for a Higgs field, the message is carried by a Higgs boson. One set of Higgs values (A, B, C, …) makes an electron 1836 times less massive than a proton; another set of Higgs values makes an electron 1836 times *more* massive than a proton. One set of Higgs values makes a proton slightly less massive than a neutron, while another set makes it three times more. One set makes photons massless but W and Z bosons massive, thereby rendering the electromagnetic interaction distinct from the weak interaction. Another set, by making all four messenger particles massless (the photon, W^+, W^-, Z^0), leaves the weak interaction indistinguishable from the electromagnetic. Every compartment of the wheel yields a different combination of masses, but in only one case—the case where all the Higgs fields are zero—will all the particles have zero mass and will all the interactions be united into one.

Spin the wheel. Rock the wheel. Let the table vibrate, as if during an earthquake. Let the ball skip over all possible values of the Higgs fields; let it bounce over the top now and then; let it enjoy the unbroken symmetry of possibilities that a state of high energy has to offer, a state just like the agitated, highly energetic world of the Big Bang. The Higgs fields, sampling all the possibilities equally, average to zero, and consequently all the particles in that world (including the messenger particles) are massless. The fundamental interactions are unified and indistinguishable, as long as the Higgs ball continues to bounce symmetrically around the wheel.

After a time, though, the ball loses energy and comes to rest. The wheel slows, and the rocking stops. The universe cools. In a game of ordinary roulette, the ball settles randomly into one of the compartments around the edge of the wheel, a point where the gravitational potential energy is at a minimum (call it zero). Friction, air resistance, and all the other dissipative forces of the real world eventually take their toll. In a game of Higgs roulette, the evolving fields likewise settle randomly into a state where the potential energy (the Higgs potential) is zero, but now there is a world of difference: although the potential energy is zero, the Higgs fields themselves are not. At whatever point the cooling universe comes to rest, somewhere around the bottom of the energy hump, that is where the values stay:

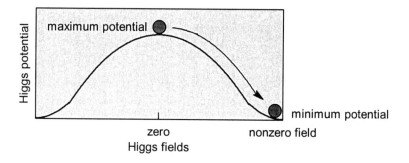

The Higgs fields remain frozen in place all throughout the universe, leaving a permanent crack in the symmetry of the forces. Henceforth the vacuum is filled uniformly with nonzero Higgs fields, everywhere with the same set of values.

Thus empowered, the Higgs fields dish out a fixed mass to any particle that happens along. Particles with different masses slog through a constant Higgs field and find their motion retarded in different ways, not unlike a shark and an octopus navigating the same body of water. Acting by means of the Higgs bosons (which, bear in mind, have not yet been detected experimentally), the frozen Higgs fields distinguish both particles and interactions. They deliver a different message to every particle, as if to say:

> *You, photon, shall be massless and transmit the electromagnetic interaction over great distances. You, W and Z bosons, shall have 86 and 97 times the mass of a proton, respectively, and carry the weak interaction over distances incredibly short.*
>
> *You, down quark, shall distinguish yourself from the up quark by having a greater mass, whereas you, neutrino, shall be nearly massless and in this way differ from your brother the electron.*

And so the story goes.

At the moment, one cannot judge conclusively whether the Higgs account of Symmetry Misplaced is verifiable history or merely legend. The Higgs mechanism makes convincing sense of the *W* and *Z* bosons (which actually do exist), but Higgs fields are not the only conceivable source of broken symmetry in the electroweak model. And although the idea fits well into the current way of thinking, the current way of thinking may change. Suspending judgment, then, we await the results of experiments at energies sufficiently high to prove or disprove the existence of the Higgs bosons.

False Vacuum: Start of Something Big

Economists tell us that monetary inflation develops when too many dollars chase too few goods and services. Cosmologists tell us that space-time inflation develops when too much energy chases too little space. The blame falls once again on a not-entirely-vacuous vacuum, a nominally empty space permeated (according to the various models) by something with a tangibility all its own: a field. A quantized "inflaton" field.

We conjure up a Higgs-like field with a wrinkle. The inflaton potential, rather than reaching a maximum when the associated field is zero, sinks instead to a local minimum. Picture a mountain range broken by a high valley into which a bouncing ball may sometimes land (or, if you like, a roulette wheel with a depression in the center of the disk):

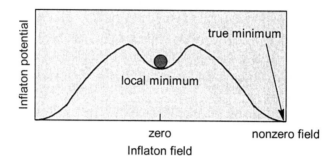

The local minimum does not offer the lowest energy possible, but it does provide a resting point that can be maintained under favorable conditions. If, for example, the overheated system undergoes a rapid enough cooling, then a ball falling into the depression might become stuck there, unable to scale the surrounding peaks. The ball (the inflaton field) remains above the true minimum until it musters the requisite energy to escape, and until that time, the vacuum has an energy higher than a proper vacuum ought to have. The space exists arguably as a "false" vacuum, and sooner or later this false vacuum will discharge its excess energy and find true repose at the bottom of the mountains. The probabilistic nature of the quantum world provides a way down. Accidents happen. Just wait.

But while it lasts, a false vacuum created by a Higgs-like field behaves in ways that challenge our everyday intuition. In some respects, it appears to be simply a concentration of energy in a confined space, like air in a balloon or paint in a spray can. In other respects, a false vacuum acts in ways that no other pressurized container can. Consider the differences, one by one:

    1. The total mass (or mass-energy, mc^2) contained in an ordinary space holds constant even as the volume is increased or decreased. If we cut the volume in half, say, we force the same number of particles to occupy half the space and hence the energy density (the energy contained per unit of volume) to

double. If we double the volume, we allow the particles to spread out into twice the space. The density falls by half:

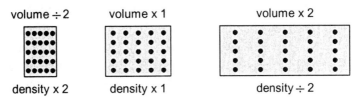

The energy contained in the false vacuum, however, arises not from a fixed amount of mass, but rather from a fixed value of the inflaton field. Since the strength of the field at each point does not depend on volume, then neither does the corresponding energy density. The energy density remains the same no matter how big or small the space:

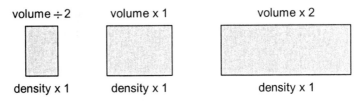

2. Mass-energy contained in an ordinary material space exerts a positive pressure. The particles push outward against the walls of the container:

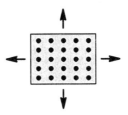

The energy contained in a false vacuum, by contrast, exerts a negative pressure. The inflaton field creates a sucking force that pulls inward:

The effect—by no means an obvious one—is predicted by the

complicated equations of general relativity, but we can construe it in a way that has at least an aura of plausibility. Suppose that Zeus, present at the creation, wishes to expand the space occupied by the false vacuum. If so, then he will have to invest energy into the system. He will have to supply enough of his own energy to overcome the suction of the false vacuum. Absent that suction (and absent Zeus's countervailing energy), the system would be unable to maintain a constant energy density in the larger space. The sucking force of the false vacuum ensures that any change in volume will be accompanied by a change in the total energy it contains.

3. Gravity is geometry, says Einstein, a local warping of spacetime caused by the presence of mass-energy. The positive pressure ordinarily exerted by matter warps the nearby space outward. Particles following the distorted tracks appear to come together, as if attracted by a gravitational field:

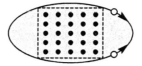

The sucking, negative pressure of a false vacuum warps the space inward. Particles following the distorted tracks appear to move apart, as if repelled by an "antigravity" field:

Therein lies the difference. A false vacuum, created when the energy of empty space is higher than it ultimately could be, produces a net gravitational repulsion rather than a gravitational attraction. On the one hand, the concentration of energy in a space—in any space—induces a positive curvature proportional to the energy density, an attractive influence. On the other hand, the sucking force of a false vacuum induces a *negative* curvature proportional to three times the energy pressure, a repulsive influence. Two competing effects thus tear the space in opposite directions, and the repulsive force wins. The space-time geometry of the false vacuum is dominated by the negative curvature induced by its negative pressure.

The false vacuum feeds on itself. The more space it has, the faster it grows. Powered by an ever growing negative pressure, the inflating universe expands the scale of space at a staggering exponential rate. *Tick.* In a moment indescribably brief, the universe grows to twice its original size. *Tick.* Four times its original size. *Tick.* Eight times. *Tick. Tick. Tick.* After ten ticks, before you know it, all distances have increased a thousandfold. After twenty ticks, a millionfold. Thirty ticks, a billionfold. Puffed up by the false vacuum, the speed of space accelerates without bound, faster and faster, until finally the expansion comes to an end. The false vacuum leaks away.

When the false vacuum dissipates, *how* the false vacuum dissipates, *whether* the false vacuum dissipates, how the false vacuum *starts*—all these questions, and more, continue to generate intense interest, and the proposed solutions are legion. The central theme of inflation, though, shines through the many variations. There is "old" inflation, which posits a deep local minimum in the inflaton potential and a quantum mechanical route to escape from it. There is "new" inflation, which starts with a shallower local minimum (even a plateau) and brings about a less precipitous exit. There is "chaotic" inflation, which allows for a variety of local minima and a variety of eventual universes. There is "eternal" inflation, which envisions the inflating false vacuum as outrunning its own decay and leaving behind a string of disconnected "pocket universes." For even though the false vacuum may have partially decayed in one place, its remnant might continue growing to points beyond:

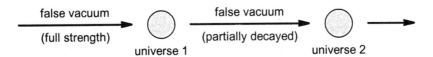

A false vacuum that decays and runs away...lives to grow another day.

The list goes on. There is "extended" inflation. There is "supernatural" inflation (not what you think it is). There is "ultimate-free-lunch" inflation, in which the total energy content of the universe is assumed to have been exactly zero right from the beginning: a quantity of algebraically positive mass-energy diminished by an equal magnitude of algebraically negative gravitational energy (the attractive potential stored in the cosmic gravitational field). "Something minus something equals nothing," asserts the equation for the ultimate free lunch, "and from nothing comes everything."

There are models, like chaotic inflation and eternal inflation, that imply a universe of universes (a "multiverse," some say), with the various members governed perhaps by different laws and different physical constants. But even then, even as part of a multiverse, we would still be isolated in our own little universe like an island in the midst of a vast ocean. Cut off from our multiverse associates, unable to communicate, we would still have to live within our own laws and draw energy from our own store of fields and particles.

There are models and models, some more speculative than others, but they all point to a universe whose formative instants were shaped by an inflationary expansion. They all look to quantum mechanics as an explanation for the origin of matter, and they all push to the limit one of nature's most basic principles of design: that big things ultimately come from small things.

Forces of Darkness

So here we are, after nearly fourteen billion years of cosmic evolution, and we come to a question every sentient being should find interesting. How will it all end?

With a Bang? Or with a Whimper? Will the forces of cosmic expansion, launched during an epoch of runaway inflation, prevail over the forces of cosmic contraction; or will the attractive pull of gravity, which draws together every bit of mass-energy in the universe, be sufficient to slow down and eventually reverse the expansion? Will the universe turn back on itself and begin to contract, like a tossed ball returning to Earth; or will it keep running on forever, like a rocket launched into space? Will the expansion slow down and finally stop, like a hockey puck succumbing at last to friction; or will it somehow manage to accelerate, drawing on a source of energy we have yet to explain?

The battle has been waged since the beginning of time, and it will not be over any time soon. But even now we may aspire to know the outcome, provided we can take accurate stock of *all* the energy and mass on hand. Because if we know both the recessional rate of the galaxies (a measure of cosmic expansion) and the cosmic mass density (a measure of gravitational contraction), then we should be able to weigh one force against the other. We should be able to predict one of three possible endings:

1. *Gravity wins.* The universe contains sufficient mass to turn back the expansion and collapse one day in a Big Crunch. *Ouch.* Cosmologists describe such a universe as "closed," with a large-scale geometry analogous to that of a sphere. Parallel light rays eventually meet.

2. *The expansion wins.* There is enough mass to slow the rate of expansion but not to stop it. The universe is "open," destined to expand without limit. Gravitational collapse is not in the cards, not even in the far distant future, and the geometry of the open universe is that of a saddle. Parallel light rays move progressively farther apart.

3. *Neither wins.* With just the right amount of mass, the attractive force of gravity exactly balances the repulsive force of expansion, and gradually—over a long, long time—the rate of expansion drops closer and closer to zero, although never actually reaching it. With that, we have the "flat" Euclidean universe predicted by inflation and supported by the latest observations. Parallel light rays remain parallel, always separated by the same distance.

Astronomers, rising to meet the challenge, continue to develop ever more accurate ways to refine their measurements. With great care, they record redshifts from distant stars. They look for gravitational distortions and anomalous motions that signal the presence of mass. They systematically inventory all the matter in the known universe, wherever and whatever it may be. And the more they do so, the more incomplete the picture seems to become. The circumstantial evidence is growing increasingly clear: there is more to the universe than meets the eye. There is more energy than we previously thought, and there is more matter than we can see.

Dark Matter

Count up all the protons, neutrons, and electrons that make up the ordinary matter of our cold universe (the chemical elements), and then count again. Something is missing. There is plainly not enough ordinary matter to explain the existence and motion of the galaxies. The stars that shine, the nebulae that glow, the quasars that broadcast radio waves—all of the electromagnetic sources in the visible universe, put together, lack the combined mass to do the job gravitationally. Something else out there, some kind of nonluminous "dark" matter, must be exerting a gravitational influence, and in great quantity, too. The protons and neutrons of ordinary matter represent but a small minority (perhaps 5%) of the mass of a universe mostly dark.

Yes, there is plenty of ordinary matter lurking in places astronomers have yet to expose: in distant gas clouds, in undiscovered galaxies, maybe in black holes. Yes, we cannot expect to track down and verify every last proton and neutron in the cosmos. But no, the dark matter is manifestly *not* composed of protons and neutrons (jointly called *baryons*, or "heavy ones"). If it were, the chemical composition of the universe would be different, since the constraints of Big Bang nucleosynthesis put strict limits on the total number of baryons handed down to us. Let the universe start out with too many protons or neutrons, and the final mixture will not contain hydrogen, deuterium, helium, and lithium in the proportions we observe today. Even if every single proton and neutron were found and tallied in the census, the collective mass would still not add up.

So what might it be? What variety of nonelectromagnetic matter might be out there, hiding under cover of darkness? Some of the dark matter, although clearly not all, may come from neutrinos. First, the universe is filled with vast numbers of them, and experiments now suggest that a neutrino indeed has a small (but not zero) mass. Second, the particles are electrically neutral and conspicuously shy in their interactions. Neutrinos leave no electromagnetic signature, and they interact microscopically only through the weak force, making them hard to detect. Their presumed mass is too small, however, to account for more than a few percent of the cosmic total, and hordes of nearly massless, highly energetic neutrinos liberated by the Big Bang would have another disqualifying feature as well: they would move too fast. Traveling at close to the speed of light, they would be unable to clump into the structures needed to produce the full array of gravitational effects.

No. Reviewing the list of obvious candidates, we cross off one particle after another. None is suited to be a major source of dark matter:

> *Protons and neutrons, no.* See above. The distribution of light elements holds down the total population of baryons. Any undiscovered excess of protons and neutrons would be betrayed by its effect on the cosmic mixture of hydrogen, deuterium, helium, and lithium.

> *Electrons, no.* There are only as many electrons as protons, and electrons are nearly 2000 times less massive. If baryons cannot make up the deficit, then neither can electrons.

> *Photons, no.* The mass-energy contributed by the cosmic microwave background adds up to just a minute portion of the total, maybe 0.005%.

> *Neutrinos, no.* Again, their mass is too low and they move too fast to constitute an appreciable portion of the dark matter. Current estimates show neutrinos contributing only between 0.1% and 5% of the total mass.

> *Anything that we definitely know to exist, no.* All told, the usual suspects furnish no more than a small fraction of the cosmic mass. Dark matter apparently arises from something that we do not yet know: something cold and sluggish enough to clump into extended structures...something heavy enough to account for all the missing mass...something stealthy enough to elude electromagnetic detection...something aloof enough to interact weakly or not at all with ordinary matter.

Whatever it is, then, dark matter appears unlikely to show up on the current roster of known particles. We have no choice but to look elsewhere, hoping to focus the search by asking the right questions. Will the mystery mass be explained, for example, by hypothetical particles that would arise in a quantum world of higher symmetry, such as those demanded by grand unified theories (which put the strong and electroweak forces on an equal footing) or "supersymmetric" theories (which interconvert fermions and bosons)? Will the lightweight but ubiquitous "axion" of some grand unified scenarios be shown actually to exist, and if it does, will the axion make a significant contribution to dark matter? Or will the supersymmetric "neutralino," another theoretical possibility, be the particle that eventually provides the key?

Why stop there? Perhaps the successful candidate will prove to be a true dark horse, an exotic species currently on nobody's short list. Or—and here is another reasonable proposal—perhaps there is really no dark matter at all. Maybe the anomalous behavior of the galaxies is attributable not to some exotic new form of matter, but rather to an exotic new law of motion. If so, we need new equations and not new particles.

We shall have to wait and see. The answers are not yet clear, and none of the proposed explanations is without problems. For now, physicists can only speculate and try to devise experiments that might bring the dark matter to light.

Dark Energy

It is enough to make even the proudest feel humble: the realization that the protons, neutrons, and electrons that we hold so dear—the building blocks of the Earth, Sun, Moon, planets, stars, and, come to think of it, ourselves—amount to very little in the grand design, a trifling few percent of the cosmic matter supply. And if such a blow to one's material self-esteem were not enough, then be prepared to be taken down yet another peg. For not only does "our" kind of matter fall woefully short of cosmic dominance, but so does "our" kind of energy as well.

Go out and tally up all the energy in the universe. Record the amount stored in the mass and gravitational field of every bit of matter (including dark matter), and once again the numbers fail to add up. "Ordinary" energy, the kind that causes gravitational attraction, makes up only about a third of the cosmic total. The remainder is "dark" energy, and nobody knows for sure what it is. We only know what it does. It causes a gravitational repulsion. Compare:

> Ordinary (attractive) energy brings matter together. Gravitational attraction causes atoms to clump into stars, galaxies, and the other large structures of the universe. Gravitational attraction makes the

universe smaller, more condensed. Gravitational attraction coun-
teracts the expansion of space that began with the Big Bang.

Dark (repulsive) energy pulls space apart and takes matter along
with it. Antigravitational repulsion inhibits the formation of stars
and galaxies. Antigravitational repulsion makes the universe bigger.
Antigravitational repulsion accelerates the expansion of space that
began with the Big Bang.

The key discovery came just in 1998, and it was a shocker. Observation of distant supernovae suggested that the universe is not behaving
like a cannonball shot into the air, which accelerates only briefly during
an initial thrust upward (analogous to the inflationary Big Bang) and
then undergoes a relentless deceleration due to gravity. The universe is
behaving instead like a rocket firing its engines continuously, picking up
speed despite the countervailing pull of gravity. Whereas the cannonball
gets one kick upward and then has nothing but inertia to keep it moving
in the same direction, the rocket (below, right) enjoys an ongoing boost
of force and acceleration:

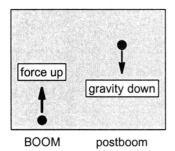 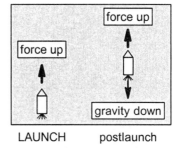

The cannonball never moves faster than its muzzle velocity once gravity
takes over. The rocket, relying on the strength of its engines, controls its
own fate. It can slow down. It can maintain a constant speed. It can go faster.

If the supernovae evidence and other indications hold up, then
apparently the universe (like the rocket) supplies its own motive force to
accelerate against gravity—not to slow down the expansion of space, but
to speed it up. The power to do so would come presumably from the so-called
dark energy of space: a vacuum energy, like the false vacuum of the
inflationary era, that produces a negative pressure and concomitant
gravitational repulsion. Left unchecked, a growing acceleration of this sort
might even one day tear the universe apart in a final, rousing "Big Rip."

The jury is still out concerning the source, strength, character, and history of the dark energy. Some researchers attribute the negative pressure to a uniform energy density that pervades all space, an effect Einstein modeled by a quantity called the cosmological constant. Others propose a different kind of quantum field ("quintessence"), one for which the energy density varies in space and time. Still others caution that the reported acceleration, although based on solid observations, needs additional confirmation. Sentient beings await the verdict with understandable interest, because the fate of the universe remains unknowable so long as the dark energy continues to be a mystery.

Higher Standards

The recipe calls for just a few simple ingredients, but they are enough to account for nearly every structure and interaction in the known universe. Three groups of fermions (twelve elementary particles in all) supply the matter, and a baker's dozen of bosons transmit the forces:

		1	2	3
FERMIONS	quarks	up down	charm strange	top bottom
	leptons	electron electron-neutrino	muon muon-neutrino	tauon tau-neutrino
BOSONS	W^+, W^-, Z^0 (weak interaction) photon (electromagnetic interaction) 8 gluons (strong interaction)		Higgs (mass)	

Most of the names are already familiar from Chapters 2 and 9, with the neutrino rechristened here as the "electron-neutrino" to distinguish it from analogous neutral particles in the other groups, or "generations." Except for mass, which increases from left to right across the chart, properties of fermions in the second and third generations are identical to those in the first.

Now, concerning the basic ingredients, note first that every generation of matter contains two quarks and two leptons ("light ones," from the Greek). Elementary in the strictest sense, the quarks and leptons are taken to be irreducibly simple particles with no internal structure of their own. They differ in their fundamental attributes and in the ways they interact:

The quarks. Each quark, in addition to having a unique mass, sports three metaphorical colors (red, green, blue) and one of six metaphorical flavors (up, down, charm, strange, top, or bottom). Quarks participate in all four fundamental interactions.

The leptons. A negatively charged electron, muon, or tauon in each generation is matched with an uncharged neutrino. Leptons, unlike quarks, are immune to the strong interaction. They all have different masses.

Add to the mix an antiparticle for each fermion (a doppelganger with the same mass but diametrically opposed charge and magnetic moment), and our recipe is complete. We have reduced the world to a short list of ingredients and instructions compact enough to fit onto an index card.

Theoreticians proudly call it the Standard Model of particle physics, and they have just cause to be proud. It works, and it works wonderfully well. Drawing on both quantum chromodynamics (the rules of the color force, as outlined in Chapter 9) and the unified electroweak theory (also Chapter 9), the Standard Model rationalizes the electromagnetic, weak, and strong interactions using a remarkably spare set of fermions and bosons. The first generation of fermions suffices to build all the neutrons, protons, atoms, and molecules of ordinary matter. The second and third generations account for the exotic but fleeting debris produced when particles collide in high-energy accelerators.

These are verifiably real particles, not simply figments of a mathematically fertile imagination. They exist. All of them, except the Higgs boson, have been detected either directly or indirectly, and their properties conform to the predictions of the Standard Model. This brief recipe, amazing to say, is almost enough to serve as the blueprint for an entire universe.

Almost, but not quite.

Under the Rug

The Standard Model, for all its successes, suffers from omissions and inconsistencies that cannot simply be wished away. Where is gravity, for one? Where is the unification between the strong and electroweak forces? Where is the dark matter? Where is the dark energy? Where is the inflaton field?

Why are there not equal amounts of matter and antimatter in the universe? Why are there fully three generations of fermions when nature apparently uses only the first to build every single proton, neutron, and electron currently in stock? Why do particles have the specific masses they do, and if a Higgs field indeed confers mass, then what is the mass of the Higgs boson itself? And speaking of bosons, why are they so different from fermions?

Why, too, if the Standard Model is otherwise so powerful, does it rely on a sleight of hand called "renormalization" to make the equations work? A quantized field at first blush is a monstrosity, a mathematical absurdity that makes no sense at all. It envisions space as filled with an infinite number of infinitesimally small vibrators, each continually bringing forth virtual quanta of the field (photons, for instance):

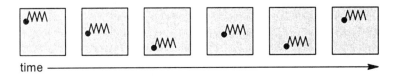

time

Here one instant and gone the next, the virtual quanta are created and annihilated under license of the energy-time uncertainty principle described in Chapter 9. Their random comings and goings break no conservation laws, but nevertheless the unending percolation of a quantized field adds up to a great embarrassment: an *infinity*, to be exact. Since the number of potential oscillators is infinite even in the tiniest region of space, then so is the calculated energy. The electron, a source of the electromagnetic field, theoretically has both infinite energy and infinite mass as it sits bathed in a cloud of virtual photons. The electronic mass measured experimentally, however, is eminently finite, approximately a billionth of a billionth of a billionth of a gram.

Enter renormalization. We redefine the mass and charge of the bare particle (as measured in the laboratory) to incorporate and thus discount the continual spray of virtual particles, as if the problematic infinity does not exist. The effect is something like taring a scale or resetting an odometer, and despite any appearance of impropriety, a renormalized field has one great practical virtue. It gives the right numbers. Although some physicists contend that the procedure merely sweeps the infinities under the rug, the numerical results so obtained are finite, well behaved, and often astonishingly accurate. Just look to the example of quantum electrodynamics, cornerstone of the Standard Model, for one of the most convincing theoretical predictions in all of science: a calculated value of 1.00115965214 for a certain magnetic property of the electron, compared with a measured value of 1.001159652188. (*But still,* you say, *the tricks we have to pull in order to make the theory work!*)

Take heart. The assorted questions and complaints are signs not so much of error, but rather of incompleteness. Too much of the Standard Model is demonstrably correct, consistent, and logically interdependent for the structure suddenly to collapse like a house of cards. What the Standard Model needs is to be upgraded to a deluxe version, not to be discarded whole. It needs to find a place as part of a broader, more comprehensive picture, valid within its own sphere yet compatible with phenomena in other domains as well.

The possibility of supersymmetry in nature, a breakdown of the wall dividing fermions from bosons, is a promising place to begin.

Supersymmetry

Is that an electron you see, or a neutrino? An up quark or a down quark? A *W* boson or a photon?

In a Higgs-less world of unbroken symmetry, conscientious observers might well disagree. They would see only massless, indistinguishable particles acting out different roles in the eye of every beholder. The particles would swap identities as ambiguously as the four corners of a square, or the six vertices of a hexagon, or the infinity of points on a circle:

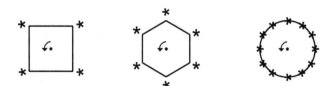

Observers, understandably confused, would be unable to tell the difference.

No one, though, would ever be so confused as to mistake an electron for a photon, or a neutrino for a *Z* boson, or a quark for a gluon, because fermions and bosons do not mix. The Standard Model has them living in separate camps, with the differences enumerated as in Chapter 9:

1. *Fermionic wave functions switch sign when identical particles are interchanged.* Bosonic wave functions stay the same.

2. *Fermions carry either a half-unit of spin or an odd multiple thereof (1/2, 3/2, 5/2, ...).* Bosons carry either integral units or zero (0, 1, 2, ...).

3. *Fermions are loners, each one occupying its own quantum state.* Bosons are joiners. They can all pile up in the same state.

4. *Fermions (electrons, neutrinos, quarks) are matter.* Bosons (photons, *W*, *Z*, gluons) are force.

An observer might be forgiven, under some circumstances, for confusing an electron with a neutrino, but never a neutrino with a photon. After all, fermions and bosons do not mix.

But suppose they do. Suppose that bosons and fermions are really twins separated at birth by an accident of broken symmetry. Suppose that under the right conditions, fermions could transform into bosons as naturally as up quarks transform into down quarks. Suppose that the present-day universe masks a higher, more inclusive symmetry, a *supersymmetry* that links fermions and bosons into a true brotherhood of matter and force.

It would be a world containing twice as many particles as we currently see, a world in which every fermion and every boson would have a heretofore unacknowledged buddy in the opposing camp: a "superpartner." For the electron (a fermion with a half-unit of spin), there would be a matching supersymmetric-electron, or *selectron* (a boson with zero spin). For the neutrino, there would be a sneutrino. For the quark, a squark.

For the photon, there would be a photino: a fermion with a spin of one-half to complement a boson with a spin of one. For the *W* boson, a wino. For the *Z* boson, a zino. For the gluon, a gluino. For every particle, a superpartner with a half-unit less spin. For every pair, a marriage made in supersymmetric heaven.

No experiment has ever produced one of these superpartners, and perhaps they do not exist—or, as many physicists believe, perhaps they are so tremendously heavy that their materialization requires huge amounts of $E = mc^2$. Supersymmetry is a legitimate prospect to explore, however, and its implications for the Standard Model are far reaching. One of the effects of having superpartners, for example, is to alleviate the infinities that otherwise plague quantum fields. Fermions and bosons tend to make opposing contributions to the quantum jitters, and the supersymmetric guarantee of "a fermion for every boson and a boson for every fermion" goes a long way toward averaging out the fluctuations. When the blip from one kind of particle is positive, the blip from the other kind is negative. The microworld becomes a calmer place. Even more, a supersymmetric universe is one where gravity and the three other forces would be able to unite into one. General relativity conjures up gravity as a device to ensure that all observers, regardless of their own acceleration, describe the same physical laws. Supersymmetry does the same thing. It transforms the coordinates of space-time in a way that guarantees the equivalence of all reference frames.

Supersymmetry thus presents itself as an important clue in the search for a comprehensive theory, but it is only a piece of the puzzle, not the whole picture. The next step calls for an even more drastic reconsideration of some long-held assumptions, and the first casualty is an idealization that dates back to the ancient Greeks. We may be forced, in the end, to abandon the notion of an elementary particle as a dimensionless point.

The Vanishing Point

It may be the last bastion of classical physics to fall: the perception of space-time as a smooth continuum, divisible without limit into ever smaller portions. Everything else in the world (energy, linear momentum, angular momentum, mass, charge, and more) comes bundled in discrete packets, yet we knowingly construct models in which the boundaries of space and time shrink to virtually nothing. No matter how tightly the borders are drawn, we assume that they can always be drawn just a little bit tighter:

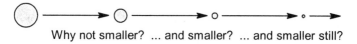

Why not smaller? ... and smaller? ... and smaller still?

Or can they? Can finite amounts of energy, mass, momentum, charge, and the various other attributes of a particle really be packed into a dimensionless point? What does it mean to vanish into a mathematical abstraction that has neither length nor height nor width? Can space-time really be chopped into arbitrarily tiny pieces?

An observer of the macroworld says yes. To Einstein, the empty space of general relativity stretches out unbroken like an elastic sheet, a trampoline on which nobody is jumping. Classically, a small piece of empty space looks qualitatively no different from a large piece. The trampoline may span the entire universe or merely a tiny dot, but in the absence of mass its flexible fabric is everywhere smooth and continuous.

An observer of the microworld sees a different picture. In the quantum mechanical universe, space is never really empty and never really quiet. A fluctuation bubbles up randomly out of the vacuum, persists for as long as the uncertainty relations allow, and then disappears, only to be replaced by another fluctuation somewhere else. "Do whatever you wish," says Heisenberg, "provided that you do it quickly and do not trespass beyond a small region of space. If you stay within the limits of the uncertainty principle, then I'll pretend not to notice."

Poof! A blip of energy, a blip of field, and then *poof!* it's gone. If the time is short, then the uncertainty in energy (Chapter 9) is large. If the space is small, then the uncertainty in momentum (Chapter 7) is large. Given enough time and space, the blips average to zero and the quantum mechanical vacuum seems to behave as decorously as the classical vacuum—but *only* if it is allowed a chance to smooth out the unending fluctuations. If we investigate the vacuum too closely, then the indeterminacy of the microworld rears its unsettled head.

Now since gravity is geometry (à la Einstein), it will be space-time itself that pays the price when a gravitational field gets the quantum jitters. Empty space, pushed to the limit of smallness, turns into a frothing cauldron of vacuum fluctuations, a messy confusion of "quantum foam." Wild variations in the gravitational field twist and rend the curvature of space-time beyond repair. From far away, we see a smooth fabric. Up close, on the scale of the Planck length, we see a fabric shredded to bits.

If space-time were truly smooth and if particles were really mathematical points, then the dogma of general relativity (*gravity is geometry*) might never be reconciled with the dogma of quantum mechanics (*gravity is a field of messenger bosons*). Macroscopic–microscopic harmony breaks down irretrievably when distances become unreasonably small and masses become unreasonably large, as they do under precisely those conditions where quantum gravity is mandated. Quantum gravity, which plays out over almost unimaginably short distances, poses problems that go unnoticed with more moderate phenomena such as water waves and electromagnetic waves. Consider some of the differences:

1. *Water waves.* Toss a stone into the still waters of a pond, and it creates a disturbance. An oscillating wave of matter spreads out in all directions, and different observers describe what happens in contrasting but complementary terms. Where a macroscopic observer sees a continuously varying wall of water, a microscopic observer sees a lumpy composite of H_2O molecules. Both descriptions make sense. There is no inherent conflict.

2. *Electromagnetic waves.* Toss an electric charge into the vacuum, and it creates a disturbance. An oscillating wave of electromagnetic field propagates outward at the speed of light, and again our observers file their reports. The macroscopic observer sees a continuously varying field, smooth and unbroken. The microscopic observer sees a lumpy mosaic of massless particles (photons), each one a discrete quantum of the electromagnetic field. As before, both descriptions make sense. There is no inherent conflict.

3. *Gravitational waves.* Toss a star into the vacuum of general relativity, and it creates a disturbance. An oscillating wave of gravitational influence propagates outward at the speed of light, and once more we ask Mr. Macroscopic Observer and Ms. Microscopic Observer to describe what they see. Mr. Macroscopic, according to general relativity, should observe a continuously varying pattern of ripples in the fabric of spacetime. Ms. Microscopic, according to quantum mechanics, should observe a lumpy composite of particles (gravitons), each one a discrete bundle of a quantized gravitational field. The quantum mechanical rules specify exactly what the hypothetical gravitational messenger must be: a massless boson with two units of spin angular momentum, twice as much as a photon.

But here our traditional quantum mechanics collapses into absurdity, because wherever quantum mechanics meets general relativity—over distances a hundred billion billion times shorter than in a nucleus—any field theory based on pointlike particles

suffers from incurable infinities. We ask nature for a renormaliz-
able quantum field that will yield a suitable graviton, and, alas,
there are no volunteers. The mechanics of point particles breaks
down in the infinitesimally small world of quantum gravity, and
the alternative picture of general relativity breaks down as well.
The havoc of quantum foam destroys the smooth fabric of space-
time that general relativity assumes as a prerequisite.

To cut the Gordian knot, we need a new set of assumptions. We need to
imagine that space-time comes packaged in discrete parcels that can be
chopped only so fine and no finer. We need to imagine that elementary
particles do not shrink all the way down to the vanishing point. We need
to picture them not merely as tiny billiard balls, but rather as tensile
structures able to manifest themselves in different ways. Because if our
new assumptions hold up, then we have a chance to understand how
everything in nature, matter and force alike, derives from the same
fundamental entity. It is an exhilarating possibility.

The new assumptions give rise to the theoretical pursuit of "string theory" (more generally called "M-theory"), and the excitement is palpable. If
string theory pans out, then we shall have a universe infinitely more simple
and infinitely more complex than anyone ever suspected.

Strings Attached

Think of it as the ultimate economy: the reduction of everything—all the
quarks, all the electrons, all the neutrinos, all the messenger bosons—to a
single building block, a teeny-weeny strand of energy vibrating in a world
of eleven dimensions, a world where what you get is not always what you
see.

Teeny-weeny, but not vanishing. The loops and membranes of string
theory are postulated to extend over a billionth of a trillionth of a
trillionth of a centimeter, a distance (the Planck length) so small that it
might never be probed directly, but nonetheless a distance greater than
zero. That slight departure is enough to ease the jitters of space-time to a
point where quantum mechanics and general relativity can coexist. The
infinities disappear.

It is a world almost incomprehensibly small. The diameter of a hydrogen nucleus is already a hundred billion billion times greater than the

Planck length, and the diameter of a hydrogen atom is a hundred thousand times greater than that. If a hypothetical string were enlarged to the width of a fingernail, then the fingernail itself would extend more than a thousand trillion trillion miles, enough for over ten billion billion round trips between the Earth and Sun. Light, traveling at a speed of 186,000 miles per second, would take a million billion years to traverse the distance. An airplane would do it in a billion trillion years. A person walking at a moderate speed would need 200 billion trillion years. A string is as small as those other numbers are big.

Yet it is big enough to have internal dimensions and to do things that a billiard ball cannot. A violin string or a drumhead, vibrating in different ways, produces an assortment of musical notes; an elementary string or membrane, likewise vibrating in different ways, produces an assortment of particles. One mode of vibration gives rise to a quark. Another gives rise to an electron. Another, a photon. Another, a gluon; and another (*mirabile dictu*), a graviton. Everything that nature needs to put together a universe comes from different facets of a single source. That, at least, is the proposal, and its promise to tie together a world of loose ends is extraordinary to contemplate.

This new harmony, moreover, need not infringe on the existing models of quantum mechanics and general relativity. On the contrary, the basic rules of the earlier, less fundamental theories become an essential part of any microcosmic symphony of strings. The quarks and leptons and gravitons of quantum mechanics are indeed *demanded* by the new theory, not merely tolerated. Therefore even if elementary particles prove ultimately not to be structureless points, as we are clearly beginning to suspect, we might still be able to reinterpret the older theoretical structures without having to tear them down entirely.

There is a catch, of course, because the conjectured strings and membranes are no ordinary violin strings and drumheads vibrating in four space-time dimensions. No, if only it would be that simple! Instead, the vibrations of M-theory occur in a supersymmetric world of eleven dimensions, ten spatial and one temporal. Fulfillment of both conditions, supersymmetry and extra dimensions, is the price exacted for the tremendous thrift and versatility needed to produce literally everything from a single piece of fabric, a one-size-fits-all "superstring."

The idea seems to come straight out of science fiction (imagine Rabi saying, "What? Three dimensions of space aren't good enough for you?"), but string theory cannot do without them. Just as a molecule with three atoms can vibrate in more ways than a molecule with only two,

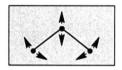   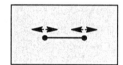

so it is with superstrings as well. They need the additional "degrees of freedom" afforded by the additional dimensions. Whether pictured as strands, loops, or membrane-like surfaces, the hypothesized superstrings must vibrate in more than three dimensions of space. If not, they are unable to deliver the full array of matter and force found in the universe.

Let nobody feel diminished for being unable to visualize a world with more than three spatial dimensions. Mathematicians have no difficulty working in abstract spaces of four dimensions, five dimensions, six dimensions, or even six hundred dimensions (they apply the same rules to all of them), but to picture a *physical* space of more than three dimensions is too much for the earthbound mind to bear. Nothing in our brains, nothing in our evolutionary history, nothing in our everyday experience gives us any clue to a world that can encompass dimensions other than north–south, east–west, and up–down. We are big creatures living in a big world, and until one lives in a box the size of the Planck length, one cannot hope to understand.

But maybe we can appreciate some of it, or at least accept the notion of a dimension so small as to go unnoticed. How many dimensions, for example, are we likely to ascribe to a high-voltage wire viewed from a distance of a hundred feet? *One*, you probably say, since the wire appears to have only length. You would be right, too, because a single number surely suffices to locate any position on the wire:

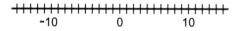

And so it seems to be, one dimension, case closed... but is that really all there is? For if we look more carefully, then a whole new world—the world of a hidden second dimension—begins to unfold all along the line. Come closer. Study the wire from a distance of one yard, or one foot, or one inch. Put yourself in the position of an ant crawling on the wire, and see that there is now width to go along with length. At every position on the line, we need a second number to specify an angle around the circumference:

Get closer still. Examine the wire with a magnifying glass or a microscope, and suddenly an even smaller world of bumps and pits comes into view. For those with eyes small enough to see, there is also a tiny dimension of up–down. Neither we nor the ant would have ever suspected it.

The ant, meanwhile, cut off from the space outside the wire, believes itself to inhabit a two-dimensional world delimited by forward–backward and clockwise–counterclockwise. Whereas a large observer has difficulty accepting an overly small dimension, a small observer likewise has difficulty accepting an overly large dimension. For the ant, anything outside the two dimensions of the wire falls into the realm of either science fiction or boldly speculative physics. Even if the creature surmises that space holds more mystery than forward–backward and clockwise–counterclockwise, it cannot easily interact with whatever world lies beyond.

Now all this fine talk of hard-to-find dimensions is not meant to depict even remotely the mind-bending hyperspace of string theory. The intention is only to make the "impossible" seem possible: that it may be *possible* for space to look different when viewed on different scales; that it may be *possible* for a dimension either to curl up so tightly or to uncurl so extensively that it escapes notice; that it may be *possible*, in short, for parts of our universe to be so large or so small as to disappear from sight.

Were string theory less elegant or the ultimate prize less seductive, scientists might be tempted to regard it as purely a mathematical recreation, an exercise unconnected with reality. But string theory is arguably a thing of rare intellectual beauty, almost unsurpassed, and its potential insights are enough to inspire a kind of rapture. It promises to be a Theory of Everything (if by "everything" you mean the reduction of all force and all matter to a single entity). It hints at a deeper understanding of the Big Bang and the inflationary universe. It holds out the prospect of other universes existing in other dimensions.

There is a strong possibility that string theory will never be tested directly. The scales of length (impossibly small) and energy (impossibly high) may prove to be forever out of the experimenter's grasp. We may never develop an instrument to show us a superstring in the way an ordinary microscope shows us a biological cell or in the way a scanning tunneling microscope shows us an atom. At best, we may have to settle for circumstantial evidence, such as might come from the detection of gravitons or superparticles. At worst, we may have to be content with nothing other than the beauty and mathematical consistency of a final, polished Theory of Everything.

For some, it may be enough. The theoretical physicist Eugene Wigner spoke of the "unreasonable effectiveness" of mathematics in describing nature, and anyone who has worked in the physical sciences would find it hard to disagree. Dirac, intrigued by a minus sign in his equations, went far out on a limb to predict the existence of a particle that nobody was looking for: the positron. He was right. The positron was discovered a few years later, a mathematical hunch come true. Einstein, commenting on the possibility that an experiment might prove general relativity wrong, ventured to say, "Then I would be sorry for the dear Lord, because the theory is correct." So strong were Einstein's feelings about truth, beauty, and mathematics that, for him, any universe worth understanding would have to be an aesthetic masterpiece. It would have to obey rules as magnificently compelling as those laid down by general relativity.

Many physicists have the same gut feeling today. Looking forward to a final string theory, all inclusive with no loose ends, they envision an interlocking structure so beautiful...so consistent...so perfect—it *must* be right.

How nice it would be to know for sure.

By the Way

Nuance and elaboration, commentary, historical and biographical notes, cross-references, a mathematical formula here and there—odds and ends to be noted, pursued further, or ignored, at the discretion of the individual reader.

CHAPTER 1 *Two critical points to bear in mind throughout: (1) never take for granted the apparent willingness of nature to yield to reason (it is a most pleasant surprise that she does), and (2) never expect to find a one-size-fits-all explanation of everything under the Sun. Different phenomena call for different models, and the secret of scientific success is to develop a sense of reasonableness and proportion—an ability to appreciate scale and scope, to ask the right questions, to make the appropriate simplifications.*

1a *our science*: Latin *scientia* "knowing," from *scire* "to know."

1b *a drop of water already holds more molecules than a galaxy contains stars*: A large galaxy might contain several hundred billion stars, far fewer than the thousand billion billion H_2O molecules in a typical drop of water (1 followed by 21 zeros).

2a *is what we see... like Earth and Moon?* Newton's classical mechanics (Chapter 4) showed that the heavens and the Earth obey the same natural laws, a refutation of Aristotelian philosophy.

2b *uncanny certainty... where random chance decides every turn*: an allusion to entropy and the second law of thermodynamics, treated in Chapter 10.

2c *particles find a balance*: a recurring theme, taken up first in Chapter 2 and invoked repeatedly thereafter.

4 *a cubbyhole... representing one possible mechanical state*: The picture is to be interpreted just as the text describes it, as a kind of filing system for certain pieces of data (at the moment, unspecified). The symbolic grid points and boxes, unlike those of an ordinary map, do not necessarily represent positions in space.

5a *the force of gravity*: referring to Newtonian gravity, an example of a "field theory" in classical mechanics (Chapters 2 and 4).

5b *the curvature of space-time*: referring to Einstein's general theory of relativity, a geometric interpretation of gravity (Chapter 5, second half).

5c *what works for Earth and Moon*: classical mechanics (Chapter 4).

5d *fails utterly for electron and proton*: The appropriate model for atoms and molecules is quantum mechanics, usually in its nonrelativistic approximation (Chapters 7 and 8, plus the first half of Chapter 9).

5e   *gluon and quark*: The Standard Model of particle physics makes use of relativistic quantum mechanics (Chapter 9, second half) to describe the atomic nucleus and its internal constituents. The model of the future may well involve string theory (Chapter 12, final section).

6   *in the chapters to follow*: The four questions, intentionally vague at this point, are all rooted in specific ideas. *Question 1*: the difference between a small system governed by an equation of motion (Chapter 4) and a large system governed by thermodynamics and statistical probability (Chapter 10). *Question 2*: the difference between classical mechanics (Chapter 4) and quantum mechanics (Chapters 7 through 9), evident in the specification of the relevant states. *Question 3*: the possibility of extreme sensitivity to initial conditions, or deterministic chaos (Chapter 11). *Question 4*: the meaning of the arrow of time, or the paradox of microscopic reversibility (Chapter 10).

7   *those questions and more*: such as ... *Chapter 2*: How is matter put together? Why do stable structures exist in the world? *Chapters 3 and 5*: How is physical law, constrained by the principle of relativity, able to tolerate the arbitrary viewpoint of every conceivable observer? Why do some things *always* change, and why do some things *never* change? *Chapter 6*: How does the electric charge make its mark on the universe large and small? *Chapter 12*: How did it all begin? Is that all there is? What is at the root of it all?

CHAPTER 2   *Lessons learned, among others: Big things come from small things. No particle is an island. Particles attract and repel, and the world hangs in the balance. The whole is more (or less) than the sum of its parts.*

8   *a particle at rest remains at rest*: the law of inertia (Newton's first law of motion). Developed at length in Chapters 3 and 4.

9   *the potential to be <u>different</u>*: The repeated use of the word "potential" is not casual, as will become clear in the course of the current chapter and also Chapter 4. To a physicist, the term signifies the "potential energy" of interaction and has a distinctly quantitative significance.

12   *shapes up in the following way*: The quantitative concept of "potential" is expressed for the first time in the graph. It will not be the last.

13   *without the tension between repulsions ... and attractions*: A star, for instance, counters the inward pull of gravity (the attraction of mass for mass) with an outward push of explosive energy (coming from the thermonuclear fusion reactions taking place within, as in a hydrogen bomb). When the nuclear fuel runs low and the fires begin to die, the attractive force of gravity becomes dominant. The star collapses on itself.

16a   *we come to know mass ... by what it does*: The view presented here and in Chapter 4 pictures gravity as a *force*, an attractive influence arising from the presence of mass. It is a phenomenological, empirical picture at best, not a fundamental explanation of how gravity works or the microscopic source from which it originates. For that, we await a proper quantum mechanical theory of gravity, a work very much in progress. See under *Higher Standards* in Chapter 12.

An especially revealing view of gravity, although still macroscopic, comes from Einstein's general theory of relativity, which treats mass as an agent that warps the contours of space and time. See *Space-Time and Gravity* in Chapter 5.

16b *millionfold, billionfold, trillionfold*: Large numbers such as these are understood according to common American usage. A million is 1 followed by six zeros. A billion is a thousand million, or 1 followed by nine zeros. A trillion is a thousand billion (a million million), or 1 followed by 12 zeros.

18a *later we shall explore that balance of forces more closely*: in the concluding section of Chapter 4 (*Though the Heavens Fall*).

18b *the germ of a possibility, the potential*: again, a deliberate use of the word "potential" to suggest how a physicist invests this everyday idea with mathematical precision. No further instances will be noted henceforth.

20a *more generally, the electromagnetic interaction*: See Chapter 6 for a fuller treatment of electric charge and the classical electromagnetic interaction.

20b *the difference lies solely in our point of view*: one of the many (and far-reaching) consequences of enforcing the principle of relativity (Chapters 3 and 5).

21a *Benjamin Franklin*: American publisher, diplomat, scientist, and inventor (1706–1790). Although some of Franklin's ideas about electricity needed modification later on, his contributions were nonetheless original and important. Along the way, he coined such terms as *positive, negative, conductor*, and *battery*—the core of the electrical vocabulary still in use today.

21b *in one camp we have the positive particles, the nuclei of atoms*: The nucleus was discovered circa 1911 by the great Ernest Rutherford (1871–1937) and two of his associates, the German physicist Hans Geiger (1882–1945) and an undergraduate named Ernest Marsden. In one of the most celebrated experiments in scientific history, they bombarded a thin metallic foil with a bulletlike beam of positively charged particles. Their principal observations were twofold: (1) Most of the incoming particles passed straight through the foil with only minimal deflection. (2) A few of the particles ricocheted back at large angles, a shocking finding that Rutherford likened to "a fifteen-inch shell bouncing off a sheet of tissue paper."

Rutherford's interpretation of the experiment gave rise to the solar-system model of the atom, the first reasonable picture of a system that would later be described quantum mechanically. The nuclear atom, in Rutherford's view, contains a hard, positively charged core at its center (the nucleus) set amidst a sea of lightweight, negatively charged electrons. The nucleus gives the atom nearly all its mass; the electrons give the atom nearly all its volume.

Rutherford, born in New Zealand, was awarded the Nobel Prize for Chemistry in 1908—not for discovering the nucleus, but for his work in radioactivity.

21c   *in the other camp we have the negative particles ... the electrons outside the nucleus*: The electron was discovered in 1897 by the English physicist J. J. Thomson (1856–1940), who was awarded the Nobel Prize for Physics in 1906. Thomson showed that a "cathode ray" (a beam produced when an electric current strips electrons from the atoms of a dilute gas) is really a stream of negatively charged particles. Measuring the deflection of the beam by electric and magnetic fields, Thomson determined the charge-to-mass ratio for the electron.

The absolute mass and absolute charge of the electron were established more than a decade later by Robert A. Millikan, who used an electric field to suspend negatively charged oil drops in midair—one drop at a time. Millikan (1868–1953), a renowned American physicist, received the Nobel Prize for Physics in 1923.

25a   *an atom of hydrogen*: Atoms are astonishingly tiny yet filled mostly with empty space. If the proton in a hydrogen atom were scaled up to the size of a golf ball, then the electron would typically be found more than a mile away.

25b   *nearly all of the visible matter in the universe derives entirely from hydrogen and helium*: Inventoried only for *visible* matter (material with an electromagnetic signature), the cosmos contains approximately 75% hydrogen and 23% helium. That amount, however, represents perhaps only 5% of the total mass of the universe, most of which comes from "dark" (nonelectromagnetic) matter. See *Forces of Darkness* in Chapter 12.

25c   *small variations in structure ... yield big differences in behavior*: the essence of chemistry.

25d   *the helium atom ... does not react*: Chemists classify helium (along with neon, argon, krypton, xenon, and radon) as a "noble gas," an element whose atoms remain aloof from other atoms. The reasons why different atoms behave in different ways are considered in Chapter 9 (concluding with *Material World*).

26a   *up and up, through uranium and beyond*: Arranged systematically according to recurring properties, the chemical elements are grouped into the rows and columns of the periodic table (a sight familiar to anybody who has ever taken a course in chemistry or physics).

26b   *after some 120 elements or more*: Only 92 elements (hydrogen through uranium) occur naturally. The rest are produced artificially in nuclear reactors and particle accelerators, sometimes in minuscule amounts that last only for an instant. The roster of new elements has grown considerably since the early 1990s.

28a   *some 50 pounds*: One pound is equivalent to 454 grams of mass and 4.448 newtons of force under terrestrial gravity. A kilogram is equivalent to 2.2 pounds.

28b   *interspersed among the protons are <u>neutrons</u>*: Lacking an electric charge, the neutron is a difficult particle to detect. The English physicist James Chadwick (1891–1974), a former student of Ernest Rutherford, discovered it in 1932 and received the Nobel Prize for Physics in 1935. (For Rutherford, see comment 21b.)

29 *the "strong" nuclear interaction*: The terms "interaction" and "force" are often used interchangeably, despite some subtle shades of meaning. Also, the shorter form "strong interaction" is synonymous with "strong nuclear interaction." Similarly for "weak interaction."

31 *it undergoes radioactive decay*: Antoine-Henri Becquerel (1852–1908), a French physicist, discovered radioactivity by accident in 1896: a photographic plate, exposed to certain uranium salts, revealed an unexpected image. Soon after, Marie and Pierre Curie discovered and isolated two new radioactive elements, polonium and radium.

Polish-born Marie Sklodowska Curie (1867–1934), who coined the term "radioactivity," shared the 1903 Nobel Prize for Physics with Becquerel and her husband, Pierre Curie (1859–1906). She went on to receive the 1911 Nobel Prize for Chemistry as well, becoming the first person to win two Nobel Prizes.

Independently of Becquerel and the Curies, Ernest Rutherford (see comment 21b) was honored for his work in radioactivity with a Nobel Prize for Chemistry in 1908.

33 *the electromagnetic force as we initially conceive it*: We shall see much later, in Chapter 9 (*Toward Unity*), that the electromagnetic and weak interactions share a common origin. Under conditions of high energy, they merge indistinguishably into a unified "electroweak" interaction.

34a *the unity of nature's design*: The brief treatment that follows is expanded upon considerably in Chapter 9 (*Toward Unity*) and parts of Chapter 12, where we are better positioned to understand the quantum mechanical origins of the strong and weak interactions. For historical and other notes, see the relevant commentary pertaining to those later chapters.

34b *the quarks are perhaps not the irreducibly simple, indivisible, end-of-the-line building blocks*: The qualification holds open the ultimate reductionism promised by string theory (Chapter 12, *Higher Standards*), which envisions quarks and all the other particles as different manifestations of a single, truly elementary source: an unimaginably tiny strand or surface of energy, vibrating in a space of more than three dimensions.

34c *weak interaction charge*: Another term is "weak isospin, third component."

36 *the "color" interaction*: revisited in Chapter 9 (*Toward Unity*), within the framework of quantum chromodynamics.

40 *a world dominated by mass*: True, but note that electromagnetic, strong, and weak forces play a role in the astrophysicist's world as well. Interactions other than gravity produce such phenomena as sunlight and sunspots, magnetic poles, stellar nucleosynthesis, neutron stars, and quasars, to cite just a few examples.

CHAPTER 3 *It seems like an eminently reasonable, almost trivial requirement to impose on a physical theory: that the results not depend on the biased perceptions of observers in arbitrary reference frames. The location of a coordinate system, the orientation of a coordinate system, the zero point of a clock, the speed at which an observer is moving—none of these things should alter the underlying laws of nature.*

Reasonable, yes. Trivial, no. By granting equal rights to all qualified observers, nature allows certain perceptions to vary in the eye of the beholder while forcing other perceptions to be the same for everybody. The consequences are profound: the blending together of space and time (the present chapter), the conservation of energy and momentum (Chapter 4), the equivalence of mass and energy (Chapter 5), and the warping of space-time to create gravity (Chapter 5).

43a *no observer is without bias*: Throughout this entire work, no real distinction is made between a passive observer (someone looking through a telescope, say) and an active observer (an experimenter, someone who deliberately manipulates nature and then takes note of what happens). Nor should the term "observer" be taken always to mean a conscious being. It might just as well designate a recording device or measuring instrument, or even something like the rings inside a tree trunk—mute testimony to events that came before.

43b *like meters and miles*: For readers unfamiliar with the units of length commonly employed in the United States, note the following conversions: (1) 1 inch = 2.54 centimeters; (2) 12 inches = 1 foot = 0.3048 meter; (3) 3 feet = 1 yard = 0.9144 meter; (4) 5280 feet = 1 mile = 0.6214 kilometer.

44 *a particle ... (it doesn't matter what)*: The particle in question might be a muon, for example, a negatively charged entity similar to an electron but some 200 times more massive. Muons are produced high above the Earth when the particles in incoming cosmic rays collide with nuclei in the atmosphere.

45a *everything ... squeezed into a dot infinitesimally small*: The idea is not as far-fetched as it sounds. The Big Bang theory, currently accepted by most cosmologists, describes the universe at its inception in just such a way: as an infinitesimally small, infinitely dense concentration of energy with no spatial extent. See *Once Upon a Time* and *Something in Nothing* in Chapter 12.

45b *it would be nothing like our present universe*: See the comment immediately above.

47 *we can conceive of all kinds of space*: The concept of a multidimensional space is enormously fruitful in mathematics, and abstract spaces spanning an arbitrary number of dimensions (often infinity) figure prominently in the theoretical models of physics. They are essential components of classical mechanics (Chapter 4), classical electrodynamics (Chapter 6), quantum mechanics (Chapters 7 through 9), and statistical mechanics (Chapter 10).

Physical spaces with more than three dimensions—that is, places for real objects to *be*—are currently of intense interest to physicists and mathematicians working on string theory (Chapter 12, *Strings Attached*).

52   *an application of high-school trigonometry*: If θ denotes the angle between reference frames *xy* and *uv*, then the coordinates are converted as follows:

$$u = x\cos\theta + y\sin\theta$$
$$v = -x\sin\theta + y\cos\theta$$

54   *if ... the universe is expanding*: Big Bang cosmology (Chapter 12).

55a   *a repulsive effect opposite to gravity*: the "dark" cosmic energy (Chapter 12).

55b   *tiny, tiny "superstrings"*: Chapter 12.

55c   *Newton's mechanical equations*: Chapter 4. *Maxwell's electromagnetic equations*: Chapter 6. *Schrödinger's quantum mechanical equations*: Chapter 8. *Einstein's equations of relativity*: Chapters 3 and 5.

56   *beginning with Galileo and Newton*: Galileo Galilei (1564–1642), the Italian scientist and mathematician, is regarded by many as the father of modern physical science. His towering contributions include the law of inertia and the law of falling bodies, foundations of classical mechanics later elaborated by Newton. Galileo's telescopic observations also led eventually to the acceptance of the Copernican view of the solar system (as centered about the Sun, not the Earth).

Isaac Newton (1642–1727), English physicist and mathematician, brought the heavens down to Earth, showing that both Moons and apples fall in precisely the same way. With his three laws of motion, Newton developed classical mechanics as a quantitative tool and established the law of universal gravitation. He also made important contributions to the study of optics and sound, and he was one of the inventors of the infinitesimal calculus in mathematics.

58   *a __stupid__ reference frame (like a roller coaster ...)*: It was Einstein's desire to reconcile observations in all possible reference frames (including accelerated ones, such as a stupid roller coaster) that led him to develop a radically new theory of gravity: general relativity, as explained in the latter part of Chapter 5. General relativity today forms the basis of modern cosmology, the study of the structure and origin of the universe as a whole. See Chapter 12.

60a   *186,000 miles per second*: 300,000 kilometers per second.

60b   *exactly the same number in all inertial reference frames*: The invariance of the speed of light was proved experimentally by Americans A. A. Michelson and E. W. Morley in 1887, who showed that the supposed "luminiferous ether"—an all-pervading invisible medium through which electromagnetic waves were presumed to propagate—did not exist. Whether or not Einstein was aware of the Michelson-Morley experiment remains in dispute, but the demonstrated absence of the ether provided a solid experimental foundation for the 1905 theory of special relativity. See comment 61 for additional remarks on the nonexistent ether.

Albert Abraham Michelson (1852–1931), one of the greatest experimental physicists of all time, was disappointed by the negative result and distrusted it until the end of his life. Among Michelson's other achievements were the accurate measurement of the speed of light and the first measurement of the size of a star. He received the Nobel Prize for Physics in 1907, the first American to do so.

60c   *as Einstein realized*: Albert Einstein (1879–1955), the celebrated German-born theoretical physicist, thoroughly revolutionized physics in the first two decades of the twentieth century. Einstein's theory of relativity (Chapters 3 and 5) shattered such basic concepts as the perception of space, time, energy, and mass. He also made important contributions to statistical mechanics and to the early theory of quantum mechanics.

Einstein received the Nobel Prize for Physics in 1921, in recognition of his quantum mechanical explanation of the photoelectric effect (comment 176c). To some, he was the world's "last classical physicist."

61   *the electromagnetic field ... requires no particular medium for its existence*: Nineteenth-century physicists could not conceive of light traveling through absolutely empty space, unsupported by any material medium. They assumed, instead, that space is filled everywhere with an invisible substance called the *luminiferous* ("light-bearing") *ether* whose sole function would be to serve as a medium for electromagnetic waves. The existence of the ether was disproved experimentally in 1887 by Michelson and Morley (comment 60b).

Had the ether existed, it would have constituted a unique reference frame, the only one in which light propagates at 186,000 miles per second. An observer at rest in the ether frame, sitting on the Throne of Zeus, could legitimately claim to be at absolute rest, and all other observers would agree.

64a   *they have a unique number, a mixture of a spatial and a temporal interval*: Suppose that observer 1 measures a spatial separation x and a temporal separation t between two events, whereas observer 2 measures separations of u and t', respectively. Despite their different perceptions of time and space, both observers record the same space-time interval, defined below. The symbol c denotes the speed of light:

$$\text{Invariant interval} = \sqrt{(ct)^2 - x^2} = \sqrt{(ct')^2 - u^2}$$

This result, which follows in part from the Pythagorean theorem, will find further application in the first half of Chapter 5.

When the temporal contribution outweighs the spatial contribution, a space-time interval is described as *timelike*. Its squared value is greater than zero. A *spacelike* interval, by contrast, has a dominant spatial contribution and a squared value less than zero. The value of a *lightlike* interval is exactly zero.

64b   *what unites us*: The German mathematician Hermann Minkowski, summing up the import of relativity, said it memorably and best: "Henceforth space by itself, and time by itself, are doomed to fade away into mere shadows, and only a kind of union of the two will preserve an independent reality." [1908 address translated in *The Principle of Relativity: A Collection of Original Papers on the Special and General Theory of Relativity* (Dover, 1952)]

65   *we ask Albert Einstein*: Einstein's credit for the theory of relativity is well deserved, but he was not the only one to grapple with these ideas. H. A. Lorentz and G. F. FitzGerald, attempting to explain the Michelson-Morley experiment (see comment 60b), proposed that lengths appear contracted to observers in motion. Lorentz then worked out the formulas that enable inertial observers to reconcile their various coordinates of space and time, and Hermann Minkowski further devised a four-dimensional space that links the three physical dimensions of space with an additional one of time. Henri Poincaré, also a major influence, independently developed a set of space-time transformations and even coined the term "principle of relativity" in the years before Einstein's first publication. It was Einstein, however, who recognized brilliantly that time, space, and simultaneity are linked physically in a way that transcends mere mathematics. His synthesis transformed relativity into one of the foundations of physical law.

Hendrik Antoon Lorentz (Dutch physicist, 1853–1928) shared the 1902 Nobel Prize for Physics with countryman Pieter Zeeman. George Francis FitzGerald (Irish physicist, 1851–1901) worked independently of Lorentz to propose the effect now known as the Lorentz-FitzGerald contraction. The German Minkowski (1864–1909) and the Frenchman Poincaré (1854–1912) were both extraordinarily creative mathematicians.

CHAPTER 4   *The mechanical ideal: to "declare the end of a thing from the beginning," an ideal embodied most nearly in the deterministic equations of classical mechanics.*

67   *we calculate, almost exactly*: The motion of the Moon, although a case study in classical mechanics, is difficult to predict with unlimited accuracy. Precise calculations are complicated by a number of small influences, including the following: (1) the gravitational effect of the Sun (which, despite being weakened by distance, is nonnegligible owing to the body's large mass); (2) gravitational effects of the planets, very weak but also very complex; (3) minor irregularities in the rotation of the Earth; and (4) tidal effects between Earth and Moon, which cause the two bodies to develop slight bulges owing to gravitational distortions. The Moon is in fact slowly receding from the Earth at the rate of about 1.5 inches per year, and the Earth's daily rotation is very slowly decreasing over time. The average period between full Moons is 29.530589 days.

68a   *the ... classical mechanics of Isaac Newton*: So great were Newton's contributions to the analysis of motion that the term "Newtonian mechanics" is now synonymous with "classical mechanics." Nevertheless, a number of other physicists, astronomers, and mathematicians played a critical role both before and after Newton (1642–1727). The following list, although highly selective, should at least convey a sense of the rich intellectual legacy that is classical mechanics.

Nicolaus Copernicus (Polish astronomer, 1473–1543) placed the Sun, not the Earth, at the center of the solar system and paved the way for ... *Tycho Brahe*

(Danish astronomer, 1546–1601), whose observations provided a basis for... *Johannes Kepler* (German astronomer, 1571–1630), who discovered the laws of planetary motion, later synthesized by Newton into a comprehensive mechanical system based on universal gravitation.

Galileo Galilei (Italian scientist and mathematician, 1564–1642) is credited with discovering the uniform acceleration of falling bodies, the parabolic trajectory of projectiles, and the law of inertia for circular motion (all of which are taken up later in the chapter). See also comment 56.

René Descartes (French mathematician and scientist, 1596–1650) extended Galileo's concept of inertia to linear motion, thus anticipating Newton's first law.

Pierre-Simon, marquis de Laplace (French physicist and mathematician, 1749–1827) applied Newton's system to a detailed study of the solar system, accounting for minor deviations of the planetary orbits.

Jean Le Rond d'Alembert (French mathematician, 1717–1783) and *Leonhard Euler* (Swiss mathematician and physicist, 1707–1783) contributed decisively to the analytical and mathematical development of mechanics...along with *Joseph-Louis Lagrange* (Italian-French mathematician, 1736–1813) and *William Rowan Hamilton* (Irish mathematician and astronomer, 1805–1865), each of whom reformulated Newton's laws in ways that—although mathematically equivalent to the original—provide tremendous new insight and analytical power.

James Prescott Joule (English physicist, 1818–1889), *William Thomson* (later *Lord Kelvin*; Scottish physicist, engineer, and mathematician, 1824–1907), and *Hermann von Helmholtz* (German scientist, 1821–1894) independently formulated and advanced the law of energy conservation.

68b   *small particles confined in small spaces*: To a physicist, the term "particle" means a body in which any internal motion and internal structure can be ignored, not necessarily something small (as in casual usage). A planet or star, for example, becomes a particle when it is viewed from such a great distance that the actual size and shape of the body are of no particular concern.

68c   *the shadowy regime of quantum mechanics*: Chapters 7 through 9.

68d   *a world of chaos*: Chapter 11.

69   *the classical, quantum, and chaotic machines of a mechanical universe*: Included in the classical (that is, unambiguously deterministic) world is also macroscopic electrodynamics, summed up in Maxwell's equations. See Chapter 6.

70a   *the instantaneous position and velocity of every particle*: The stated conditions are appropriate for classical mechanics, elaborated later in the chapter.

70b   *the rise and fall of a wave*: appropriate for both electrodynamics (Chapter 6) and nonrelativistic quantum mechanics, often called wave mechanics (Chapter 8).

72   *compressed into a single point*: The technical term is a point in "phase space."

76   *paid for in full*: A body near Earth has gravitational potential energy *mgh*, where *m* is the mass, *h* is the height above the surface, and *g* is the acceleration

due to terrestrial gravity, equal to 9.8 meters (32 feet) per second per second. Moving with velocity v, the body has a kinetic energy of $\frac{1}{2}mv^2$.

Were it not for friction, the sum of kinetic and potential energy would remain rigorously constant. In the real world, however, part of the kinetic energy shows up as heat. See Chapter 10 (*The First Law: Work and Heat*).

77 *for worlds penetrating into atomic nuclei and subatomic particles*: The relevant models, electroweak theory and quantum chromodynamics, are described in Chapter 9 (*Toward Unity*) and various parts of Chapter 12.

83 *"mass"... is the source of gravity*: We must leave it at that, without even a hint of how the force of gravity emanates mysteriously from this attribute we call mass. A classical, macroscopic theory like Newton's law of gravitation can do no more than tell us *what* happens, not *why*. The model describes wonderfully well how the force of gravity varies with mass and distance, and it gives us the mathematical tools to calculate trajectories—with great accuracy, too—but a classical explanation cannot tell us anything about how mass acts microscopically as a source of gravity. For that kind of understanding, we need a quantum theory of gravity (currently unavailable, although many physicists are encouraged by the progress being made in string theory; see *Higher Standards* in Chapter 12).

84a *such wondrous sights*: One of the earliest known experiments of this sort was reported in 1586 by Simon Stevin, a Flemish mathematician. Not long after, Galileo stated that a light object and a heavy object dropped simultaneously from the Leaning Tower of Pisa would hit the ground with less than a handbreadth between them, thus idealizing the outcome. Whether or not he actually performed the experiment is an open question.

84b *Einstein, with a wholly new world view*: general relativity.

84c *Newton's equation of motion promises to supply the third*: Promises are not always kept, at least not the promise of an exact solution. Even to solve the equations for only three interacting bodies (let alone more than that) is already a mathematical impossibility, and so in systems such as Earth-Moon-Sun we must invariably resort to simplifications and approximations. The expedients are often quite reasonable, but nevertheless they represent departures from the mathematical straight-and-narrow.

86a *free will... disappears in a clockwork universe*: Laplace went so far as to imagine an omniscient observer, someone who (if the laws of nature were truly deterministic) would be able to know the past and future of everything in the universe. To do so, one would have to specify a position and velocity for each particle, along with all the forces acting on them—no simple task, but theoretically and philosophically possible in a Newtonian world.

Quantum mechanics, of course, yanks out the mainspring of a clockwork universe, but Laplace was wrong for other reasons as well. Even in a classical world, the jittery hand of deterministic chaos (Chapter 11) vitiates any promise of mechanical omniscience. (For a note on Pierre-Simon Laplace, see comment 68a.)

86b   *in Newton's world*: In general, references to Newton imply a system innocent of both quantum mechanics and chaos. The term "Newtonian physics" also excludes Einsteinian relativity, although relativistic effects do not impair the deterministic nature of classical mechanics.

87   *a swift, strong change in momentum demands a ... greater force*: The quantity (force)×(time), called *impulse*, determines the change in momentum effected by a constant force acting over a given interval. If the time is halved, then the force must be doubled to produce the same change in momentum.

90   *for every manifestation of symmetry*: The principle is contained in *Noether's theorem*, formulated by the German mathematician Amalie Emmy Noether (1882–1935).

91   *angular momentum*: For a classical particle of mass m moving in a circle of radius r at a linear velocity v, the angular momentum has magnitude mvr. Its direction is perpendicular to the plane of motion.

92   *a twofold change in velocity requires a fourfold change in kinetic energy*: Kinetic energy is proportional to mass times the square of velocity.

94   *yet still the solar system endures*: Notwithstanding the apparent stability of Earth, Sun, Moon, and planets, the orbits are indeed slowly changing with time. Frictional forces, for one, rob the bodies of energy and thereby alter their trajectories, as evidenced by the eventual reentry of all artificial satellites. According to general relativity, too, any mass in motion is expected to radiate energy by the emission of gravitational waves. The effect is analogous to the emission of electromagnetic waves by a moving charge, but of a vastly smaller magnitude. See comment 239b.

96   *details of the orbit, including the exact shape*: The degree of ellipticity varies considerably. Some planetary orbits, such as those of Venus and Neptune are nearly circular; others, such as those of Mercury and Pluto, are highly distorted. For Earth, a relatively mild deviation from circularity causes the orbital distance to vary by only 3.4% during the course of a year.

CHAPTER 5   *Newtonian mechanics is revisited in the light of Einsteinian relativity, and the repercussions shake physics to its core. Special relativity, by placing all inertial observers on the same footing, leads to the equivalence of mass and energy: $E = mc^2$. General relativity, by granting the same rights to observers even in accelerated reference frames, leads to a revolutionary new theory of gravitation: a force-free warping of space-time in the presence of mass.*

98   *early in the twentieth century*: The special theory of relativity was published in 1905; the general theory, in 1916.

100a   *matter and antimatter*: See comment 294a.

100b   *ripples in space-time*: Treated later in the present chapter and also at the end of Chapters 9 and 12, "gravitational waves" are the mechanism by which mass transmits its influence throughout space-time—analogously to the way that electric charge transmits its influence by means of electromagnetic waves (Chapter 6, *Riding the Wave*). Although such ripples have proved (so far) too weak to detect directly, they are a key component of general relativity. Like electromagnetic waves, gravitational waves are expected to travel at the speed of light. See comment 239b.

100c *black holes and wormholes*: Black holes are discussed briefly in the closing paragraphs of this chapter and in comment 292b below. Wormholes, mentioned here only in passing, are hypothetical "shortcuts" in the landscape of space-time. A wormhole is pictured as providing a short corridor between distant points in the universe, similar to the way an underwater tunnel links two shores otherwise accessible only by a long overland route.

100d *the Big Bang*: theory describing the origin of the universe (Chapter 12).

102 *the geometry of a right triangle*: the Pythagorean theorem, expressed here as $x^2 + y^2 = r^2$.

103a *in three dimensions, the general recipe would amount to this*: The two-dimensional case has been cited above (comment 52) as requiring the application of four factors deriving from a single angle of rotation, θ. These four factors (represented below by the subscripted symbol a) correspond to the four fixed angles between axes u and x, axes u and y, axes v and x, and axes v and y:

$$u = x\cos\theta + y\sin\theta = a_{ux}x + a_{uy}y$$
$$v = -x\sin\theta + y\cos\theta = a_{vx}x + a_{vy}y$$

In three dimensions, there are nine angles to consider: between u and x, between u and v, between u and z, between v and x, between v and y, between v and z, between w and x, between w and y, and between w and z. The transformation then involves application of nine trigonometric factors:

$$u = a_{ux}x + a_{uy}y + a_{uz}z$$
$$v = a_{vx}x + a_{vy}y + a_{vz}z$$
$$w = a_{wx}x + a_{wy}y + a_{wz}z$$

103b *each observer records an invariant length*: as ensured by the generalized Pythagorean relationship, $r^2 = x^2 + y^2 + z^2 = u^2 + v^2 + w^2$.

104 *velocity (V)*: An uppercase symbol is used here to distinguish V from the lowercase v reserved for axes uv.

106a *a high-speed particle like a muon*: See comment 44.

106b *"proper time"... the time recorded by a clock locally at rest*: A "proper length" is defined analogously for a reference frame in which two events are simultaneous. Observers in relative motion record a dilated time and a contracted length in place of the "proper" values.

108 *a "minus" sign enters somewhere into the mix*: See comment 64a.

109 *space and time... function together without entirely sacrificing their individual identities*: The term "space-time" is hyphenated throughout this work,

a usage consistent with earlier convention but increasingly giving way to "spacetime." The hyphen, reminiscent of a minus sign, is a salutary reminder that space and time are not blurred as indistinguishably as sometimes suggested.

112   *gamma depends on the relative velocity of the two reference frames*: The formula, with the velocity denoted by V, is as follows:

$$\gamma = \frac{1}{\sqrt{1 - \dfrac{V^2}{c^2}}}$$

116   *it rolls to the bottom of the hill and begins to climb up the other side*: The argument here is abstract, but note that Galileo actually did experiments very much like the one described. He allowed balls to roll down one inclined plane and climb up another, making careful measurements of time and distance.

117   *a certain kind of uranium nucleus breaks into pieces*: The isotope is uranium-235, made from 92 protons and 143 neutrons. Nuclear fission is treated in Chapter 2 under *The Weak Nuclear Force*, where it is contrasted with beta decay.

118   *the balls cover 16 feet in the first second*: The acceleration due to terrestrial gravity, g, is 32 feet per second per second, and the distance covered during a free fall of t seconds is equal to $\frac{1}{2}gt^2$. Hence the observed values are 16 feet, 48 feet, and 80 feet during each of the first three seconds.

Systematic measurements of falling bodies were first made by Galileo, who discovered the relationship between distance and time-squared in the early 1600s.

120   *the unfortunate passengers have no sense of gravity*: Despite the mass of the nearby Earth, the freely falling passengers judge themselves to be weightless. No "force" of gravity pulls their feet to the floor, because both the elevator and its occupants are falling toward the center of the Earth at the same rate.

It is a common misconception that passengers in a spacecraft are weightless because of the absence of any large mass nearby. Quite the contrary, however, since space travelers typically find themselves only marginally farther away from massive objects like the Sun, Earth, and Moon. Orbiting at 100 miles above the Earth's surface, for example, a capsule is subjected to just a 5% reduction of terrestrial gravity—the difference between the inverse square of 4000 miles (the distance from surface to center of the planet) and 4100 miles (the slightly greater distance from orbit). What makes the astronauts *locally* weightless is the common acceleration that they share with their vehicle, the same effect that makes free fall in an elevator so exhilarating, at least for a time.

121a   *Einstein gave it the status of a founding principle*: The idea is to *postulate* that something is true (here, that gravitational force is indistinguishable from accelerated motion) and develop a theory based on the assumption. If experimental results agree with theoretical predictions, then the original postulate is correct. We can use it confidently to make other predictions and prove other assertions. If not, we devise a new theory.

This same postulate-based reasoning applies to special relativity as well, which flows entirely from two assumptions: (1) The laws of physics are the same in all inertial frames. (2) The speed of light is invariant for all inertial observers, regardless of their state of motion.

121b   *Einstein ... the cornerstone of his general theory of relativity*: Had Einstein not lived, any number of scientists would undoubtedly have deduced the special theory of relativity. Lorentz, Poincaré, and Minkowski were already close (see comment 65), and indeed these three had previously worked out the key equations. Most historians of science agree, though, that the *general* theory of relativity was a unique intellectual triumph, and were it not for Einstein we might still be waiting for such a theory today. Few discoveries in science and mathematics are coupled as closely to a single person as general relativity is to Einstein.

123   *the deviation ... near the Sun*: In 1919, three years after the publication of general relativity, British astronomer Arthur S. Eddington (1882–1944) put the theory to the test. Measuring the positions of stars during a solar eclipse, he observed deviations in line with those predicted by Einstein's equations. Einstein became an instant world celebrity, the most famous scientist of his day.

Considered objectively, Eddington's measurements were uncertain enough to counsel caution, but the scientific community was eager to accept general relativity as true, if only because it was so mathematically compelling and aesthetically beautiful. The warm reception differed strikingly from the one afforded to Michelson and Morley, whose technically *exquisite* measurements in support of special relativity faced greater skepticism. Many physicists, including Michelson himself (see comment 60b), were loath to renounce the ether and embrace the topsy-turvy rethinking of time and space that followed. Scientists are not without their prejudices.

124   *clocks may appear to run slower*: Imagine a rotating carousel outfitted with two clocks, one at the center and another along the edge:

To observer 1, positioned above the disk, the clock along the edge appears to be in motion. It runs slow compared with a clock outside. The clock at the center, coincident with the axis of rotation, appears stationary. It keeps proper time.

Now consider the perspective of an observer 2 stationed at the center of the carousel. To observer 2, both clocks look to be at rest. The clock in the center stays in place, clearly, and the clock on the edge similarly remains fixed along a radius as the carousel turns. It seems not to move, since observer 2 rotates along with the carousel. But if the principle of equivalence is valid, then both observers must record the same difference between the two clocks.

And so they do. Observer 1, looking from the outside in, sees a rotating disk and attributes the effects to accelerated motion. Observer 2, looking from the inside out, attributes the same asynchrony to the centripetal force suffered by the clock at the edge but not by the clock at the center (like the pull one feels when swinging a tethered ball overhead).

Similarly, lengths along the direction of rotation (tangent to the circle) appear contracted while lengths measured along the radius remain undisturbed. The one observer attributes the geometric distortion to a force acting within a stationary reference frame, whereas the other observer attributes the same distortion to the overall acceleration of the disk. Both have an equally valid claim.

127   *a concession to our three-dimensional minds*: Even professionals who specialize in relativity concede the impossibility of visualizing a *flat* four-dimensional space, let alone a warped one.

130a   *the astrophysicist's neutron stars and black holes*: See comment 292b.

130b   *there are a few small deviations that only general relativity can explain*: For example, a slight turning of the orbit of Mercury that amounts to 43 seconds of arc per century (just over one-hundredth of a degree).

131   *they took us to the Moon and back*: Newtonian physics was used to plot the course of the Apollo missions, including the improvised rescue of Apollo 13.

CHAPTER 6   *A look at the other great domain of classical physics, the macroscopic and deterministic world of electric charge. Here the four Maxwell equations unite electricity with magnetism and engender the electromagnetic wave as the fruit of the union. It will be the last stop before quantum mechanics.*

133a   *huge numbers of electrons are stripped from molecules of H_2O*: The violence of a thunderstorm is formidable, with a typical lightning flash able to generate an electric current as high as 20,000 amperes.

133b   *the beta decay of a neutron*: The process, governed by the weak interaction, produces a proton, an electron, and an antineutrino. See Chapter 2 (*The Weak Nuclear Force*) and Chapter 9 (*Toward Unity*).

133c   *as long as they live*: Relativistic quantum mechanics allows for the creation and annihilation of particles, as described in Chapter 12 (*Half Empty, or Half Full?*). Mass and energy are interconverted according to $E = mc^2$.

Note also that electric charge is conserved even during interactions such as beta decay, where particles of one kind turn into particles of another.

135a   *each interaction adds to the total*: a statement of the "superposition principle" for classical fields.

135b   *specified in full by Coulomb's law*: The electrostatic force exerted by charges q_1 and q_2 is proportional to $q_1 q_2 / r^2$, where r is the distance between the particles. Like charges repel (the sign of the force is positive), and unlike charges attract (the sign of the force is negative).

The law of electrostatic force is named for Charles-Augustin de Coulomb (1736–1806), the French physicist who formulated it.

135c *so does Newton's force of gravitation*: The force is proportional to $m_1 m_2 / r^2$, where m_1 and m_2 are the masses and r is the distance between them.

137 *spacing between "lines of force"*: Michael Faraday introduced lines of force as a way to visualize the physical effects of a field without having to use mathematical equations. Developed in the 1830s, the concept remains useful to this day. (For a note on Faraday, see comment 147b.)

138 *a magnetic dipole cannot be pulled apart into isolated north and south monopoles*: Quantum mechanics does not forbid the existence of magnetic monopoles, but for reasons still unclear they do not seem to occur in the universe—or, better to say, no magnetic monopole has yet been found. See also comment 300d.

143 *an atomic electron does produce a magnetic dipole*: The reference is to "spin angular momentum," introduced in Chapter 8 (*Bridging Two Worlds*) and elaborated in Chapter 9 (*In a Spin: Fermions, Bosons, and the Pauli Principle*).

145 *James Clerk Maxwell, a giant of nineteenth-century science*: The Scottish physicist and mathematician is ranked with Newton and Einstein for the creativity and impact of his work. In addition to synthesizing the classical theory of electromagnetism, Maxwell (1831–1879) made major contributions to thermodynamics and statistical mechanics (Chapter 10), notably the relationship governing heat, temperature, and the motion of microscopic particles.

147a *Gauss's law for electricity*: German mathematician Carl Friedrich Gauss (1777–1855), considered one of the most brilliant of all time, made important contributions to physics and astronomy in addition to mathematics. In collaboration with the German physicist Wilhelm Eduard Weber (1804–1891), his research in electricity and magnetism bore practical fruit in the invention of the telegraph. Gauss's mathematical work also played a part in developing the principle of energy conservation.

147b *Faraday's law of electromagnetic induction*: Self-taught, the physicist and chemist Michael Faraday (1791–1867) overcame class barriers in nineteenth-century England to become one of the greatest experimental scientists in history. His painstaking research laid much of the groundwork for the comprehensive electromagnetic theory later put forth by Maxwell.

147c *Maxwell–Ampère's law for the magnetic field*: André-Marie Ampère (French physicist and mathematician, 1775–1836) formulated Ampère's law, which specifies the magnetic force arising between two electric currents.

Before Ampère, it was Hans Christian Oersted (Danish physicist and chemist, 1777–1851) who discovered—serendipitously—a relationship between electricity and magnetism. During a lecture in 1820, he observed that an electric current passing through a wire caused a nearby compass needle to deflect. The French physicists Jean-Baptiste Biot (1774–1862) and Félix Savart (1791–1841) confirmed the effect and formulated a mathematical equation to describe it.

148   *Maxwell's equations are relativistically correct as they stand*: Einstein realized that one of two theories—Maxwellian electrodynamics or Newtonian mechanics—was inconsistent with relativity. It turned out to be Newtonian mechanics that needed modification at high velocities, along with fundamental notions of space and time. Maxwell's electromagnetic theory remained unscathed.

150   *we trust in Maxwell's equations to supply the correct description*: Some two decades after Maxwell predicted the existence of electromagnetic waves other than visible light, the German physicist Heinrich Hertz (1857–1894) discovered them. Hertz produced radio waves in the laboratory and showed that these signals behave in ways similar to ordinary light. Measuring the speed and wavelength of the oscillations, he confirmed the validity of Maxwell's theory.

In 1895 another German physicist, Wilhelm Conrad Röntgen, discovered electromagnetic waves in the form of X rays—by accident, while studying the effects of electricity on a gas at low pressure (so-called cathode rays). Important in its own right, Röntgen's work also paved the way for J. J. Thomson to discover the electron in 1897 (see comment 21c). Röntgen (1845–1923) was awarded the first Nobel Prize for Physics, in 1901.

151   *in a vacuum*: The speed of light is slightly less than c in air, notably more so in some other materials.

153   *a wavelength of 3 meters*: One meter is equal to 39.37 inches, or 3.28 feet.

CHAPTER 7   *The first of three chapters on quantum mechanics. A discussion of Heisenberg's uncertainty principle and the wave–particle duality sets the stage for more to come.*

164   *and things smaller still*: protons and neutrons, for example, and their constituent quarks.

169   *it is a process called "diffraction"*: For diffraction and interference to be significant, a wave must pass through an opening with dimensions comparable to the wavelength.

170   *as if 1 + 1 = 4*: meaning that the wave *amplitudes* add, but the energies do not.

175   *they carry energy ... linear momentum ... angular momentum*: but not mass. A photon has zero rest mass and travels perpetually at the speed of light. See comment 229.

176a   *an invariant quantity called "Planck's constant"*: The German theoretical physicist Max Planck (1858–1947) was the first to show that energy is quantized. He was awarded the Nobel Prize for Physics in 1918.

176b   *it has units of (energy) ×(time)—or, equivalently, (momentum) ×(length)*: The dimensions are those of "action," an important quantity in both classical and quantum mechanics.

176c   *the size of that lump depends only on the frequency*: The first evidence that light had a particle-like aspect came from the photoelectric effect, explained by Einstein in 1905. Here a beam of light strikes the surface of a metal, delivering sufficient energy to eject some of the metallic electrons and thereby cause an electric current to flow. The onset of the current, however, depends not on the intensity of the light (which, if light were acting as a wave, would be proportional to the square of the amplitude) but instead on its frequency. Electromagnetic energy comes packaged in discrete bundles (photons), each of them proportional to frequency, and the current is triggered one particle at a time. If a photon lacks the requisite energy, then it cannot dislodge an electron. Below a certain threshold frequency, even the most intense beam will not liberate a single electron.

178a *electrons and protons and neutrons ... fall victim both to wavelike interference and to the capricious rule of probability*: In 1924, the French physicist L.-V. de Broglie suggested that particles might have some of the properties of waves, namely a characteristic wavelength inversely proportional to momentum. Experimental proof came just three years later, when U. S. physicists C. J. Davisson and L. H. Germer and, independently, the English physicist G. P. Thomson demonstrated that electrons indeed undergo diffraction. It was a dramatic confirmation of the validity of the new quantum mechanics.

Louis-Victor de Broglie (1892–1987), who made his proposal in a doctoral thesis, received the Nobel Prize for Physics in 1929. Clinton J. Davisson (1881–1958) and George Paget Thomson (1892–1975) shared the prize in 1937.

178b *nevertheless, they do*: Richard Feynman, one of the inventors of quantum electrodynamics, expresses the feelings of many physicists when he describes the electron diffraction experiment in the following terms: "... a phenomenon which is impossible, *absolutely* impossible, to explain in any classical way, and which has in it the heart of quantum mechanics. In reality, it contains the *only* mystery. We cannot make the mystery go away by 'explaining' how it works. We will just *tell* you how it works .... No one will give you any deeper representation of the situation." [R. P. Feynman, R. B. Leighton, and M. Sands, *The Feynman Lectures on Physics, Volume III (Quantum Mechanics)*, Addison-Wesley, 1965]

180a *it is called Heisenberg's uncertainty principle*: The German physicist Werner Heisenberg (1901–1976) proposed the uncertainty principle in 1927, shortly after his formulation of matrix mechanics (an original form of quantum mechanics in which observable properties are represented as arrays of numbers). He was awarded the Nobel Prize for Physics in 1932.

180b *a number roughly the size of Planck's constant*: A more exact formulation of the uncertainty principle states that the product of the uncertainties in position and momentum ($\Delta x\,\Delta p$) is no smaller than $\hbar/2$, where $\hbar$ (read "h-bar") is Planck's constant divided by 2π.

183 *the names change, but the mathematics does not*: The mathematics is called Fourier analysis, after the French mathematician Jean-Baptiste-Joseph Fourier (1768–1830). Fourier analysis, used widely throughout engineering and the physical sciences, decomposes an arbitrarily complex waveform into independent components. Originally applied to the study of heat, it is a technique of extraordinary power and generality.

CHAPTER 8   *The quantum mechanical state—the wave function—takes shape, and the rules for its interpretation and evolution are developed.*

188a   *the same generic equations common to all such disturbances*: The point has been made before, but it is well worth emphasizing. *There are far more phenomena in nature to describe than there are equations to describe them.* Remarkably, the same mathematics (albeit with varying interpretations of the symbols) often applies to a host of different processes.

188b   *the mathematics of an electron is the mathematics of a wave*: This statement reflects the point of view known as "wave mechanics," a technique introduced in 1926 by Erwin Schrödinger on the basis of de Broglie's wave–particle duality (see comment 178a). Wave mechanics is not the only way to formulate a quantum theory, but it is widely taught and applied throughout the physical sciences. The computational facility and physical interpretation of wave mechanics make it especially convenient for practical applications.

An alternative approach, the more abstract "matrix mechanics," was developed nearly simultaneously by Werner Heisenberg, Max Born, and Pascual Jordan. Matrix mechanics and wave mechanics were later shown to be equivalent, first by Schrödinger and then by Dirac (who merged the two methods into a particularly elegant and powerful system, sometimes called "transformation theory.")

Theoretical physicists Erwin Schrödinger (Austria, 1887–1961) and Paul Adrien Maurice Dirac (England, 1902–1984) shared the Nobel Prize for Physics in 1933. Heisenberg (Germany, 1901–1976) received the prize in 1932, and Max Born (Germany, 1882–1970) was a cowinner in 1954.

190   *the form of each mode ... and the extent to which that mode contributes to the total vibration*: a description of Fourier analysis (see comment 183).

192   *it tells us the statistical probability of finding the particle*: The probabilistic interpretation of a wave function originates with the German physicist Max Born (1882–1970), who also contributed to the development of matrix mechanics (see comment 188b). Born received a belated share of the Nobel Prize for Physics in 1954.

193   *according to Heisenberg*: the uncertainty principle.

200   *the laws of quantum mechanics must also withstand any overall turning of the reference frame*: alluding to the notion of "gauge symmetry," a key feature of relativistic quantum mechanics. See the latter part of Chapter 9.

201   *it represents all there is to know*: Einstein, for one, never accepted this assertion, and to the end of his life argued that quantum mechanics was an incomplete theory—that its probabilistic formulation was only apparent, masking an underlying reality still to be exposed.

The crux of the problem deals with the seeming disconnection between cause and effect in quantum mechanics, where *identical* states measured in identical ways can yield different results. Opposing this view are the "hidden-variable" theories advanced by David Bohm and others, which postulate that supposedly identical states are not really identical at all. They differ initially by the values of certain hidden variables, which (if we knew them) would account deterministically for the subsequent divergence in measured properties.

The question remains controversial, although experiments undertaken since 1980 tend to support the indeterminate view of the microworld demanded by quantum mechanics. Hidden variables, moreover, introduce special forces and anomalies that strike most physicists as unattractive at best, contrived at worst.

202a   *an abstract space*: called "Hilbert space," named for German mathematician David Hilbert (1862–1943). It is what mathematicians call a *complex* space, made up of numbers that are part real and part imaginary (they incorporate the so-called imaginary unit, $i = \sqrt{-1}$).

The vector formulation described here is based on the approach taken by Dirac, who unified the wave mechanics of Schrödinger with the matrix mechanics of Heisenberg, Born, and Jordan. See comment 188b.

202b   *not until later, not until an observer intrudes with a measurement, must the system choose either component 1 or component 2*: The statement is in keeping with the orthodox view of quantum mechanics, known as the "Copenhagen interpretation" (after the views of influential Danish physicist Niels Bohr and his associates). Most practicing scientists accept the Copenhagen interpretation and use it in their daily work; few are happy with it.

Bohr's interpretation is an affirmation of the philosophy of logical positivism: that science should concern itself only with those quantities that are experimentally accessible, renouncing all others. According to the Copenhagen interpretation, the indeterminacy inherent in a wave function is a part of nature that must be accepted at face value. The equations tell us precisely what numbers are likely to be measured, and that information—nothing else—is all there is to know.

Nobody disputes the accuracy or the predictive power of quantum mechanics, but the idea of a wave function "choosing" between components 1 and 2 sticks in the philosophical craw of many scientists. The Copenhagen interpretation effectively shuts off debate by asserting that "if we can't measure it, then we have no right to discuss it." The issue of how a superposition of components collapses into just one is thus left unsolved, the so-called measurement problem of quantum mechanics.

Attempts to resolve the measurement problem have included various hidden-variable theories (see comment 201) and also "many-universe" theories. According to the latter, the system never really chooses just *one* component, but instead embraces all of them simultaneously—each existing in a parallel universe inaccessible to the rest. Newer approaches, some undertaken only since the 1990s, have shown considerable promise in recent years.

205   *quantum mechanics ... had to be formulated by working backward from classical mechanics*: Newton's equations, however, do not provide a mathematically

convenient starting point. Quantum mechanics is based on the alternative *Lagrangian* and *Hamiltonian* formulations of classical mechanics, both of which are fully equivalent to the Newtonian system but which use different, more easily manipulated equations.

207a   *the proton and neutron, too*: A neutron, although neutral overall, contains within it three electrically charged quarks and thus acquires a magnetic moment.

207b   *Niels Bohr, one of the founders of quantum mechanics*: The famed Danish physicist (1885–1962) was the first to devise a way to quantize the energy and angular momentum of a hydrogen atom, circa 1913. He received the Nobel Prize for Physics in 1922.

The Bohr atom, although superseded by the quantum mechanics of the 1920s, proved to be an important step on the road to a more comprehensive theory. Bohr himself, along with his many colleagues at the Institute for Theoretical Physics in Copenhagen (including Heisenberg), played a major role in developing quantum mechanics and its interpretation. See also comment 202b.

209   *for atoms, the Schrödinger equation*: Austrian physicist Erwin Schrödinger (1887–1961) shared the 1933 Nobel Prize for Physics. See also comment 188b.

Note that the Schrödinger equation is a nonrelativistic approximation, valid for low velocities only. It gives excellent results for most chemical applications.

CHAPTER 9   *In symmetry there is force. First, the indistinguishability of quantum mechanical particles divides the world into bosons and fermions, force and matter. Interchange symmetry imposes an exclusionary influence on fermions that gives atoms their size and hardness, thereby creating the chemical differences that make life possible. Second, the symmetry of a wave function with respect to a shift in phase—a symmetry enforced locally and relativistically all throughout space-time—stitches together the quantum mechanical universe. If two observers, separated in time and space, are to perceive a rotated wave function as doing essentially the same thing, then nature must supply a force field that communicates the difference in phase. A quantized field of messenger particles, bosons, arises to guarantee the local symmetry.*

211a   *only those processes that conserve energy ... linear momentum ... angular momentum*: See *Constancy and Change* in Chapter 4.

211b   *the space-time interval must be invariant*: See *Relativity and Invariance* in Chapter 3 and *Four of a Kind: Space-Time* in Chapter 5.

211c   *a facet of nature so beautiful*: Aesthetic judgments are purely subjective, of course, but many theoretical physicists confess to finding great beauty in the symmetry of physical law.

215   *the governing "Pauli principle"*: named for the Austrian-born theoretical physicist Wolfgang Pauli (1900–1958), who formulated the interchange principle in the mid 1920s. Pauli was awarded the Nobel Prize for Physics in 1945.

216a *the physicist S. N. Bose*: In 1924–1925, the Indian physicist and mathematician Satyendra Nath Bose (1894–1974) collaborated with Einstein on a quantum statistical theory of the particles now known as bosons. Particles in this class are said to obey Bose-Einstein statistics.

216b *spin angular momentum*: See *Bridging Two Worlds* in Chapter 8.

216c *in honor of Enrico Fermi*: The Italian-born Fermi (1901–1954), together with P. A. M. Dirac (see comment 188b), developed a theory of such particles in 1926–1927. Fermions are said to obey Fermi-Dirac statistics, as opposed to Bose-Einstein statistics (see comment 216a).

Fermi is well known for his many contributions to nuclear physics, which include engineering the first controlled chain reaction triggered by nuclear fission. He received the Nobel Prize for Physics in 1938.

216d *they "exclude" one another, as if pushed away by a force*: It is precisely this force of exclusion—which has its roots in the symmetry properties of interchangeable fermions—that gives atoms their size and makes matter incompressible. The fermionic constituents of atoms resist being squeezed into arbitrarily small volumes, as explained in the upcoming section *Material World*.

The exclusionary force suffered by electrons in atoms traces back ultimately to the electromagnetic force, one of the four fundamental interactions.

216e *no two of them can simultaneously occupy the same quantum state*: This particular statement, applying only to fermions, is often called the "Pauli exclusion principle." It is a consequence of the more general "Pauli principle" described previously, which asserts that wave functions are multiplied by either 1 or −1 upon interchange of identical particles. See also the preceding comment.

219 *it depends on how the electrons are configured*: The whole of chemistry is contained in this statement.

223a *like the American Congressman*: the late Thomas P. ("Tip") O'Neill, Jr., Speaker of the U.S. House of Representatives from 1977 through 1986.

223b *Einstein's macroscopic conception of gravity*: general relativity (Chapter 5).

227a *physicists call the operation a "gauge" transformation*: The word "gauge," which connotes a standard of measurement (a railroad gauge, for example), was chosen by Hermann Weyl in 1918. Investigating the relationship between local symmetry and the classical electromagnetic field, Weyl was reminded of the standard distance between rails—and although the theory was flawed, the term stuck. It has no deeper meaning except as a label, plainly inapt, for a certain kind of phase transformation. Both classical and quantum mechanical electromagnetic fields obey a form of gauge symmetry.

German-born Hermann Weyl (1885–1955), a mathematician and physicist, made important contributions to quantum mechanics and relativity theory.

227b *local gauge symmetry, coupled with special relativity*: The modern approach to gauge invariance originates with C. N. Yang (b. 1922) and Robert L. Mills (1927–1999), who formulated a theory of the strong interaction in 1954.

Their work on "non-Abelian gauge transformations" exposed a deep connection between local symmetry and quantum fields. Theories built on this foundation are often referred to as Yang-Mills theories.

Chen Ning Yang, a Chinese-born American theoretician, shared the 1957 Nobel Prize for Physics with Tsung-Dao Lee.

229   *with zero rest mass a photon travels unceasingly and untiringly at the speed of light*: A massless particle has a vanishing rest energy, arising from $m = 0$ in the equation $E = mc^2 = 0$. The relativistic energy resides entirely in the momentum portion (p) of the four-dimensional energy-momentum vector, and hence a massless particle must move continually at the speed of light to maintain a fixed energy pc. See *Conservation of Momentum-Energy* in Chapter 5 for a discussion of relativistic energy and momentum.

230a   *greater than Planck's constant, h, or roughly so*: See comment 180b.

230b   *so long as the product $\Delta E \, \Delta t$ remains less than approximately h*: more precisely, $\hbar/2$ (as in comment 180b).

231   *quantum electrodynamics, which serves as a model for the rest*: In 1926, Dirac (see comment 188b) was the first to devise a relativistic quantum theory of the electron. Quantum electrodynamics, familiarly called QED, was subsequently refined and brought to fruition in the late 1940s by the independent efforts of three theoretical physicists: Richard P. Feynman (U. S., 1918–1988), Julian S. Schwinger (U. S., 1918–1994), and Tomonaga Shin'ichiro (Japan, 1906–1979).

Dirac received the Nobel Prize for Physics with Erwin Schrödinger in 1933. Feynman, Schwinger, and Shin'ichiro shared the prize in 1965.

232a   *a down quark emits a W particle and turns into an up quark, thus converting a neutron into a proton*: Conservation of electric charge requires the boson here to be negative (W^-, with a charge of -1). Before the event, the total charge coming from the neutron is zero. Afterward, the combined charge of the proton ($+1$) and the W^- boson (-1) is still zero. The short-lived W^-, a virtual particle, gives way immediately to an electron (-1) and antineutrino (0).

Positive or negative W bosons are exchanged in processes where the participating fermions switch charges. Neutral Z^0 bosons are exchanged in processes where all fermionic charges remain the same.

232b   *as in the related charged-current process*: The drawing is meant to suggest a "Feynman diagram," a graphical representation of particle interactions developed by Richard Feynman (see comment 231). Physicists use such diagrams to help visualize complex processes without having to resort to equations.

233a   *the W and Z bosons are indeed heavy particles*: approximately 86 and 97 times, respectively, more massive than a proton.

233b   *they have only a brief time to hop from fermion to fermion*: perhaps as little as a tenth of a trillionth of a trillionth of a second (a decimal point, 24 zeros, and then a 1).

234a *their apparent divergence arises ... from broken symmetry*: The idea of spontaneous symmetry breaking, in particular the Higgs mechanism, is elaborated in Chapter 12 (*Something in Nothing*).

234b *the "electroweak" interaction*: The electroweak theory was developed independently by theoretical physicists Steven Weinberg (U. S., b. 1933), Abdus Salam (India-Pakistan, 1926–1996), and Sheldon L. Glashow (U. S., b. 1932), cowinners of the 1979 Nobel Prize for Physics.

234c *quarks*: Beginning in the 1960s, physicists tried to rationalize the discovery of ever more "elementary" particles by proposing an internal structure for protons and neutrons. Most prominent was the Eightfold Way developed by Murray Gell-Mann of the United States and, independently, Yuval Ne'eman of Israel. A similar approach was taken as well by the American physicist George Zweig.

Gell-Mann subsequently introduced the quark as a fundamental building block, with three quarks needed to construct both a proton and a neutron. His theory predicted further that quarks come in three distinct types known today as flavors (of which there are actually six, as noted in Chapter 2, Chapter 12, and in the present Chapter 9). Gell-Mann (b. 1929) won the Nobel Prize for Physics in 1969.

The term "quark" is a whimsical usage inspired by Gell-Mann's reading of a line from James Joyce's *Finnegans Wake*: "Three quarks for Muster Mark!" The vowel sound is pronounced variously as in either "quart" or "mark." Eightfold Way is an allusion to Buddhist philosophy.

234d *the fancifully named attribute of "color"*: In the mid 1960s, Oscar Greenberg of the United States and Yoichiro Nambu of Japan proposed a new property to resolve certain anomalies in the theoretical behavior of quarks. The new property allowed for three distinct states, which Gell-Mann dubbed "color."

235a *the nonliteral colors red, green, blue*: sometimes specified (just as nonliterally) as red, yellow, blue.

235b *equal weights of red, green, and blue quarks produce chromatically neutral (colorless) protons and neutrons*: Quarks are fermions, endowed with a half-unit of spin angular momentum. The attribute of color, by distinguishing quarks with identical flavors, prevents them from violating the Pauli exclusion principle. The two up quarks in a proton, for instance, would occupy identical states and consequently be unable to coexist were it not for their contrasting colors. The different colors do for quarks in a proton or neutron what different spin states (up and down) do for the electrons in an atom, as described earlier in the chapter.

235c *in electromagnetic interactions, a one-dimensional phase rotation calls forth a field of photons*: Physicists classify the underlying symmetry of the electromagnetic interaction as $U(1)$. The more complicated symmetry of the color interaction is designated $SU(3)$.

235d *a field of "gluons"*: Again, it was Gell-Mann who coined the whimsical term (see also comments 234c and 234d). The gauge bosons "glue" the quarks into protons, neutrons, and mesons.

236a   *quantum "chromodynamics," our evolving theory of the strong field*: Many hands have played a role in developing the model, but the first breakthrough came in the early 1970s with the discovery of asymptotic freedom (see comment 237a) by David J. Gross, Frank Wilczek, and H. David Politzer.

Gross (b. 1941), Wilczek (b. 1951), and Politzer (b. 1949) were awarded the Nobel Prize for Physics in 2004.

236b   *a richer, more complex quantum field born of a more complex symmetry*: SU(3), as mentioned in comment 235c. The weak interaction falls under the heading SU(2), whereas the electromagnetic interaction by itself is classified as U(1). The unified electroweak interaction combines the symmetries U(1) and SU(2).

237a   *unlike electromagnetic attractions and repulsions, quark–quark interactions become stronger with distance*: and, conversely, they vanish entirely when the separation shrinks to zero. The technical term is "asymptotic freedom."

237b   *the quarks are unable to escape*: Since free quarks do not occur naturally and cannot be produced artificially, experimental confirmation of their existence is necessarily indirect. Available evidence, however, tends to support the reality of quarks and gluons.

238   *certain kinds of mesons supply the glue that holds together a nucleus*: Japanese physicist Hideki Yukawa (1907–1981), winner of the 1949 Nobel Prize, predicted the existence of mesons and described how they might implement the strong interactions between protons and neutrons in a nucleus.

239a   *"grand unification" of the quantum fields*: taken up in Chapter 12, in a cosmological context.

239b   *indirect evidence tends to confirm their existence*: In 1974, American astrophysicists Joseph H. Taylor, Jr., and Russell A. Hulse discovered a closely coupled pair of neutron stars that emitted radio pulses in a highly distinctive pattern. Regular variation of the radio signals, along with a slow decay in the orbits, is consistent with the emission of gravitational waves. The evidence is indirect, but persuasive.

Taylor (b. 1941) and Hulse (b. 1950) shared the 1993 Nobel Prize for Physics. For neutron stars, see comment 292b.

240   *a form of "string theory"*: See the concluding sections of Chapter 12.

CHAPTER 10   *The microscopic laws of motion offer no sense of time, but in the world at large, time marches on. Macroscopic systems progress irreversibly from one state of equilibrium to another, driven forward by the statistical imperatives of large numbers and overwhelming odds. Energy becomes increasingly dispersed and harder to extract, even as the total amount remains constant. The entropy of the universe grows and grows.*

246a   *conditions under which all gases seem to behave alike*: The reference is to the theoretically "ideal" (or "perfect") gas. Gases at sufficiently low density—typically under conditions of high temperature, low pressure, and large volume—

all tend to behave in the same way, regardless of chemical composition. The atoms or molecules in a sparsely populated gas act more like noninteracting particles than chemically distinct entities, and the macroscopic properties they engender are modeled by a single generic equation.

246b   *the model stands on its own*: Even if we were to discover one day that atoms do not really exist (an unlikely prospect, admittedly), the apparatus of thermodynamics would not be affected in the slightest. Bulk thermodynamics depends in no way on the microscopic structure of a system.

247   *understood microscopically, its pressure arises from the impacts of individual molecules*: Statistical analysis of particles in a gas was first undertaken by James Clerk Maxwell and Ludwig Boltzmann in the latter half of the nineteenth century.

For notes on Maxwell and Boltzmann, see comments 145 and 267b, respectively.

251   *accept a better offer and embark on a new history*: In a more technical work, such an equilibrium would be considered a "metastable" state rather than a "stable" state. It exists locally in a high valley of potential, stable only with respect to small disturbances. By contrast, the valley of *lowest* possible potential is uniquely stable in a global sense.

254a   *uncountably many more ways*: Strictly speaking, the ways can be counted quite straightforwardly using combinatorial arithmetic. The number is so large, however, that one might just as well dispense with any calculation and accept the value as infinite.

254b   *a fixed amount of global energy*: The total amount of energy in the universe, according to some cosmologists, may well be zero: equal parts positive and negative. Under the "free-lunch" scenario of cosmic origins, the positive contribution of rest energy and kinetic energy is balanced exactly by a negative contribution arising from gravitational energy (which is always attractive).

256   *some of the energy … appears in the form of heat*: Before the nineteenth century, heat was erroneously thought to be a liquid ("caloric") that literally flowed from one body to another. That kind of thinking began to change in 1798 when the physicist Benjamin Thompson, working as a military advisor to the King of Prussia, noted a relationship between the heat produced during the boring of a cannon and the amount of mechanical work done. Some four decades later, the English physicist James Prescott Joule measured the mechanical equivalent of heat experimentally, thus proving that work and heat were interchangeable forms of the same basic quantity: energy. They could be measured in the same units.

Benjamin Thompson, Count Rumford (1753–1814), was an American-born British physicist and a founder of the Royal Institution of Great Britain. James Prescott Joule (1818–1889) was an English physicist who played a major role in the development of thermodynamics. The standard unit of energy, the joule, is named for him.

257a   *the "first law of thermodynamics," a restatement of the law of energy conservation*: The law has a long history, and many mathematicians, scientists, and engineers have contributed to its formulation. For some of the key names, see comments 68a and 256.

257b   *Mike sees microscopic energy going into <u>translational</u> motion*: Unlike the three other forms listed, translational motion over comparatively long (but still microscopic) distances can usually be described by classical mechanics. The energy levels are spaced together so closely as to appear continuous.

261   *the second law ... recognizes that work readily dissipates into heat, but heat does not readily turn back into work*: The second law of thermodynamics is stated macroscopically in a number of ways, all of them equivalent. The German physicist Rudolf Clausius (1822–1888) formulated the law by asserting that "heat cannot of itself pass from a colder to a hotter body" or, alternatively, "the energy of the universe is constant; the entropy of the universe tends to a maximum." In the Kelvin-Planck formulation, the assertion is that "a heat engine operating cyclically cannot convert heat into work without some other effect on its surroundings." As a consequence, it is impossible to build a perpetual motion machine by extracting heat from one source (such as air or water) and converting 100% of it into mechanical energy.

The second law and the concept of entropy owe much to the earlier studies of Sadi Carnot (1796–1832), a French engineer and physicist who made a detailed analysis of cyclical steam engines. Carnot understood that heat can be converted into mechanical work (by having a thermal reservoir fall from a higher temperature to a lower temperature), just as mechanical work can be converted into heat.

For brief notes on Lord Kelvin (William Thomson) and Max Planck, see comments 68a and 176a, respectively.

262   *a quantity called entropy*: Its dimensions are heat (energy) divided by temperature.

264   *it cannot go home again*: James Clerk Maxwell likened the second law of thermodynamics to the statement, "If you throw a tumblerful of water into the sea, you cannot get the same tumblerful of water out again." You can get *another* tumblerful of water out again—even one that looks superficially and macro-scopically just like the original—but not *the* tumblerful you first tossed.

267a   *the system falls into the probabilistic nirvana of equilibrium, and there it stays*: If one waits long enough, certainly, then eventually a nonequilibrium microstate—even something wildly improbable, like a configuration with all the particles packed into a corner—is bound to occur. The objection is more philosophical than realistic, though, because such waiting times can be long indeed. For example, with only 100 particles (a vanishingly small system by macroscopic standards), an all-on-the-left microstate represents just one of roughly 1,000,000,000,000,000,000,000,000,000,000 alternatives. Even if each microstate endures for only a trillionth of a second, the system would still need over 30 billion years to run through the entire set.

267b *the statistical imperative is to increase the microscopic disorder present in the universe and ... maximize the global entropy*: The statistical interpretation of entropy and the second law is due to the Austrian physicist Ludwig Boltzmann (1844–1906), who developed a mathematical relationship between macroscopic entropy and the number of microstates. The equation (in its original form, "$S = k \log W$") is engraved on Boltzmann's tomb in Vienna.

CHAPTER 11 *Where the simple becomes complex, and where determinism carried too far leads to chaos.*

269 *chaotic systems obey rules of the road as rigorously as Earth and Moon*: A great number of people, in many fields, have contributed to the elucidation of such rules over the last forty years. The following is only a partial list: Michael Berry, Mitchell Feigenbaum, Albert Libchaber, Edward Lorenz (see comment 279), Benoit Mandelbrot, Robert May, David Ruelle, Robert Shaw, Stephen Smale, James Yorke. For an interpretive history of chaos through the mid 1980s, see James Gleick's *Chaos: Making a New Science* (Viking, 1987).

274 *this "logistic difference equation" ... becomes the paradigm*: It was a biologist, Robert May, who used this equation—long known, but woefully underappreciated—to encourage a broader community to study the effects of chaos and complexity. In what he called a "messianic" paper (a 1976 review article in the journal *Nature*), May urged scientists in all fields to consider how chaos might affect their world view. He argued persuasively that students should discover the reality of nonlinear systems early in their education, not least by playing with this simple equation and discovering its secrets for themselves. ["Simple Mathematical Models with Very Complicated Dynamics," *Nature*, Volume 261 (1976), pp. 459–467]

277 *the length of the repeating cycle soon doubles from two to four, and then from four to eight*: called "period doubling."

278 *the chaotic outcome varies disproportionately*: The statement can only hint at the rich mathematical structure that goes into the apparent hodgepodge of a chaotic system. The period doubling that precedes the onset of chaos, for instance, is but a single example of the many rules and patterns to be found.

Even within a chaotic regime, for all its complexity, there are often regions of surprising regularity, periodicity, and stability. Simple cycles may suddenly appear and disappear with small changes in the controlling parameters. Chaos may vanish, only to reemerge after a new sequence of period doubling. Much can happen, and chaotic behavior is usually so rich that the mind cannot take it all in at once. Sometimes only the visual enlightenment of computer graphics can make the picture plain.

279 *the "butterfly effect"*: The term is associated with a weather model developed in the early 1960s by Edward Lorenz, a research meteorologist at MIT. Lorenz's computer model, which showed that extreme sensitivity to initial conditions can impede long-range forecasting, included a two-lobed graph reminiscent of a seagull or butterfly. The tongue-in-cheek image of a wing-flapping, chaos-inducing, havoc-wreaking butterfly came later.

281   *we fool ourselves*: Contrary to some accounts, physicists and mathematicians were well aware of the complexity of nonlinear systems long before chaos exploded as an interdisciplinary phenomenon in the late 1970s and early 1980s. In 1908, for example, Henri Poincaré (comment 65) described what is now known as extreme sensitivity to initial conditions: "a small error in the former [the initial conditions] will produce an enormous error in the latter. Prediction becomes impossible...." [*Science and Method* (Dover reprint, 1952)] Richard Feynman, in Volume III of the influential *Feynman Lectures on Physics* (see comments 178b and 231) described the determinism-in-name-only of nonlinear classical systems in similar fashion: "...given an arbitrary accuracy, no matter how precise, one can find a time long enough that we cannot make predictions for that long a time. Now...this length of time is not very large." And there were many other workaday scientists, too, most of whom did not write books but were aware of these ideas nonetheless.

What cannot be denied, however, is that nonlinear complexity was long passed over in favor of idealized model systems that could be solved exactly. Nonlinear systems were often dismissed as being just too difficult to tackle, without any realization that structure and rules might lie buried in the mire of complexity and chaos. In that sense, the emergence of nonlinear dynamics in the last few decades has been a development of enormous significance, a veritable revolution.

Still, there is scant reason to write off science before 1975 as hopelessly unrealistic. The simplified linear approach, although disparaged by some, has had its moments. Among other things, it has produced the entire body of knowledge and understanding sketched out in the preceding ten chapters.

282   *Prigogine...speaks of the difference between being and becoming*: Ilya Prigogine, *From Being to Becoming* (W. H. Freeman, 1980). Prigogine (1917–2003), a Belgian physical chemist born in Russia, received the 1977 Nobel Prize for Chemistry.

283   *chance thus becomes an essential part of natural law*: Prigogine (see comment 282) makes an eloquent case for a reformulation of physical law on a nondeterministic foundation. His ideas, although not accepted by all, are universally respected.

For a nonmathematical account of Prigogine's approach, see his 1997 book *The End of Certainty: Time, Chaos, and the New Laws of Nature* (Free Press).

CHAPTER 12   *The more we know, the more we wonder. Our excursion into physical law concludes with a peek over the horizon, a glimpse into the knowable unknown. Three principal issues are considered: (1) the origin of the universe, (2) the cosmic dark matter and dark energy, (3) the ultimate structure of matter on the smallest of scales.*

284 *Nobel physicist I. I. Rabi*: Isidor Isaac Rabi (U. S., 1898–1988) received the Nobel Prize for Physics in 1944. The line is oft quoted.

286 *tantamount to a Theory of Everything*: The label, sometimes shortened to TOE, has gained currency in the popular press in connection with string theory. See the final sections of this chapter.

288a *the Hebrews believed instead*: More specifically, it was the Rabbis of the Talmudic era (third through sixth centuries) and medieval Jewish philosophers such as Maimonides (1135–1204) and Nachmanides (1194–1270). Commenting on the text of Genesis, they insisted upon *creation ex nihilo* as a matter of faith.

288b *the Doppler effect, a simple property of waves*: The phenomenon is named for its original discoverer, the Austrian physicist Christian Doppler (1803–1853).

In 1848, six years after Doppler's publication, the French physicist Armand-Hippolyte-Louis Fizeau (1819–1896) independently described the same effect. The phenomenon, particularly when applied to light, is sometimes referred to as the Doppler-Fizeau effect.

Both Doppler and Fizeau showed how the wavelength of starlight would be altered owing to the star's motion relative to the Earth. Together, their explanation anticipated by over eighty years the broader observations of Edwin Hubble (comment 289). Fizeau, whose measurements of the speed of light failed to confirm the existence of the luminiferous ether, was also a predecessor of A. A. Michelson (comment 60b).

289 *his name is Hubble*: The reference is to Edwin Powell Hubble (1889–1953), the American astronomer who discovered the intergalactic redshift described in the paragraphs immediately following.

290 *general relativity demands it*: Chapter 5. See also comment 320a.

291a *a Big Bang of creation*: The term was coined, mockingly, during a radio interview by British astronomer Fred Hoyle, who never intended it to stick. Hoyle (1915–2001) advocated instead the "steady-state" model of cosmology, which calls for the continuous creation of matter as the universe expands. A relatively small amount of new matter every ten billion years or so would be sufficient to maintain a constant density over the vast reaches of space, and this approach to cosmic management was considered reasonable by many. The steady-state theory, fully consistent with general relativity, enjoyed wide support until later evidence effectively disproved it.

The Big Bang, also justifiable according to general relativity, was originally proposed in the 1920s by the Russian physicist Alexander Friedmann (1888–1925) and the Belgian astronomer Georges Lemaître (1894–1966). The modern version was first worked out in the 1940s by the Russian-born American physicist George Gamow (1904–1968) and his colleagues.

291b   *a Nobel Prize for astrophysicists Arno Penzias and Robert Wilson in 1978*: Arno A. Penzias (Germany, b. 1933) and Robert W. Wilson (b. 1936) were young researchers at the Bell Telephone Laboratories in Holmdel, New Jersey, when they made their discovery in the mid 1960s. They shared half of the 1978 Nobel Prize for Physics, the other half going to Soviet physicist P. L. Kapitsa for unrelated work.

291c   *electromagnetic ripples emanating from a primeval disturbance*: Penzias and Wilson discovered the microwave background radiation by accident, stumbling upon an unwelcome source of noise that interfered with measurements they were making for other purposes. At nearby Princeton University, however, physicists Robert H. Dicke (1916–1997), P. James E. Peebles (b. 1935), and associates were planning explicitly to look for microwave radiation as a remnant of the Big Bang. They did not detect the radiation themselves, but it was the Princeton group that explained the significance of the Penzias-Wilson discovery. Both groups published their work simultaneously in 1965.

292a   *only 2.7 degrees above absolute zero, spread evenly throughout the universe with a variation of only 0.001%*: The results quoted are from the Cosmic Background Explorer satellite (COBE), reported in January of 1993. The characteristic temperature was determined to be 2.726 degrees, uniform to within one part in a hundred thousand.

A follow-up satellite, the Wilkinson Microwave Anisotropy Probe (WMAP), has improved on the already stunning results of COBE. Launched June 30, 2001, WMAP was designed to provide data with 100 times the resolution and 30 times the sensitivity of COBE.

292b   *the cataclysmic explosion of a "supernova"*: For most of its life, a star radiates energy as a consequence of nuclear fusion: the progressive building-up of heavy nuclei by melding together lighter nuclei (such as hydrogen and helium) at extremely high temperature. The outward radiative pressure counterbalances the inward pull of gravity, enabling the star to maintain its volume. Beyond iron, however, a nucleus with 26 protons, fusion becomes an unprofitable business. The energy consumed exceeds the energy produced, and the star begins to lose its internal support. Heavy elements sink into an inner core around which the lighter elements are wrapped like layers of an onion, and then, as the available nuclear fuel burns away, the radiated energy becomes insufficient to overcome gravity. The iron core begins to collapse, squeezing the protons in its nuclei and the electrons outside into an enormously dense, rapidly spinning kernel of neutrons. Separate nuclei and electrons mostly cease to exist, and the result is a "neutron star" with a mass of one or two Suns compressed into a radius of perhaps 12 miles. A spoonful of the neutron core might weigh 50 billion tons under terrestrial gravity.

Given sufficient mass (at least eight Suns), a neutron star can later explode into a supernova when lighter material from its outer layers—unsupported from within—collapses and rebounds off the inner core. The shock wave that follows is tremendous. Energy produced by the collapse radiates outward, blasting away part of the outer layers and fusing together even heavier elements in the enormously high temperatures that obtain. Going out in a blaze of glory, the supernova burns with exceptional brightness for a brief time, and then the star is no more. It leaves behind a legacy of heavy elements, including all those found within and around us on Earth.

With even more original mass, the neutron core may evolve into a black hole. Rather than bouncing off the core, matter falling in from the outer layers is captured and imprisoned by the black hole.

293a   *approximately 13.5 billion years ago*: The calculated age of the universe continues to be refined in the light of increasingly accurate measuring techniques. As of October, 2003, for example, the value was 13.7 ± 0.2 billion years; three months later, it was 13.5 ± 0.2 billion years.

Changes in the reported values are becoming steadily smaller. Earlier estimates ranged from 10 to 20 billion years.

293b   *an infinitesimally small and infinitely hot point*: Standard general relativity, which construes space-time as mathematically real (it uses ordinary rational and irrational numbers like 1, −1.4, and √3), requires either a singularity at time zero or else a preexistent, eternal universe. There are less conventional alternatives, though, which can do away with the singularity. In *A Brief History of Time* (Bantam, 1988), Stephen Hawking describes one involving imaginary time (based on the "imaginary" number $i = \sqrt{-1}$, the square root of −1).

String theory, a very different approach, hints at an especially intriguing resolution. By imposing a small but finite size on the universe even at its smallest, a successful string theory would likewise remove the singularity and possibly rob time itself of any presumed zero point. See "The Myth of the Beginning of Time" by Gabriele Veneziano, in *Scientific American*, May, 2004, pp. 54–65.

294a   *evil-twin particles of "antimatter"*: A particle and its antiparticle have the same mass but opposite electric charges, magnetic moments, and other fundamental attributes. Brought together, they annihilate to produce pure energy according to $E = mc^2$.

Except for their rarity, antiparticles are not particularly unusual or exotic forms of matter. A positron, for example, is simply a particle with exactly the same mass as an electron but a charge of +1 unit rather than −1 unit. Positrons and other antiparticles are produced naturally during certain radioactive processes and artificially in particle accelerators every day.

294b   *materializing ... out of an intangible store of energy*: as allowed by $E = mc^2$.

294c   *were it not for a slight excess of matter over antimatter in the beginning*: one of the great unanswered questions in particle physics and cosmology.

295a   *almost four hundred thousand years later*: 380,000, according to data available in 2004. The number is variously estimated at between 300,000 and 400,000 years, with more recent values tending toward the longer end. The caution noted in comment 293a applies.

295b   *the universe became transparent to electromagnetic radiation*: an event that cosmologists call "decoupling."

296   *the improbability of it all makes one think*: One of the possible explanations, unacceptable to many cosmologists, draws on the "anthropic principle." Its premise is simple: Had events unfolded any differently, we (conscious beings) would not be here today. Not being here, we would be unable to speculate about cosmic origins and marvel about the improbability of it all.

297   *the model of cosmological inflation*: The idea of inflation as arising from a vacuum energy (see the section *Something in Nothing*) goes back to Willem de Sitter (1872–1934), a Dutch mathematician and astronomer who proposed it in 1917. The theoretical physicist Alan Guth (b. 1947) developed a fully realized model of the inflationary universe in 1980–1981, and Andrei Linde (b. 1948) introduced "chaotic inflation" in 1982.

The inflationary universe, like many other scientific breakthroughs, has numerous coauthors, and any attempt to give credit to all is fraught with peril. The following list of names is meant only to highlight some of the major contributors to inflation and related subjects. It is in no way intended to be complete: Andreas Albrecht, Erast Gliner, James Hartle, Stephen Hawking, James Peebles, David Schramm, Lee Smolin, Alexei Starobinsky, Paul Steinhardt, Edward Tryon, Michael Turner, Alexander Vilenkin.

299   *increased first by a factor of two, then four, then eight... and so on*: The story is told of King Shirham, who wanted to reward his grand vizier, Sissa Ben Dahir, for inventing the game of chess. Vizier Sissa, who surely understood the power of exponential expansion, responded with what struck the king as a peculiarly modest request: a single grain of wheat on the first square of the chessboard, two grains on the second, four grains on the third, eight grains on the fourth, and so on, a straightforward doubling from square to square.

The numbers grew more rapidly than the king imagined. Over a million grains were required for the twenty-first square, a billion for the thirty-first, a trillion for the forty-first, a quadrillion for the fifty-first, a quintillion for the sixty-first. All told, the amount would have been 18,446,744,073,709,551,615 grains—more than all the wheat in India.

300a   *space would have increased by some thirty or forty powers of ten*: or more. Some versions of the inflationary theory call for a much greater expansion.

300b   *the initial impetus provided by inflation has been opposed inexorably by gravity ever since*: But see also the recent evidence of cosmic acceleration, discussed under the heading *Dark Energy* and commented on further below.

300c   *inflation solves the vexing horizon problem*: As an alternative to inflation, some physicists (notably João Magueijo) advocate the varying-speed-of-light theory. Under this scenario, electromagnetic radiation in the early universe is presumed to have propagated faster than it does today—fast enough to establish thermal communication among all the particles, without the need for any exponential stretching of space. The idea is controversial. See João Magueijo, *Faster than the Speed of Light* (Perseus, 2003); and "Plan B for the Cosmos," in *Scientific American*, January, 2001, pp. 58–59.

300d *inflation ... accounts for both the large-scale homogeneity and local inhomogeneity of the observable universe*: Inflationary models also may explain the absence of magnetic monopoles (see *Magnetism: Poles Apart* in Chapter 6 and comment 138 above). Moreover, by postulating a primordial inflaton field (discussed later in the chapter, culminating in the section *False Vacuum*), the inflationary hypothesis provides a source for the original expansion—albeit not the dark energy.

301 *one very specific value, <u>exactly</u>, with scarcely any room for error*: the "flatness problem" of cosmology. One second after the Bang, absent inflation, the value of a parameter called "omega" would have to have fallen fortuitously between 0.999999999999999 and 1.000000000000001 if the universe is to exhibit the properties it does today. With inflation, there is no such restriction.

304 *physicists call them "Higgs fields"*: after Peter Higgs (b. 1929), a theoretician at the University of Edinburgh who proposed such fields in 1964. Other physicists also expressed similar ideas at the time.

309 *particles ... slog through a constant Higgs field*: The postulated field is not only constant, but *scalar* as well. In contrast to a vector field (an electric field, for example), which has an independent value in each of the three principal directions, a scalar Higgs field delivers just a single number—the same—at each point. The constant scalar field provides no sense of direction and thus does nothing to impair the inherent isotropy, or directionlessness, of space.

313a *a quantum mechanical route to escape from it*: called "tunneling." Trapped in a valley, a classical particle must stay put if it lacks the energy to leap over a nearby mountaintop. A quantum mechanical wave function, strangely enough, always has some probability (even if very small) to tunnel through the mountainside and leak away.

313b *"supernatural" inflation (not what you think it is)*: not magical or divine, but rather a supersymmetric variation of a model called "natural" inflation.

316a *the nebulae that glow*: Nebulae (singular "nebula") are diffuse and vastly extended clouds of interstellar dust and gas, mostly hydrogen.

316b *the quasars that broadcast radio waves*: A quasar is a very bright, very distant celestial object that emits radio waves in addition to visible light.

317a *protons and neutrons, no*: Researchers sometimes use the term "massive compact halo object" and its unfortunate acronym MACHO to denote candidates for baryonic dark matter.

317b *neutrinos, no*: The term "weakly interacting massive particle" (chosen for the sake of another precious acronym, WIMP) is used to denote dark matter that interacts only by gravity and the weak interaction. Neutrinos, if they indeed have a mass, would be WIMPs according to this terminology.

317c   *"supersymmetric" theories (which interconvert fermions and bosons)*: treated later in the chapter, under the heading *Supersymmetry*.

318   *if so, we need new equations and not new particles*: See, for example, "Does Dark Matter Really Exist?" by Mordehai Milgrom, in *Scientific American*, August, 2002, pp. 42–52.

319a   *the key discovery came just in 1998*: Two teams independently reported similar findings. Saul Perlmutter of Lawrence Berkeley National Laboratory (California) headed the Supernova Cosmology Project. Brian Schmidt of the Mount Stromlo and Siding Spring Observatories (Australia) headed the High-Z Supernova Search Team.

319b   *observation of distant supernovae*: These objects are classified as Type Ia supernovae, not the Type II supernovae described in comment 292b. Whereas a Type II supernova comes about when a star collapses owing to loss of nuclear fuel, a Type I supernova forms when a small, dense star exerts its gravity to capture mass from a larger and more diffuse companion star nearby. The supernova explosion that accompanies the subsequent collapse, a nuclear bomb of sorts, burns with a uniform intensity that astronomers use as a "standard candle" to measure distance.

The 1998 measurements showed that several dozen Type Ia supernovae appear dimmer than they would if the universe were either expanding at a constant rate or decelerating. Subsequent observation of even more distant supernovae has strengthened the case for acceleration.

320a   *an effect Einstein modeled by a quantity called the cosmological constant*: For Einstein, who assumed that the universe was static, the spatial flux demanded by general relativity could mean only one thing: a problem with the equations. To counter any contraction brought about by gravity, he therefore introduced— *ad hoc*—a vacuum energy that would allow space to expand. By proper adjustment of an antigravitational "cosmological constant," it became possible to stabilize the theoretical universe of general relativity.

Not long afterward, Hubble's measurements showed that the universe was indeed not static, but rather that it was expanding. Einstein, faced with observational proof of a dynamic space-time, soon came to regret the cosmological constant, calling it his "greatest blunder."

His only mistake may have been to doubt himself. Today, with reports of a vast dark energy at work, more and more scientists believe that the cosmological constant represents a very real and very important force in the universe.

320b   *"quintessence"*: For an introduction, see "The Quintessential Universe" by Jeremiah P. Ostriker and Paul J. Steinhardt, in *Scientific American*, January, 2001, pp. 46–53.

320c   *so long as the dark energy continues to be a mystery*: Two articles in the February, 2004, issue of *Scientific American* provide additional perspective and detail concerning cosmic acceleration.

In "From Slowdown to Speedup" (pp. 62–67), Adam G. Riess and Michael S.

Turner discuss evidence that the acceleration has not been in effect continuously since the Big Bang—that, initially, the expansion of space had been slowing down (as expected) before it suddenly began to accelerate.

In "Out of the Darkness" (pp. 68–75), Georgi Dvali points to string theory as a possible explanation for the acceleration. As discussed in the final pages of this chapter, superstrings are believed to vibrate in spatial dimensions beyond the usual three of length, width, and height. Some of these "extra" dimensions are presumably quite small and thus easy to overlook, while others may be infinitely large and similarly easy to overlook (although for different reasons). If gravitons, the hypothesized messengers of the gravitational interaction, were occasionally to leak into these large but hidden dimensions, they might cause a warping of space-time and an accompanying acceleration. "Gravitational leakage" of this sort would also account for the extraordinary weakness of gravity compared with the other fundamental forces.

320d   *three groups of fermions ... and a baker's dozen of bosons*: For readers who wish to refresh their memory concerning fermions and bosons, see Chapter 9 (under the heading *In a Spin: Fermions, Bosons, and the Pauli Principle*).

320e   *quarks and ... leptons*: Baryons ("heavy ones" such as protons and neutrons) are particles built from three quarks, whereas the various leptons (which contain no quarks) have the electron and neutrino as prototypes. Each class obeys its own conservation law, stated as follows: (1) The total number of baryons in the universe is constant. If one baryon disappears (say a neutron undergoing beta decay), then another one must arise to take its place (the proton that it becomes). (2) The total number of leptons in the universe is constant. If a lepton appears out of nowhere (say the electron produced during neutron decay), then an antilepton must simultaneously appear in order to cancel it out (the antineutrino produced alongside).

325   *an idealization that dates back to the ancient Greeks*: Democritus (ca. 460–ca. 370 B.C.) was one of the first to articulate an atomistic theory of nature. He pictured the physical world as arising from an infinite number of infinitesimally small atoms moving in an infinitely large void. The atoms were assumed to be indivisible, incompressible, and indestructible point particles.

326   *a messy confusion of "quantum foam"*: The name was coined by John Archibald Wheeler (b. 1911), one of the leading physicists, teachers, and mentors of the twentieth century.

328a   *in the infinitesimally small world of quantum gravity ... general relativity breaks down*: Special relativity may break down as well, once a certain threshold of length and energy is crossed. Theoretician Giovanni Amelino-Camelia and others argue that Einsteinian special relativity (including $E = mc^2$) will have to be modified to describe phenomena on the Planck scale. The theories under development, still controversial, are classified as "doubly special relativity (DSR)" for their postulation of two impassable limits: (1) a maximum speed, c (as Einstein said); and (2) a minimum length or maximum energy (something new). Advocates suggest that DSR might explain the dark energy and also provide a better picture of cosmic origins than inflation, in addition to playing a role in quantum gravity. Some say DSR will rewrite modern physics. [*New Scientist*, 8 February 2003, pp. 28–32]

328b   *more generally called "M-theory"*: Before the mid 1990s, string theorists suffered from an embarrassment of riches: five variations on a theme, fully five different versions of a ten-dimensional string theory (incorporating one dimension of time, three ordinary dimensions of space, and six additional "curled-up" dimensions of space, exceedingly small). Then, in a landmark 1995 lecture, Edward Witten showed that all five variations are equivalent and can be subsumed under a broader model formulated in eleven dimensions (one temporal, ten spatial). The new approach, M-theory, uses membranes of two dimensions and higher in addition to the one-dimensional strings of earlier theories.

M-theory is very much a work in progress, with its full implications yet to be elaborated. Those in the know remain vague about the meaning of "M," suggesting variously that it denotes matrix, membrane, mystery, magic, or mother.

Physicist and mathematician Edward Witten (b. 1951) has been one of the most influential figures in string theory since the 1980s. He was awarded the prestigious Fields Medal in 1990, often considered the equivalent of a Nobel Prize for mathematics.

328c   *vibrating in a world of eleven dimensions*: The idea of spatial dimensions existing beyond the usual three dates back to 1919, when Polish mathematician Theodor Kaluza formulated a unified theory of Einsteinian gravity (general relativity) and Maxwellian electromagnetism in five-dimensional space-time. Seven years later, Swedish mathematician Oskar Klein refined and clarified the concept of a tightly curled extra dimension of space.

The modern approach to string theory began with attempts by Gabriele Veneziano in the late 1960s and Yoichiro Nambu in the early 1970s. Michael Duff, Michael Green, Brian Greene, David Gross, Joël Scherk, John Schwarz, Nathan Seiberg, Andrew Strominger, Cumrun Vafa, Edward Witten, and many others have all made important contributions.

331   *hyperspace*: space of more than three dimensions.

332a   *Eugene Wigner spoke of the "unreasonable effectiveness" of mathematics*: The Hungarian-born American physicist Eugene Paul (Jenó Pál) Wigner (1902–1995) was awarded half the 1963 Nobel Prize for Physics. The other half went to J. Hans D. Jensen and Maria Goeppert Mayer for unrelated work.

For "unreasonable effectiveness," see *Communications in Pure and Applied Mathematics*, Volume 13 (1960), pp. 1–14.

332b   *"I would be sorry for the dear Lord"*: as quoted in R. Clark, *Einstein: The Life and Times* (Avon Books, 1984).

Glossary

This glossary is intended to be self-contained to the extent possible, with only a minimum of explicit cross-references. Most of the terms that appear in a particular entry are defined elsewhere in entries of their own.

absolute zero   The coldest temperature possible, at which all motion except zero-point vibration comes to a stop: $-273.15°C$ (equivalently, $-459.67°F$).

acceleration   A vector representing the rate at which velocity changes, taking into account both magnitude and direction.

alpha decay   Emission of an alpha particle by a radioactive nucleus. See also *beta decay*; *gamma ray*; *radioactivity*.

alpha particle   A nucleus of helium-4, consisting of two protons and two neutrons tightly bound by the strong interaction. Alpha particles are emitted by certain radioactive nuclei. See also *alpha decay*; *beta particle*; *gamma ray*.

amplitude   The maximum displacement attained by a wave or other oscillation.

angular momentum   A vector associated with the turning of a body or its motion about a point. Angular momentum, the rotational counterpart of linear momentum, obeys a strict conservation law in both classical and quantum mechanics. (1) For a particle of mass m moving in a circle of radius r at linear velocity v, the classical angular momentum has magnitude mvr. Its direction is perpendicular to the plane of motion. (2) In quantum mechanics, angular momentum is represented by a mathematical operator designed to correspond with the classical quantity. See also *correspondence principle*.

anisotropic   Having different properties along different axes, thus affected by direction, angle, or orientation. Contrasted with *isotropic*.

annihilation   Conversion of mass into pure energy, usually occurring when a particle meets its antiparticle. The amount of energy produced is governed by the equation $E = mc^2$. Contrasted with *creation*. See also *antimatter*; *mass-energy*.

antielectron   A positron (the antiparticle of an electron). See *antimatter*; *positron*.

antimatter   Matter composed of antiparticles, such as antielectrons (positrons), antiprotons, antineutrons, antineutrinos, antimesons, and antiquarks. A particle and its antiparticle have the same mass and spin but opposite electric charge, magnetic moment, and other internal attributes. They annihilate on contact and disappear in a burst of energy.

antineutrino   The antiparticle of a neutrino, produced (along with an electron) during the beta decay of a neutron into a proton. See also *antimatter*.

atom   A composite particle, electrically neutral, consisting of a positive nucleus (with a charge of $+Z$) set amidst Z electrons. The atomic number Z, equal to the number of protons in the nucleus, determines the chemical behavior of the atom. In isolation, an atom acts as the smallest unit of a chemical element capable of exhibiting the properties of that element. See also *molecule*.

atomic number   The number of protons in the nucleus of an atom.

axion   A hypothetical particle low in both mass and energy, predicted by some grand unified theories. Axions, if they exist, are possible candidates for dark matter.

baryon   A fermion constructed from three quarks and subject to all four forces, including the strong force. Examples: protons, neutrons, and their antiparticles. See also *lepton*; *meson*.

beta decay   Radioactive processes that bring about the interconversion of protons and neutrons, together with the emission of electrons, positrons, neutrinos, and antineutrinos. Beta decay is mediated by the weak interaction, which effects a change in quark flavor. See also *alpha decay*; *beta particle*; *gamma ray*.

beta particle   An ordinary electron produced during the transformation of a neutron into a proton. See also *alpha particle*; *beta decay*; *gamma ray*.

Big Bang   Model of cosmic origins that pictures the nascent universe as an infinitesimally small, infinitely dense concentration of energy out of which space-time abruptly began to emerge. An initial burst of inflation is usually postulated as a mechanism for driving the expansion.

black hole   An object with a gravitational field so intense that nothing (not even light) can escape from it. Particles that come too close to a black hole are captured and imprisoned.

boson   Any elementary particle that possesses either zero spin angular momentum or an integral unit (0, 1, 2, ...). Unlike fermions, two or more bosons are not prohibited by the Pauli principle from occupying the same quantum state. Certain kinds of bosons, passed between fermions, mediate the four fundamental interactions. Examples: photons, W, Z, gluons. See also *messenger particles*.

brane   An extended structure in M-theory. A string is a one-brane; a membrane is a two-brane; and, in general, a p-brane is an object in p dimensions.

butterfly effect   Extreme sensitivity to initial conditions, a key concept in chaos theory. A small change in the initial state of a system produces a disproportionately large effect on the final state after a sufficient lapse of time.

chaos   Unpredictable, seemingly random behavior in a system that nominally obeys deterministic laws: a consequence of nonlinear feedback. Sometimes called *deterministic chaos*. See also *butterfly effect*; *determinism*; *period doubling*.

charge   (1) In general: an attribute that enables a particle to participate in one of the four fundamental interactions. Examples: electric charge, strong interaction charge (color, def. 1). (2) Used without qualification: shorthand for *electric charge*.

charged-current process   A weak interaction mediated by *W* bosons, resulting in new electric charges for the participating fermions. Example: collision of a neutrino and neutron to yield an electron and proton. Compare with *neutral-current process*.

classical electrodynamics (classical electromagnetism)   Study of electric and magnetic fields and their effects on charged particles and currents, valid under conditions where the fields appear continuous. The classical electromagnetic field, arising from a huge number of photons, betrays no obvious grain or particle-like aspect. Contrasted with *quantum electrodynamics*.

classical mechanics   Study of force and motion for particles having definite positions and momenta. Also called Newtonian mechanics, classical mechanics is a thoroughly deterministic model rooted in Galilean relativity and Newton's three laws of motion. Since classical quantities are measurable simultaneously with unlimited accuracy, the system follows a well-defined sequence of position and momentum (a path). Contrasted with *quantum mechanics*. See also *first law of motion*; *second law of motion*; *third law of motion*; *relativity*.

classical physics   Physics exclusive of quantum mechanics, often synonymous with classical mechanics and electrodynamics. Quantum mechanics and Einsteinian relativity are usually classified as modern (or twentieth-century) physics, although relativity by itself is a classical theory.

closed universe   Cosmological model in which the large-scale curvature of space-time is positive, resulting in a non-Euclidean geometry analogous to that of a sphere. There is a both a temporal aspect and a spatial aspect: (1) In a *temporally closed* universe, which arises from a sufficiently large mass density, the governing rule is "what goes up must come down." Gravitational attraction eventually halts and reverses the expansion of space, causing the universe ultimately to collapse back to an infinitesimal point. The ensuing "Big Crunch" puts an end to time. (2) A *spatially closed* universe exists as a finite volume with no boundary, its space curved back onto itself under the influence of gravity. A light ray launched anywhere on the surface eventually returns to its starting point. Compare with *flat universe*; *open universe*.

color   (1) In quantum chromodynamics: the attribute of matter that enables a particle to participate in the strong force. Endowed with three distinguishing "colors," quarks emit and absorb massless bosons (gluons) that mediate the strong force. Color in quantum chromodynamics is analogous to electric charge in quantum electrodynamics. (2) In ordinary usage: the brain's subjective interpretation of the energy of visible electromagnetic radiation.

commuting operators   Operations that can be performed in arbitrary sequence (either *A* first and then *B*, or *B* first and then *A*), with no discernible difference in outcome. Example: rotating the hour hand of a clock. Turning the hand one hour clockwise and then two hours counterclockwise gives the same net result (one hour counterclockwise) as turning it two hours counterclockwise and then one hour clockwise.

In quantum mechanics, commuting operators represent observables that can be measured simultaneously with unlimited accuracy, unconstrained by the uncertainty principle. Quantum mechanical properties that cannot be determined simultaneously, such as position and momentum, are governed by the uncertainty principle and are represented instead by noncommuting operators. Different sequences of measurement yield different results, because the first operation produces an uncontrollable disturbance that prejudices the second.

conservation (conservation law)   Constancy of a particular quantity throughout any changes a system may undergo. The total amount of a conserved quantity (it might be energy, linear momentum, angular momentum, or charge, among others) is redistributed into different channels but is neither augmented nor diminished. Any increase in one place is matched by a decrease somewhere else.

Conservation does not necessarily imply invariance. For a quantity to be invariant (the speed of light, for example), its value must be the same in all reference frames. The numerical value of a conserved quantity, however, may differ from frame to frame even though it remains unchanged in each one.

correspondence principle   A guiding principle of quantum mechanics, enunciated by Niels Bohr: The laws and equations of quantum mechanics reduce to those of classical mechanics under conditions where Planck's constant may be considered negligible. The classical limit is approached when particle wavelengths are small, system dimensions are large, and the number of quanta is high. See also *wave–particle duality*.

cosmic microwave background radiation   The relic heat of the Big Bang, manifested today as a low-level electromagnetic field pervading all space. The radiation, characterized by a temperature just 2.7 degrees above absolute zero, falls primarily in the microwave region of the electromagnetic spectrum. It is uniform to within 1 part in 100,000.

cosmological constant   A parameter that Einstein added *ad hoc* to his equations of general relativity, simulating a force that would enable the universe to expand and thereby act against gravity. Its effect is similar to the false vacuum in the inflationary model of the Big Bang and also, possibly, to the cosmic dark energy recently discovered.

cosmology   Study of the origin, structure, and evolution of the universe.

Coulomb's law   A relationship describing the dependence of electrostatic force on charge and distance: The force between two point charges is (1) directed along the line between the two particles, (2) repulsive between like charges and attractive between unlike charges, (3) directly proportional to the product of the charges, and (4) inversely proportional to the square of the distance between them. Compare with *gravitation, law of universal*.

creation   The materialization of mass from energy, as allowed by $E = mc^2$. Contrasted with *annihilation*. See also *mass-energy*.

crystal   A solid that exhibits translational symmetry. The atoms of a crystal fall into an ordered arrangement that repeats itself over long distances.

current   A flow of liquid, electric charge, or other material. See also *electric current*.

dark energy   The still unexplained driving force responsible for cosmic acceleration, believed by some to be a vacuum energy of undetermined origin. Evidence for the acceleration of the universe (a speeding-up of the expansion of space) has been gathering since 1998, and dark energy is believed to make up over two-thirds of the cosmic total. See also *cosmological constant*.

dark matter   Mass of undetermined origin that has no electromagnetic signature, thus detectable mostly by gravitational effects. Dark matter is believed to make up 95% of the total cosmic mass.

degree of freedom   A basic, independent way in which a body can move or a system can change, one of a limited number of possibilities. The total is constrained by the number and kind of internal parts in the system. Example: A point particle can move independently in three perpendicular directions, none of which affects the progress in any other. The system thus enjoys three degrees of freedom for translation.

density   The ratio of a particular quantity (mass or energy, for example) to the volume in which it is contained. Packing a large amount of something into a small volume produces a high density.

determinism   A characteristic of certain scientific theories: that systems evolve regularly and predictably in accordance with strict laws, proceeding uniquely from cause to effect. The condition of a system at any given time serves as the proximate cause for its condition in the instant that follows. There are no surprises. Compare with *chaos*.

deuterium   A nonradioactive isotope of hydrogen that contains one proton and one neutron in its nucleus. Most of the deuterium existing in the universe today is believed to have been formed shortly after the Big Bang. Also called *heavy hydrogen*, or *hydrogen-2*. See also *tritium*.

diffraction   Interference by waves as they recombine after spreading past an obstacle. The effect is manifested in a pattern of alternating strong and weak bands.

dimensions   (1) Construed narrowly: the three coordinates of space (x, y, z) or the four coordinates of space-time (t, x, y, z), with each coordinate corresponding to an independent axis. (2) Construed broadly, for abstract spaces of all kinds: the independent components from which an arbitrary object can be built. Examples: the three perpendicular directions of ordinary space (as in def. 1), the eigenstates of a quantum mechanical system, the standing waves of a vibrating string. Each independent component is a dimension of the system. None of them can be built up by combining any of the others, but all of them—together—can be combined to produce any object desired. (3) Units associated with particular quantities, such as mass, length, or time.

direct proportionality   A relationship between x and y such that $y = kx$ for some fixed number k. If x is increased, then y must be increased by the same percentage to maintain the proportionality. Compare with *inverse proportionality*.

Doppler effect   Perceived changes in the wavelength and frequency of radiation emitted by a moving source. See also *redshift*.

eigenstate   An allowed condition of a quantum mechanical system, characterized by a specific value (an *eigenvalue*) of an observable property.

eigenvalue   See under *eigenstate*.

Einsteinian relativity   See under *relativity*.

electric charge   The attribute of matter that enables a particle to participate in the electromagnetic interaction. In classical theory (Maxwellian electromagnetism), an electrically charged particle produces and interacts with continuous electric and magnetic fields. In relativistic quantum mechanics, a charged particle emits and absorbs discrete photons.

Electric charge comes in two opposing varieties (positive and negative) and is both quantized and conserved. Except for quarks, which have fractional values, the net charge on a system is a multiple of the fundamental charge of a proton or electron.

electric current   A flow of electric charge, expressed as the rate at which a stated amount of charge passes a given point.

electric dipole   A pair of electric charges separated in space. The two charges have the same magnitude but opposite signs. Compare with *magnetic dipole*.

electric field   The influence emanating either from a static electric charge or from a variable magnetic field, manifested classically as a force acting on a stationary charge brought into the region. A vector quantity, the electric field has both magnitude and direction at each point in space. Its strength is measured as the force exerted on a unit of charge. See also *field; gravitational field*.

electric potential   The potential energy acquired by a unit electric charge at every position in an electric field. Electric potential is a scalar quantity, specified in full by a single number at each point. See also *field; gravitational potential*.

electromagnetic field   The interconnected electric and magnetic fields produced by an electric charge in motion, described classically by Maxwell's equations.

electromagnetic induction, law of   See under *Maxwell's equations*.

electromagnetic interaction   The attraction and repulsion of electrically charged particles, second strongest of the four fundamental interactions. In classical physics, electromagnetic force and potential are described fully by the four Maxwell equations. In quantum electrodynamics, the interaction is viewed as arising from an exchange of photons.

electromagnetic radiation   Energy emitted by an electromagnetic wave.

electromagnetic spectrum   The full (infinite) range of wavelength and frequency spanned by electromagnetic waves, arranged in order of increasing energy: radio, microwave, infrared, visible, ultraviolet, X ray, gamma.

electromagnetic wave   A disturbance of the electromagnetic field that varies in space and time, produced by the movement of charged particles and accounted for classically by Maxwell's equations. An electromagnetic wave consists of electric and magnetic fields oscillating at right angles both to each other and to the direction of travel. Energy is delivered straight ahead, along the line of motion.

electron   An elementary particle with an electric charge of -1, a spin of $\frac{1}{2}$, and a mass 1836 times less than that of a proton. A basic component of all atoms, the electron is considered a structureless point with no extension. A kilogram of them (weighing 2.2 pounds on Earth) would contain over a million trillion trillion particles.

electrostatics   Study of the energy and forces between electrically charged particles at rest. See also *Coulomb's law*; *electric field*; *electric potential*.

electroweak interaction   Unified force that results when the electromagnetic and weak interactions become indistinguishable at sufficiently high energy. The electroweak interaction is mediated by four messenger bosons: the photon, W^{+}, W^{-}, and Z^{0}. See also *quantum chromodynamics*; *quantum electrodynamics*.

element   A material, one of more than a hundred basic chemical substances, containing only atoms with the same number of protons in their nuclei. The various isotopes of an element have different numbers of neutrons, but not protons.

emergent phenomena   Properties and patterns that are not explainable by reducing a complex system to its simplest parts; the antithesis of *reductionism*. Example: A colony of ants behaves in ways that cannot be predicted from the behavior of each ant taken in isolation.

energy   The ability to do work, either to change a state of motion (kinetic energy) or to alter the potential of a particle in a field (potential energy). Energy is redistributed and interconverted among many different forms, but is neither created nor destroyed. The total amount of energy in the universe or in any isolated system is strictly conserved. See also *entropy*; *mass-energy*.

entropy   A thermodynamic quantity involving the ratio of heat to temperature, associated microscopically with the number of channels into which energy is dispersed. The broader the dispersal of energy, the higher is the entropy. See also *second law of thermodynamics*.

equation of motion   A mathematical relationship that lays out the past and future of a system evolving under a particular influence. Specification of both the initial state and the prevailing influence is needed to write down (and attempt to solve) an equation of motion. Examples: Newton's second law of motion, the Schrödinger equation, Maxwell's electromagnetic wave equation.

equation of state   A mathematical relationship that interconnects the macroscopic variables of a system (such as the pressure, volume, temperature, and quantity of a gas). An equation of state might specify, for example, the extent to which pressure changes when volume is doubled and temperature is halved.

equilibrium   A self-sustaining balance of forces, manifested by a particular set of unchanging properties. Compare with *steady state*. See also *macrostate*; *microstate*.

equivalence, principle of   One of the fundamental postulates underlying Einstein's general theory of relativity: that the effects of acceleration cannot be distinguished from a gravitational force. Valid over small regions of space-time, the equivalence principle posits that the laws of physics are the same for observers in both a gravitational field and an accelerated reference frame.

Euclidean geometry   Study of lines, points, angles, surfaces, and solids in a flat space, based on a set of axioms proposed by the Greek mathematician Euclid (ca. 300 B.C.). Among the most prominent features of Euclidean geometry are the following three: (1) Parallel lines never meet. (2) The sum of the three angles in a triangle is equal to $180°$. (3) The ratio of the circumference of a circle to its diameter is equal to π. See also *non-Euclidean geometry*.

exclusion principle   See under *Pauli principle*.

exponential growth   Repeated doubling over a fixed interval: 1, 2, 4, 8, 16, ….

false vacuum   A vacuum that exists temporarily in a state with an energy density greater than the lowest one possible, as if trapped in a high valley or poised on top of a hill. In contrast to a classical system, a quantum mechanical false vacuum has a certain probability to decay to a lower, more stable state and thereby discharge its excess energy. Such *metastable states* (see under *stable equilibrium*) are suggested as possible mechanisms for cosmic inflation.

feedback   The redirection of some portion of a system's output to its input.

fermion   Any elementary particle that possesses one-half unit of spin angular momentum or an odd multiple thereof (1/2, 3/2, 5/2, …). The Pauli principle forbids two or more fermions from occupying exactly the same quantum state. Examples: electrons, neutrinos, quarks, protons, neutrons; in general, fundamental particles of matter rather than force. Compare with *boson*.

field   An influence or effect that pervades space and time, arising from a particular source but acting independently of that source. A *vector field* has both magnitude and direction at every point and is expressed by as many components as there are dimensions. A *scalar field* has only magnitude and is expressed by just a single number at each point. Examples of vector fields: electric force, gravitational force. Examples of scalar fields: electric potential, gravitational potential, Higgs field, temperature.

A field is represented mathematically by the values assigned to it at infinitesimally small points. Classical and quantum fields require different treatment and interpretation.

first law of motion   The principle of inertia, originally understood by Galileo and later adopted by Newton as the first of his three laws: A body at rest tends to remain at rest unless acted on by an outside force. A body moving in a straight line at constant speed continues to move in the same direction at the same speed unless acted on by an outside force. See also *classical mechanics*.

first law of thermodynamics   A statement of the principle of energy conservation, applied specifically to macroscopic systems: The change in energy for a system is equal to the heat transferred and the work done.

fission, nuclear   Disintegration of a heavy nucleus into two lighter nuclei of comparable mass, accompanied by the release of converted mass-energy and by-products such as neutrons or alpha particles. Compare with *fusion*.

Uncontrolled nuclear fission provided the explosive power of the atomic bombs dropped on Hiroshima and Nagasaki. Controlled nuclear fission is used commercially to produce energy in nuclear reactors.

flatness problem   Inability of the original Big Bang theory (without inflation) to explain the evolution of a flat universe. Conditions at time zero would have had to be almost unbelievably precise to bring about this one special outcome.

flat universe   Cosmological model in which the large-scale geometry of space-time is Euclidean (flat), balanced on a knife's edge between an open configuration (negative curvature) and a closed configuration (positive curvature). A flat universe is destined to expand forever while undergoing a gradual, unending deceleration—provided there is no dark energy. Compare with *closed universe*; *open universe*.

flavor   An attribute with six values to distinguish different kinds of quarks: up, down, charm, strange, top, bottom. During a weak interaction such as beta decay, the flavor of a quark changes (for example, from up to down).

fluctuation   An uncontrolled and random change from one condition to another, often sudden.

force   (1) In classical physics: a vector that changes the velocity of a massive object; a directed push or pull. The strength of a force follows the slope of a change in potential. The steeper the change, the stronger the force. (2) In general: often synonymous with *interaction*.

four-vector (four-dimensional vector)   In relativity theory, a vector consisting of four components: three corresponding to a spatial portion and one to a temporal portion. The four components are perceived differently by observers in relative motion, but the magnitude of the four-dimensional vector remains invariant. Spatial and temporal contributions to the vector's length combine with opposite signs. Example: the momentum-energy vector. See also *space-time interval*.

frequency   The number of cycles completed by a wave or other oscillation during a stated interval of time.

friction   Dispersal of energy in the form of heat, usually caused by the rubbing of one object against another. Energy lost to friction is difficult to recover in a form suitable for doing useful work.

fundamental forces (fundamental interactions)   Four primal agencies that influence matter, in descending order of strength: the strong force, the electromagnetic force, the weak force, gravity.

fusion, nuclear   Formation of a heavier nucleus by the melding together of two or more lighter nuclei. The process, which occurs naturally in the stars, requires exceedingly high temperature and pressure. Compare with *fission*.

Uncontrolled nuclear fusion serves as the mechanism for the thermonuclear bomb, or hydrogen bomb. Controlled fusion, a formidable engineering challenge, is envisioned as a possible source of abundant and clean energy.

Galilean relativity   See under *relativity*.

gamma ray   Electromagnetic radiation of very high energy, produced either by radioactivity or by annihilation. See also *alpha particle*; *beta particle*; *electromagnetic spectrum*.

gas   A highly disordered fluid state of matter, lacking even the local order (clusters of particles) characteristic of a liquid. The energetic particles of a gas fill the space of whatever container they occupy. Example: steam. Compare with *solid*; *liquid*; *plasma*.

gauge invariance   See under *local symmetry*.

general relativity (general theory of relativity)   Einstein's theory of relativity as generalized to an accelerated (noninertial) reference frame, by necessity a theory of gravity as well. General relativity derives from the principle of equivalence ("gravity cannot be distinguished from acceleration"), and its conclusion is to equate gravity with geometry. Rather than submit to a force, a gravitating object freely follows the contours of a space-time warped by the presence of mass. See also *mass-energy*; *relativity*; *special relativity*.

geodesic   The straightest line possible in a curved space or space-time. Example: a great circle on a sphere.

gluon   Any of eight messenger particles that simultaneously mediate and participate in the strong interaction (specifically, the color interaction), hence a quantized excitation of the strong force field. Gluons are massless bosons that carry the "color charges" red, green, blue. See also *quantum chromodynamics*.

grand unification   Hoped-for marriage of the electroweak theory (which accounts for the electromagnetic and weak interactions) and quantum chromodynamics (which accounts for the strong interaction). If grand unification proves successful, then three of the four fundamental forces will be understood as sharing a common origin. See also *superunification*.

gravitation, law of universal   A relationship discovered by Isaac Newton: The gravitational force between two point masses is (1) directed along the line between the two particles, (2) always attractive, (3) directly proportional to the product of the masses, and (4) inversely proportional to the square of the distance between them. Compare with *Coulomb's law*.

gravitational field The influence emanating from the mass of an object, manifested classically as a force acting on another mass brought into the region. A vector, the gravitational field has both a magnitude and a direction at each point in space. Its strength is measured as the force exerted on a unit of mass. See also *electric field*; *field*.

gravitational potential The potential energy acquired by a unit mass at every position in a gravitational field. Gravitational potential is a scalar quantity, specified in full by a single number at each point. See also *electric potential*; *field*.

gravitational wave Oscillations in space-time caused by changes in local curvature and the movement of mass, a consequence of general relativity. Gravitational waves, traveling at the speed of light, broadcast gravitational influence to distant points, analogously to the way that electromagnetic waves broadcast electromagnetic influence.

graviton A massless boson with two units of spin angular momentum, predicted on theoretical grounds to be the messenger particle of a quantized gravitational field.

gravity (gravitational interaction) The attraction of mass for mass, weakest of the four fundamental interactions. In the Newtonian picture, gravity is viewed as an inverse-square-law force transmitted by a gravitational field. In Einsteinian general relativity, gravity is an inertial effect created by the warping of space-time in the presence of mass-energy and pressure.

harmonic One of a set of standing waves, as on a vibrating string: The first harmonic is the vibration of lowest energy, the fundamental mode. It has zero nodes between the endpoints. Next in line, with one node, is the second harmonic (or first overtone). After that, with two nodes, comes the third harmonic (or second overtone); and after that, with three nodes, comes the fourth harmonic, and then the fifth, the sixth, the seventh, and so forth.

heat (1) Microscopic picture: the dispersal of energy into the random, disordered motions of atoms and molecules. (2) Macroscopic picture: a change in energy manifested by a change in temperature, not by the bulk displacement of a body. Compare with *work*. See also *thermodynamics*.

heavy hydrogen Deuterium.

Heisenberg uncertainty principle See *uncertainty principle*.

Higgs field Any of various scalar fields (see under *field*) hypothesized to have nonzero values in the vacuum, despite a minimum energy density. The Higgs fields are believed to have settled into a random (but permanently fixed) set of values when the early universe cooled, causing them to undergo spontaneous symmetry breaking. Interaction with the frozen Higgs fields imparts a characteristic mass to particles of different types.

horizon problem   Inability of the original Big Bang theory (without inflation) to explain the present-day homogeneity of the cosmos. Particles in the early universe would have had insufficient opportunity to interact and reach thermal equilibrium.

Hubble's law   An observation due to Edwin Hubble: that galaxies are receding from view at speeds proportional to their current distances from Earth.

hydrogen   Simplest and lightest of the chemical elements, an atom with a nucleus consisting only of a single proton. All the hydrogen present today is believed to have formed within minutes after the Big Bang. See also *deuterium*; *tritium*.

hyperspace   A space that has more than three dimensions, making it problematic to draw or picture in the mind's eye.

inertia   The tendency of a body to persist in its current state of motion unless acted on by a force. Absent a push or pull, objects at rest remain at rest. Objects traveling in a straight line at constant speed continue to do so without speeding up or slowing down. See also *first law of motion*; *inertial reference frame*.

inertial reference frame   An unaccelerated frame of reference, one in which the law of inertia is enforced. The coordinate system is viewed as either at rest or traveling in a straight line at constant speed. See also *inertia*; *reference frame*.

inflation   Exponential expansion of the universe, hypothesized to have occurred during the first instants of cosmic evolution (and possibly continuing, in some form, to the present day). Numerous models of the "inflationary universe" have been proposed since 1980. See also *Big Bang*; *false vacuum*; *inflaton*.

inflaton   Postulated quantum field thought to be responsible for cosmic inflation, arising from a false vacuum of some sort. See also *Big Bang*.

infrared radiation   Electromagnetic radiation with energy intermediate between microwaves and visible light. Infrared photons correspond to the vibrational frequencies of many molecules. See also *electromagnetic spectrum*.

interaction   The influence exerted by one particle on another, expressed either as force or potential energy.

interference   Combination of two waves with the same frequency, leading to either reinforcement or diminution of the joint intensity. Waves that combine with the same phase interfere constructively, whereas waves that combine a half-cycle out of phase (180°) interfere destructively. At points in between, the interference is partial.

invariance   The constancy and identity of a particular quantity in all reference frames, perceived as exactly the same number by each observer. Examples: distance in ordinary space; the invariant interval in space-time. Counterexample: individual coordinates of space and time. Compare with *conservation*. See also *relativity*.

invariant interval   See *space-time interval*.

inverse proportionality A relationship between x and y such that $y = k/x$ for some fixed number k. If x is increased, then y must be decreased reciprocally to maintain the proportionality. Compare with *direct proportionality*.

inverse square law A relationship whereby one quantity is inversely proportional to the square of another, such that $y = k/x^2$ for some fixed number k. Examples: Coulomb's law of electrostatic force, Newton's law of gravitational force. The force, varying in each case as $1/r^2$, quadruples when the distance r between the particles is halved.

ion An electrically charged atom or molecule, formed by addition or removal of electrons from the neutral structure. Positive ions, deficient in electrons, are called *cations*. Negative ions, having a surplus of electrons, are called *anions*.

isotopes Atoms of the same element that differ only in the number of neutrons contained in their nuclei. Example: hydrogen (one proton), deuterium (one proton and one neutron), tritium (one proton and two neutrons). The three isotopes are all forms of hydrogen. They have similar chemical properties but different masses.

isotropic Having the same properties along different axes, thus invariant to direction, angle, or orientation. Contrasted with *anisotropic*.

kinetic energy Energy invested in the motion of a body, equal to $\frac{1}{2}mv^2$ for a mass m moving with a velocity of magnitude v. See also *potential energy*.

lattice A periodic arrangement of points in space.

law See specific law. In scientific usage, the term denotes a pattern or relationship (often expressed mathematically) that is observed consistently under stated conditions.

lepton A fermion subject to gravity and the weak force but not to the strong force. Leptons that have electric charges interact electromagnetically as well. Examples: electrons, neutrinos, and their antiparticles. See also *baryon*; *meson*.

light Electromagnetic radiation, especially (but not limited to) those wavelengths visible to the human eye.

linear Describing a relationship in which a quantity y is directly proportional to the first power of another quantity x. Example: $y = mx$, where m is a fixed number (not $y = mx^2$ or $y = 1/x$ or $y =$ anything else). Contrasted with *nonlinear*.

linear momentum See *momentum*.

liquid A fluid state of matter that occupies a definite volume but lacks a definite shape. Particles in a liquid, more energetic than those in a solid, interact to form local clusters but are unable to maintain long-range order. Example: water. Compare with *solid*; *gas*; *plasma*.

local symmetry Symmetry implies that a particular operation leaves an object unchanged (for example, rotation of a square by 90°). If a symmetry is honored only when observers at every point in space-time execute the same operation simultaneously, then it is said to be a *global* symmetry. Everybody must cooperate and perform the identical rotation or translation or phase shift (or whatever else) if the symmetry is to be preserved. A *local* symmetry, by contrast, is one for which the operation can be executed at any arbitrary time or place without prejudice to other observers. A combination of local symmetry and special relativity leads to *gauge invariance* and the existence of a quantized field.

MACHO   [MAssive Compact Halo Object] Presumed dark matter composed of baryons (neutrons and protons) but difficult or impossible to detect by electromagnetic means. Example: a black hole. Compare with *WIMP*.

macrostate   The minimal set of variables needed to characterize the state and properties of a macroscopic system. Example: The values of pressure, volume, and temperature are sufficient to represent the macrostate of a dilute gas. Compare with *microstate*.

macroworld   Aspects of nature large enough to be observed with the unaided eye and understood in the context of classical physics. Compare with *microworld*.

magnetic dipole   The simplest known source of a magnetic field, arising either from a loop of electric current or from the two fixed poles of a bar magnet. The strength of a magnetic dipole is called its *magnetic dipole moment*, or simply *magnetic moment*. Compare with *electric dipole*; *magnetic monopole*.

magnetic field   The influence manifested by either an electric charge in motion, a changing electric field, or a magnetized body, understood classically as a force acting on another magnet or electric current. A vector quantity, the magnetic field has both magnitude and direction at each point in space. Compare with *electric field*.

magnetic moment   See under *magnetic dipole*.

magnetic monopole   The magnetic equivalent of electric charge: an isolated north or south pole of a magnet, existing as an independent entity. Although theoretically admissible, no magnetic monopole has yet been observed or produced.

magnitude   The scalar size of a quantity expressed without regard to its direction or algebraic sign. Example: distance between two points, as opposed to distance and orientation taken together.

mass   (1) Informally: the quantity of matter resident in a body. Twice the volume of a substance contains twice the mass. (2) Inertial mass: a measure of the resistance a body offers to a change in motion, as defined by Newton's second law. The greater the mass, the greater is the force needed to produce a given acceleration. (3) Gravitational mass: a measure of a body's susceptibility to gravity, as defined by Newton's law of universal gravitation (and, according to general relativity, indistinguishable from inertial mass). (4) In special relativity: the source of a body's rest energy, defined by $E = mc^2$. (5) In general relativity: an agency that warps the curvature of space-time.

mass-energy   A consequence of special relativity: that mass is a form of congealed energy able to be tapped and converted into other forms. The mass-energy equation ($E = mc^2$) states that mass and energy are essentially the same quantity, differing only by a conversion factor. The relationship is analogous to the one between miles and kilometers, both of which express the same measure—distance—in different units. See also *annihilation*; *creation*; *rest energy*.

massless particle   A particle with a vanishing rest energy, characterized by zero mass in the equation $E = mc^2 = 0$. With its relativistic energy invested entirely in momentum (p), the particle travels perpetually at the speed of light and has total energy pc.

matter   The tangible stuff of the universe: anything that has mass and takes up space (although $E = mc^2$ and the wave–particle duality blur the distinction between tangible matter and intangible energy).

Maxwell's equations   Four mathematical relationships that describe the classical electromagnetic field and electromagnetic wave: (1) Gauss's law for electricity, equivalent to Coulomb's law of electrostatic force. (2) Gauss's law for magnetism, a statement that magnetic monopoles do not exist. (3) Faraday's law of electromagnetic induction, which specifies how a changing magnetic field gives rise to an electric field. (4) Maxwell-Ampère's law for the magnetic field, which combines Ampère's law (the relationship between an electric current and a magnetic field) with Maxwell's displacement current (the induction of a magnetic field by a changing electric field).

mechanical variables (state variables)   A set of numbers sufficient to specify the state of a system at any instant. Supplied with all the mechanical variables at some initial time t_1, a suitable equation of motion will predict their values at a later time t_2. Examples: (1) The position and momentum of every particle (classical mechanics). (2) The components of a wave function (quantum mechanics). (3) Macroscopic quantities that remain constant in equilibrium, such as pressure, volume, and temperature (thermodynamics). See also *state (def. 1)*.

mechanics   Study of the motion of objects in relation to force and energy.

membrane   In M-theory, a two-dimensional surface (also called a *two-brane*).

meson   One of various bosons built from a quark and an antiquark. Mesons respond to the strong force and help mediate the interactions of protons and neutrons in atomic nuclei. See also *baryon*; *lepton*.

messenger particles   Bosons that mediate an interaction between eligible fermions, such as the following: (1) photons, which transmit the electromagnetic interaction from one electrically charged particle to another; (2) W and Z bosons, which carry the weak interaction; and (3) eight species of gluon, which mediate the color force between quarks (thus giving rise to the strong interaction). The hypothesized graviton and Higgs boson are messenger particles as well.

microstate   A microscopic distribution of energy and position consistent with a given set of macroscopic properties, usually one of many such distributions possible. Compare with *macrostate*.

microwave background radiation   See *cosmic microwave background radiation*.

microwave radiation   Electromagnetic radiation with energy intermediate between radio waves and infrared. Microwave photons trigger the rotation of molecules and reorient the spins of electrons. See also *electromagnetic spectrum*.

microworld   Aspects of nature that involve small sizes and small transfers of energy, usually understood within the framework of quantum physics. Examples: molecules, atoms, electrons, protons, neutrons, quarks. Compare with *macroworld*.

molecule   A stable, electrically neutral combination of two or more atoms, held together electromagnetically in accordance with the laws of quantum mechanics. Nuclei and electrons are shared collectively by the entire molecule, and the identities of the original atoms are lost. A molecule has chemical properties different from any of its atomic constituents.

The molecule stands out as an example of how nature makes much out of little. Millions of different molecules arise from only several dozen distinct atoms, the chemical elements.

momentum (linear momentum)   A vector associated with the linear motion of an object, subject to a strict conservation law. (1) In classical mechanics, the momentum of a body with mass m moving at velocity v is equal to mv. Its direction is along the line of motion. (2) In quantum mechanics, momentum is represented by a mathematical operator designed to correspond with the classical quantity. See also *angular momentum*; *correspondence principle*.

M-theory   Proposed superunification that assigns eleven dimensions to space-time and replaces point particles with superstrings and branes (membranes in two or more dimensions). Advocates hope that M-theory will provide a comprehensive and integrated understanding of all four fundamental forces, reconciling quantum mechanics with general relativity. Introduced only in the mid 1990s, the theory has already unified five competing string theories under a common umbrella.

multiverse   A "universe of universes," of which ours is but one part. The idea has support in some circles, but the existence of a multiverse remains unproven.

muon   An elementary particle nearly identical to the electron but with a mass approximately 200 times greater. See also *Standard Model of particle physics*.

muon-neutrino   See under *neutrino* and *Standard Model of particle physics*.

neutral   (1) Pertaining to electric charge or some other fundamental attribute of matter: having a net value of zero. (2) In chemistry: neither acidic nor basic.

neutral-current process   A weak interaction mediated by a Z boson, resulting in no change of electric charge for any of the participating fermions. Example: a neutrino rebounding (intact) off a neutron. Compare with *charged-current process*.

neutralino   Hypothetical supersymmetric particle with an electric charge of zero, advanced by some as a candidate for dark matter. The neutralino, a combination of two or more superpartners (belonging to the photon, Z boson, and perhaps others), is believed to be the lightest supersymmetric species possible and hence unable to decay into even lighter particles.

neutrino An elementary particle with an electric charge of 0, a spin of ½, and little or no mass (probably less than a hundred-thousandth that of an electron). The neutrino, a lepton, plays a role in beta decay and certain other weak interactions. It does not participate in either the electromagnetic or strong interaction.

According to the Standard Model of particle physics, a neutrino-like particle accompanies an electron-like particle in each of the three generations. The electron-neutrino (the common neutrino) is paired with the electron in the first generation. The muon-neutrino is paired with the muon in the second generation. The tau-neutrino is paired with the tauon in the third generation.

neutron A particle with an electric charge of 0, a spin of ½, and a mass slightly greater than that of a proton; one of the principal components of an atomic nucleus. The neutron, although often treated as elementary, is actually a composite particle built from one up quark and two down quarks.

neutron star Late stage in the life of a star, reached when portions of the outer layers collapse under the influence of gravity to form a dense inner core of neutrons. The pressure is so extreme that protons in the various atomic nuclei merge with electrons outside to form free neutrons. Under certain conditions, a neutron star can later evolve into a supernova.

Newtonian mechanics Classical mechanics, especially as relating to Newton's laws of motion and Galilean relativity. Contrasted with *quantum mechanics.*

Newton's laws of motion Three principles upon which classical mechanics is built. See *first law of motion*; *second law of motion*; *third law of motion.*

node A point, line, or surface of a standing wave where the disturbance is zero.

non-Euclidean geometry Study of lines, points, angles, surfaces, and solids in a curved space (such as a sphere or saddle, among other possibilities). Familiar features of Euclidean geometry are no longer valid and must be modified to suit the particular space. For example: (1) Parallel lines may either diverge or converge, depending on the spatial curvature. (2) The sum of the three angles in a triangle may be greater or less than 180°. (3) The ratio of the circumference of a circle to its diameter may be greater or less than π.

nonlinear Describing any relationship in which a quantity y is not directly proportional to the first power of another quantity x. Example: $y = mx^2$, where m is a fixed number. Contrasted with *linear.*

nucleon A proton or neutron in a nucleus.

nucleosynthesis Production of a nucleus heavier than hydrogen, typically by fusion or radioactivity. Formation of the very lightest elements is believed to have been completed a few minutes after the Big Bang. Synthesis of heavy elements has been taking place in the stars ever since.

nucleus The pit of an atom: a very small, very dense, positively charged structure that gives the atom nearly all its mass. A given nucleus contains a fixed number of protons and neutrons bound together by the strong force (and opposed, to some extent, by electrostatic repulsions between protons).

observer   Idealized witness or measuring device that records relevant details of a system or event. The term need not refer to an actual human being.

open universe   Cosmological model in which the large-scale curvature of space-time is negative, resulting in a non-Euclidean geometry analogous to that of a saddle (hyperbolic). There is a both a temporal aspect and a spatial aspect: (1) A *temporally open* universe has insufficient gravitational mass to halt and reverse the outward expansion. Space and time continue to grow forever. (2) A *spatially open* universe is both infinite and unbounded. Parallel light rays eventually diverge. Compare with *closed universe; flat universe.*

operator   A mathematical recipe for transforming one object into another, used in quantum mechanics to effect a change of state and to simulate measurement. Example: a rotation operator, which turns a coordinate system through a specified angle about an axis tilted in a specified direction. See also *commuting operators.*

orbit   The path that one body follows under the influence of another.

orbital angular momentum   Angular momentum acquired by a particle moving about a point. Examples: (1) Classical: the revolution of a planet around the Sun. (2) Quantum mechanical: the indeterminate motion of an electron about an atomic nucleus. Compare with *spin angular momentum.*

particle   (1) A body for which any internal structure and internal motion can be ignored. The decision whether to treat an object as a particle depends on the phenomenon under investigation. (2) Informally: a small bit or fragment.

particle accelerator   A very large, very complex apparatus that uses electromagnetic fields to accelerate particles to extremely high energies. By examining the debris generated in high-energy collisions, physicists are able to glean information about the nature of matter and interactions.

Pauli principle   A fundamental law of quantum mechanics, enunciated by Wolfgang Pauli: (1) The wave function for a system of identical bosons remains the same (it is *symmetric*) when all the particles are interchanged. (2) The wave function for a system of identical fermions reverses sign (it is *antisymmetric*) when all the particles are interchanged.

The more limited *Pauli exclusion principle* follows from statement (2) above: identical fermions may not share exactly the same quantum state. In an atom, a single energy level can accommodate no more than two electrons, and only if they carry opposing spins. The exclusion principle thus determines the structure and properties of all the chemical elements. In addition, it mandates that quarks come in three different "colors" to differentiate species having the same flavor and spin.

period   The time required to complete one cycle of a wave or other oscillation.

period doubling   A phenomenon observed in certain nonlinear systems, often preceding the onset of chaos: the successive doubling of the length of a repeating cycle, brought about by a steady change in a controlling parameter. When the parameter first attains some threshold value, the system alternates cyclically between two outcomes. When the parameter passes the next threshold, the cycle grows to four outcomes; and then eight, and then sixteen, and so forth.

periodic table   A tabular arrangement of the chemical elements, organized into rows and columns to display recurring properties.

phase   (1) A measure of the progress of a repeating cycle, expressed as a fraction of a completed oscillation. A cycle begins at $0°$, reaches its midpoint at $180°$, and finishes at $360°$ (to begin anew). (2) Synonymous with *state (def. 2)*: a bulk state of matter, such as solid, liquid, gas, or plasma.

phase space   Abstract coordinate system in which each mechanical variable is assigned an independent axis. A classical state occupies a single, precisely defined point that has within it the position and momentum of every particle.

phase transition   Pertaining to *phase (def. 2)*: a change in the state of matter, such as the melting of a solid or the boiling of a liquid.

photon   A quantized excitation of the electromagnetic field, displaying properties of both wave and particle. An individual photon has a wavelength and frequency (like a wave), but it also delivers energy, momentum, and angular momentum (like a particle). As a massless messenger boson, a photon moves unceasingly at the speed of light and mediates the electromagnetic interaction between charged particles. See also *electroweak interaction*; *quantum electrodynamics*.

Planck scale (energy, length, time)   Rough estimate of conditions under which superunification of all four fundamental interactions takes place; a basic parameter in string theory, quantum gravity, and models of the early universe. The numbers are obtained by combining Planck's constant with the speed of light and the proportionality constant in Newton's law of universal gravitation.

Approximately the size of a superstring, the Planck length (about a billionth of a trillionth of a trillionth of a centimeter) represents a scale below which the turbulence of quantum foam wreaks havoc in space-time. The Planck energy (equivalent to 10 billion billion proton masses) represents the energy needed to probe a region that small. The Planck time (a ten-millionth of a trillionth of a trillionth of a trillionth of a second) represents the age of the universe when the cosmic radius extended only as far as the Planck length.

Planck's constant   A fundamental constant associated with the size of a quantum, having dimensions of (energy)×(time) or, equivalently, (momentum)×(length). Classical mechanics becomes increasingly accurate in systems where Planck's constant appears negligibly small. See also *correspondence principle*.

plasma   A state of matter similar to a gas but composed of electrically charged species rather than neutral atoms or molecules. The oppositely charged particles, present in roughly equal numbers, enable a plasma to conduct electricity. Most matter in the universe exists in this form. Example: the unbound electrons and nuclei in a typical star.

polarization   Of a wave: the direction in which the amplitude is displaced.

positron   The antiparticle of an electron, having the same mass but a positive charge of equal magnitude. Also called *antielectron*.

potential   (1) Energy invested in the position of a body in a field, distinguished from kinetic energy (the energy of motion). See also *electric potential*; *gravitational potential*. (2) In a more general sense, a potential is a measure of an imbalance in some quantity at two points. Example: Temperature acts as a potential for the transfer of heat. Thermal energy (heat) flows from a body at higher temperature to a body at lower temperature.

potential energy   See under *potential (def. 1)*.

pressure   The ratio of force to area. The same force applied over a small area exerts more pressure than it does over a large area.

principle   See specific principle.

probability   The statistical likelihood that an event will occur, calculated as the quotient A/T: the number of actual occurrences (A) divided by the total number of events possible (T).

proper time   In special relativity, the shortest possible time between two events: the space-time interval measured by an observer in a reference frame where the events occur in the same place. A clock at rest reads its own proper time.

proton   A particle with an electric charge of +1, a spin of ½, and a mass 1836 times greater than that of an electron; along with the neutron, a principal component of an atomic nucleus. The proton, although often treated as elementary, is actually a composite particle built from two up quarks and a down quark.

A kilogram of protons (weighing 2.2 pounds on Earth) would contain nearly a thousand trillion trillion particles. A thousand trillion of them, placed side by side, would form a line of only 2 meters (6.6 feet).

Pythagorean theorem   The square of the hypotenuse of a right triangle (length c) is equal to the sum of the squares of the two sides (lengths a and b): $c^2 = a^2 + b^2$.

quantum chromodynamics   Theory of the strong-force color interaction, incorporating quantum mechanics and special relativity. The model describes the strong force as arising from an exchange of colored gluons (massless messenger bosons) between colored quarks. See also *electroweak interaction*; *quantum electrodynamics*.

quantum electrodynamics   Theory of the electromagnetic interaction, incorporating quantum mechanics and special relativity. The model describes the electromagnetic force as arising from an exchange of uncharged photons (massless messenger bosons) between electrically charged particles. See also *electroweak interaction*; *quantum chromodynamics*.

quantum foam   A consequence of applying the uncertainty principle to point particles: violent fluctuations in the fabric of space-time on an infinitesimally small scale. Also called *space-time foam*.

quantum mechanics   Study of the energy and motion of particles too small and having too little mass to obey classical mechanics. Instead of well-defined positions and momenta, quantum mechanics uses (1) a probabilistic wave function to specify the state of a system, and (2) mathematical operators to represent the act of measurement. Evolution of a state is described by a suitable equation of motion, the solution of which yields discrete eigenstates and eigenvalues. The uncertainty principle, the superposition principle, the Pauli principle, the correspondence principle, and wave–particle duality form the basis of quantum mechanics.

The Schrödinger equation, a nonrelativistic approximation, provides a satisfactory description of most atoms and molecules (and thus most of chemistry). Classical treatment of the electromagnetic field also helps simplify the analysis of such systems, with little loss of accuracy. Analysis of the strong and weak interactions, however, demands a quantized field that meets the requirements of special relativity. See also *electroweak interaction*; *quantum chromodynamics*; *quantum electrodynamics*.

quark   An elementary particle that possesses the quantum mechanical attributes of both flavor and color, thus a participant in all four fundamental interactions. Quarks have fractional electric charges ($-1/3$ or $+2/3$), a spin of $\frac{1}{2}$, and varying masses. They occur only in tightly bound groups of two (as in a meson) or three (as in a proton or neutron), and they are the simplest known constituents of any composite particle that responds to the strong force—all matter except leptons. See also *Standard Model of particle physics*.

radiation   The emission of particles or energy. Examples: electromagnetic radiation, thermal radiation (heat), radioactivity.

radioactivity   The emission of particles or energy from an atomic nucleus, typically in the form of alpha particles, beta particles, gamma rays, protons, neutrons, and neutrinos. Also called *radioactive decay*.

radio wave   The least energetic radiation of the electromagnetic spectrum, characterized by wavelengths greater than 1 meter. Some radio waves have just enough energy to reorient the spin angular momentum of an atomic nucleus.

redshift   A consequence of the Doppler effect: an apparent stretching of the wavelength when a wave source is traveling away from an observer. The term takes its name from the properties of visible light, where wavelengths increase (and frequencies decrease) as the spectrum shifts from blue to green to red. A corresponding "blueshift" occurs when the source moves toward—not away from—the observer.

Redshifted light from distant galaxies provided the first evidence of an expanding universe. See also *Big Bang*; *cosmic microwave background radiation*.

reference frame   An observational framework for locating an event in space and time, consisting of a coordinate system and a set of clocks. See also *inertial reference frame*.

relativity   A theoretical framework built upon the *principle of relativity*: that the laws of nature must be the same for all observers, regardless of their particular reference frames. Observers recognize some quantities as invariant (the magnitude of a vector, for example) and others as variable (the individual components), and they apply coordinate transformations to reconcile any apparent differences.

One set of transformations, comprising *Galilean relativity*, is an approximation valid at low velocities, where the speed of light appears effectively infinite. Time exists independently of space in the Galilean limit. Another set of transformations, comprising *Einsteinian relativity*, blends time with space and is applicable at all velocities. Its effects become most pronounced at speeds approaching the velocity of light. See also *general relativity*; *invariance*; *special relativity*.

renormalization   A mathematical procedure to remove certain anomalies (infinities) from a quantized field, without which the theory would yield nonsense.

rest energy   Potential energy ($E = mc^2$) resident in the mass of a body at rest in an inertial reference frame, a consequence of special relativity. See also *mass-energy*.

rotation   The turning of an object around an axis or central point, independent of any translation or vibration. See also *rotational symmetry*.

rotational symmetry   Invariance to rotation. An object has rotational symmetry if it retains the same appearance and properties after being turned through a specified angle about a specified axis. See also *translational symmetry*.

scalar   A quantity possessing magnitude but not direction, represented by a single number. Scalars have the same value in all rotated reference frames. Examples: distance, speed, temperature. Contrasted with *vector*.

Schrödinger equation   A nonrelativistic quantum mechanical equation of motion, applicable to particles traveling at velocities low compared with the speed of light. Given the masses and potential energies of the particles, the Schrödinger equation determines the allowed wave functions and energies. It is applied with great success to the study of atoms and molecules.

second law of motion   One of Newton's three laws: force is equal to mass times acceleration. The larger the mass, the smaller is the acceleration produced by a given force. Equivalently stated: the force impressed is equal to the rate at which momentum changes with time. See also *classical mechanics*.

second law of thermodynamics   A principle pertaining not to the conservation of energy (as does the first law), but rather to the dispersal of energy: The universe moves inexorably toward greater global disorder, generating additional entropy with every spontaneous change. Energy, dribbling into more and more channels, becomes increasingly difficult to extract even though its total remains constant. All isolated systems run down and eventually come to equilibrium.

solid A dense, tightly knit state of matter in which atoms and molecules lack the energy to undertake large-scale translations and are limited instead to rotations and low-amplitude vibrations. The structure of a solid, constrained to have a fixed shape and volume, usually exhibits a considerable amount of microscopic symmetry and order. Example: ice. Compare with *liquid*; *gas*; *plasma*.

space (1) A coordinate system encompassing the physical dimensions of length, width, and height, thus a set of possible places for things to be and events to occur. Specification of coordinates along any three independent axes is sufficient to pinpoint an arbitrary position. Flat spaces are conveniently modeled by rectilinear axes, whereas curved spaces are better handled with coordinates that conform to a particular geometry (such as radius, latitude, and longitude to mark locations on a sphere). See also *space-time*; *time*. (2) In general, especially in mathematics: a coordinate system of any dimension, constructed to correspond with the independent components of an object.

space-time A four-dimensional coordinate system defined in connection with a given inertial reference frame. Three spatial coordinates (x, y, z) and one temporal coordinate (t) pinpoint an event in space-time. See also *relativity*; *space*; *space-time interval*; *time*.

space-time interval An invariant measure of the separation between events in space-time, having the same value in all inertial reference frames. The space-time interval, a scalar that combines a spatial contribution with a temporal contribution, is analogous to distance in ordinary space (which remains unchanged despite any rotation of the coordinate system).

sparticle See under *superpartner*.

special relativity (special theory of relativity) Einstein's theory of relativity as restricted to an inertial reference frame, deduced from two postulates: (1) The laws of physics are the same for observers in all inertial frames. (2) The speed of light is the same in all inertial frames. Among the consequences of special relativity are a blending together of space and time, a rethinking of the concept of simultaneity, and the realization that mass is a form of energy. See also *general relativity*; *mass-energy*; *relativity*; *space-time*; *space-time interval*.

speed A scalar representing the rate at which distance varies with time. Speed is the magnitude of velocity, a vector.

spin angular momentum (spin) In quantum mechanics: an angular momentum intrinsic to a particle itself, not arising from any overt motion. Quantum mechanical spin has no direct classical counterpart, and descriptions of a "spinning" electron as a spinning top are purely metaphorical. Compare with *orbital angular momentum*.

spontaneous symmetry breaking The difference between equal opportunity and unequal outcomes: the failure, owing to a random accident, of a system to realize an exact symmetry inherent in a particular law. Examples: (1) The freezing of a liquid with its molecules oriented in a specific (but randomly acquired) direction. The outcome is just one of an infinity of equally likely alternatives. (2) The random differentiation of the electron and neutrino caused by interaction with a Higgs field, resulting in the acquisition of different masses. Until a particular value of the field is established, the two particles are massless and indistinguishable.

stable equilibrium   A state resting at a point where the potential is at a minimum, as in a valley (flat). The system relaxes back to equilibrium after a small disturbance.

If the valley of equilibrium is the lowest one possible, then the equilibrium is *globally stable*. If the system has access to valleys lower in potential, then the equilibrium is *locally stable*, or *metastable*. Contrasted with *unstable equilibrium*.

Standard Model of particle physics   Framework for understanding three of the four fundamental interactions, a synthesis of quantum chromodynamics and the unified electroweak force. The model, developed in the 1970s, has been a great success, but nearly all physicists agree that it is not a final theory. Possible extensions include the incorporation of supersymmetry and the replacement of point particles by superstrings.

The Standard Model classifies matter particles (fermions) as either quarks or leptons and assigns them to three generations. The first generation, accounting for all ordinary matter in the known universe, contains the up and down quarks, the electron, and the neutrino. The second generation, which covers particles born in high-energy processes, contains the charm and strange quarks, the muon, and the muon-neutrino. The third generation, similarly, contains the top and bottom quarks, the tauon, and the tau-neutrino. Particles in the second and third generations are analogous to those in the first, distinguishable only by their greater masses. Each generation accommodates two quarks of contrasting flavors and two leptons that have the attributes of electron and neutrino.

Force particles (bosons) are assigned to each of the three fundamental forces treated by the model: the photon to mediate the electromagnetic interaction, the W and Z particles to mediate the weak interaction, and the gluons to mediate the strong interaction. In addition, the hypothesized Higgs boson imparts mass to the various particles.

standing wave   A mode of vibration that develops in a confined space, oscillating up and down but unable to move forward. The wave "stands" in place and is punctuated by a pattern of stationary nodes. See also *harmonic*.

state   (1) The condition of a system, specified in full by a particular set of mechanical variables. (2) A state of matter: an aggregation of atoms, molecules, or ions shaped by interparticle forces and often macroscopic in extent. Examples: solid, liquid, gas, plasma. Synonymous with *phase (def. 2)*.

statistical mechanics   The use of statistical analysis, rather than an equation of motion, to study a large number of particles. With a big enough system, the probabilities of certain occurrences become overwhelmingly high and the statistics increasingly accurate. The approach helps identify the microscopic origin of macroscopic properties. Compare with *mechanics*; *thermodynamics*.

steady state A condition in which the macroscopic properties of a system appear constant, although not as in a true, self-sustaining equilibrium arising from an internal balance of forces. A nonequilibrium steady state must be maintained actively from without, by connection to an external source or sink of material and energy. Example: living systems.

string theory A class of superunified theories in which point particles are replaced by superstrings vibrating in a space-time of greater than four dimensions. String theory, still under development, hopes to reconcile quantum mechanics with general relativity and thereby achieve an integrated understanding of all four fundamental forces. See also *M-theory*.

strong (nuclear) interaction The strongest of the fundamental forces, responsible for the binding of quarks into protons and neutrons and, secondarily, for the binding of protons and neutrons into atomic nuclei. Short-range and independent of mass or electric charge, the strong interaction derives from the color force between quarks. It is mediated by the exchange of gluons.

supernova Catastrophic explosion of a dying star, accompanied by the release of heavy elements built up during prior nucleosynthesis. A supernova is so bright that it can sometimes outshine an entire galaxy during the brief time that it burns. See also *neutron star*.

superpartner A consequence of supersymmetry: an as-yet-undiscovered counterpart for every known particle, differing in spin by one-half unit. For every fermion with spin-$\frac{1}{2}$, supersymmetry demands a partner boson with spin-0. For every boson with spin-1, a partner fermion with spin-$\frac{1}{2}$.

Bosonic superpartners are generally named by prefixing an *s* to the name of the fermion, turning a spin-$\frac{1}{2}$ *particle* into a spin-0 *sparticle*. Fermionic superpartners of known bosons generally add the suffix *ino* or replace the suffix *on* with *ino*. Examples: electron/selectron, quark/squark, proton/sproton, neutron/sneutron, photon/photino, *W*/wino, *Z*/zino, gluon/gluino.

superposition Combination of independent components, such as in the vector addition of three perpendicular directions in space. In classical electromagnetism, superposition refers to the addition of fields from different sources. In quantum mechanics, superposition denotes the combination of independent eigenstates to represent a complete wave function (or wave vector).

superstring Teeny-weeny building block of string theory: a strand of energy on the Planck scale. A superstring is endowed with supersymmetry and is assumed to vibrate in more than four space-time dimensions. See also *M-theory*.

supersymmetry A proposed symmetry that transforms fermions into bosons (and bosons into fermions) under certain conditions. See also *superpartner*; *superstring*.

superunification Hoped-for marriage of a grand unified theory (which would merge the electroweak and strong interactions into one) and quantum gravity (which would provide a microscopic understanding of the gravitational field). If superunification is realized, then all four fundamental interactions will be understood as sharing a common origin.

Neither a grand unified theory nor a theory of quantum gravity has yet been developed. Many physicists regard M-theory as the best chance for achieving both.

surroundings   In thermodynamics, the portion of the universe that interacts with or affects a system. Example: If an observer treats a volume of water as the system (the portion of direct interest), then the ambient atmosphere and the glass containing the water would constitute the surroundings.

symmetry   The ability of a system to appear unchanged despite undergoing some nominal transformation. An object is symmetric under a particular operation if an observer (with eyes closed) cannot tell whether or not the operation was performed. Example: A square, which looks the same when rotated in a plane by 90°, 180°, 270°, or 360°, has fourfold rotational symmetry. See also *rotational symmetry*; *translational symmetry*.

symmetry breaking   See *spontaneous symmetry breaking*.

system   (1) In general: any structure or process of interest to an observer. (2) In thermodynamics: the portion of the universe under direct investigation, considered separately from its surroundings.

tau-neutrino   See under *neutrino* and *Standard Model of particle physics*.

tauon   See under *Standard Model of particle physics*.

temperature   (1) Interpreted macroscopically: a measure of the tendency for heat to flow between two points, from a hot body (at a higher temperature) to a cold body (at a lower temperature). (2) Interpreted microscopically: a measure of the average speed (or kinetic energy) of the particles in a system.

thermal equilibrium   A self-sustaining condition in which the temperature of a system stays uniform throughout, maintained internally by the exchange of energy among microscopic particles. There is no macroscopic flow of heat between any two points. See also *equilibrium*.

thermodynamics   Macroscopic treatment of the deployment, interconversion, and dispersal of energy in its many forms, focusing on work (mechanical energy) and heat (thermal energy). Compare with *statistical mechanics*. See also *first law of thermodynamics*; *second law of thermodynamics*.

third law of motion   One of Newton's three laws, related to the conservation of momentum: for every action, there is an equal and opposite reaction. Example: When a ball bounces off the ground, the force delivered by the ball to the ground is equal in magnitude and opposite in direction to the force delivered by the ground to the ball. The ball pushes against the ground, and the ground pushes back. See also *classical mechanics*.

time   A cyclical sequence of standardized events (such as the ticks of a clock) to which other sequences can be referred, thus a way to quantify the notions of before and after. The theory of relativity demonstrates that perceptions of time cannot be separated from perceptions of space, particularly when velocities approach the speed of light. Observers in relative motion differ in their assessment of simultaneity. See also *space-time*.

torque   A force that produces rotation or twisting.

translation   Motion in a straight line: the displacement of an object from one point in space to another, independent of any rotation or vibration. See also *translational symmetry*.

translational symmetry   Invariance to spatial translation. An object has translational symmetry if it retains the same appearance and properties after being displaced through a specified distance in a specified direction. See also *rotational symmetry*.

tritium   A radioactive isotope of hydrogen containing one proton and two neutrons in its nucleus. See also *deuterium*.

turbulence   Fluid flow characterized by erratic variations in speed and direction at each point.

ultraviolet radiation   Electromagnetic radiation with energy intermediate between visible light and X rays. Ultraviolet photons typically are able to excite the electrons of atoms and molecules. See also *electromagnetic spectrum*.

uncertainty principle   A fundamental law of quantum mechanics, enunciated by Werner Heisenberg: The position and momentum of a particle cannot be measured simultaneously to unlimited accuracy. The more precisely one quantity is known, the more uncertain (or indeterminate) is the other. A similar relationship holds for energy and time, as well as for various other observables. Also called *Heisenberg uncertainty principle* and *(Heisenberg) indeterminacy principle*. See also *commuting operators*; *operator*.

universe   (1) In general, everything: the cosmos, the world, all phenomena known and unknown. (2) In thermodynamics: the totality of system and surroundings.

unstable equilibrium   A state resting at a point where the potential (although flat) is at a local maximum, as on top of a hill. A slight disturbance will cause the system to "roll down" to a state of lower potential, thus destroying the equilibrium. Contrasted with *stable equilibrium*.

vacuum   Space empty of matter but typically permeated by fields and energy.

vector   A quantity having both magnitude and direction, specified by one number (a component) along each axis in a coordinate system. The individual components take on different values when the reference frame is rotated, but the vector's magnitude (a scalar) remains the same. Examples: velocity, momentum, force, acceleration. Contrasted with *scalar*.

velocity   A vector that combines speed and direction. See also *acceleration*.

vibration   The periodic displacement and return of a body to its starting point, independent of any overall translation or rotation.

virtual particle   A short-lived creation of the vacuum, materializing out of pure energy and existing for a time before disappearing. The energy ($E = mc^2$) that finances the materialization is "borrowed" temporarily and must be repaid. The term of the loan is set by Heisenberg's uncertainty principle for energy and time. See also *mass-energy*.

visible electromagnetic radiation   Familiarly, light: the spectrum of electromagnetic radiation visible to the human eye and brain, comprising the hues red, orange, yellow, green, blue, and violet. With energies intermediate between infrared and ultraviolet radiation, photons in this range are able to excite the outlying (and therefore most loosely bound) electrons of certain atoms and molecules. They are responsible for the perception of color. See also *electromagnetic spectrum*; *light*.

volume   Capacity: the amount of space taken up by an object in three dimensions, distinguished from area (the amount of space taken up in two dimensions) and length (the amount taken up in one dimension). The concept of volume can be extended to a space with more than three dimensions.

wave   A periodic disturbance in space and time, either fixed in place (a standing wave) or moving (a traveling wave). See also *amplitude*; *frequency*; *period*; *phase*; *polarization*; *wavelength*.

wave function   Mathematical representation of a quantum mechanical state, obtained as a solution to an appropriate equation of motion (such as the Schrödinger equation). The square of a wave function establishes the probability of observing a system in its various eigenstates. Sometimes called *wave vector*.

wavelength   The distance between equivalent points on consecutive cycles of a wave (such as successive crests or troughs).

wave–particle duality   A principle of quantum mechanics, first recognized by Louis de Broglie: that the characteristics of waves and particles are intertwined and fundamentally unified. Matter can have a wavelength and undergo interference like a wave, just as light can deliver energy and momentum like a particle.

***W* boson**   Either of two massive bosons that help mediate the weak interaction (one with a charge of −1, the other with a charge of +1). A third massive boson, the neutral Z^0, completes the set. Representing individual quanta of the weak force field, the three particles are collectively called *weak gauge bosons* or *intermediate vector bosons*. See also *electroweak interaction*; *messenger particles*.

weak (nuclear) interaction   One of the four fundamental interactions, manifested principally as various forms of beta decay (such as the conversion of a neutron into a proton). Mediated by the massive *W* and *Z* bosons, the weak force is of exceptionally short range. Its action generally brings about a change in quark flavor.

WIMP   [Weakly Interacting Massive Particle] Presumed dark matter not composed of protons and neutrons, believed to interact only through the weak force and gravity. Example: a neutrino with mass. Compare with *MACHO*.

work   The application of force over distance, resulting in the displacement of a body and requiring a change in energy. Interpreted microscopically, work arises from the ordered, coherent movement of large numbers of particles. Compare with *heat*. See also *thermodynamics*.

wormhole   A "neck" or "tunnel" in the geometry of space-time, providing a shortcut between distant points. Wormholes, although predicted by general relativity, have not yet been discovered.

X ray   Highly energetic electromagnetic radiation, second in strength only to gamma rays. Photons in the X-ray range have enough energy to excite the innermost (and therefore most tightly bound) electrons of many atoms. See also *electromagnetic spectrum*.

Z boson   A massive and electrically neutral messenger boson, one of the mediators of the weak interaction; a quantum of the weak force field. Also called *weak gauge boson* or *intermediate vector boson*. See also *W boson*.

zero-point energy (zero-point vibration)   A consequence of the uncertainty principle: that a particle can never stand absolutely still, lest it betray its position and momentum simultaneously. Even in a state of minimum energy—and even at absolute zero, the coldest temperature possible—a particle executes a slight jittery motion.

Further Reading

The list below is highly selective, more a personal library than a survey of the many worthy books available to the scientifically curious nonscientist. Except where noted, the works cited require little or no mathematical background.

Beyond books, look for articles and news about current research in sources such as New Scientist, Science News, Scientific American, *and* Sky and Telescope, *to name just a few of the leading periodicals accessible to a general reader.*

Abbott, Edwin A. *Flatland: A Romance of Many Dimensions.* 6th ed. New York: Dover, 1952. Classic fantasy of life in a two-dimensional world, not only entertaining but also useful for coping with four-dimensional space-time...and worse. Written by a nineteenth-century English headmaster with scholarly interests in literature and theology.

Aczel, Amir D. *God's Equation: Einstein, Relativity, and the Expanding Universe.* New York: Four Walls Eight Windows, 1999. Cosmology and general relativity, viewed through the prism of Einstein the man.

Adair, Robert K. *The Great Design: Particles, Fields, and Creation.* New York: Oxford University Press, 1987. Treats many of the same subjects as the present work (classical physics, thermodynamics, relativity, quantum mechanics, cosmology), but with greater mathematical rigor. Will be appreciated by readers with some prior background in physics and math, at least at the high-school level.

Asimov, Isaac. *Understanding Physics. Vol. 1: Motion, Sound, and Heat. Vol. 2: Light, Magnetism, and Electricity. Vol. 3: The Electron, Proton, and Neutron.* Dorset Press, 1988. A carefully planned course in physics for laypeople, originally published in 1966 and later reprinted as three volumes in one. Volumes 1 and 2 will enable the interested reader to go beyond the classical physics offered in the present Chapters 3, 4, 5, 6, and 10. Volume 3, although still useful, does not reflect advances since 1964. High-school math a prerequisite.

Atkins, P. W. *Atoms, Electrons, and Change.* New York: Scientific American Library, 1991. A leading authority describes the quantum mechanics of atoms and molecules—chemistry, in other words—with unmatched clarity. Like all books in the Scientific American Library, this volume is a pleasure to look at as well as to read.

————. *Galileo's Finger: The Ten Great Ideas of Science.* Oxford and New York: Oxford University Press, 2003. General principles of scientific theory, with topics similar to the present work but also including chapters on biology, geology, and mathematics. Written in an informal, almost intimate voice.

————. *Molecules*. New York: Scientific American Library, 1987. Beautifully illustrated excursion into a small world.

Ball, Philip. *Designing the Molecular World: Chemistry at the Frontier.* Princeton, New Jersey: Princeton University Press, 1994. Recent advances in chemistry and materials science, for readers who wish to explore a territory barely hinted at in the present work.

Bondi, Hermann. *Relativity and Common Sense: A New Approach to Einstein.* Garden City, New York: Anchor Books, 1964. Patient, careful introduction to special relativity, an account for the layperson with much to recommend it.

de Duve, Christian. *Vital Dust: Life As a Cosmic Imperative.* New York: BasicBooks, 1995. The endlessly fascinating world of biology is ignored in the present work. Readers who wish to learn something about the molecular basis and origin of life can do no better than read *Vital Dust*. A serious book suitable for the educated general reader.

————. *Life Evolving: Molecules, Mind, and Meaning.* New York: Oxford University Press, 2002. A follow-up to the author's *Vital Dust*, written at a somewhat broader level—and with a certain amount of personal musing and reflection as well.

Davies, P. C. W., and J. Brown, eds. *Superstrings: A Theory of Everything?* Cambridge, England: Cambridge University Press, 1988. Informative interviews with prominent physicists pro and con, preceded by an introduction covering relativity, quantum mechanics, symmetry, and supersymmetry. A snapshot of a then-new field, frozen in the mid to late 1980s. The basic material is still valid, but string theory has progressed greatly since publication.

Dodd, J. E. *The Ideas of Particle Physics: An Introduction for Scientists.* Cambridge, England: Cambridge University Press, 1984. As the subtitle makes clear, strictly for scientists with the appropriate mathematical background. A prepared and motivated reader will find the ideas of the present Chapter 9 fleshed out considerably in this book.

Einstein, Albert. *Relativity: The Special and the General Theory.* Translated by Robert W. Lawson. New York: Crown Publishers, 1961. There have been numerous popular accounts of relativity, but many readers still prefer to go back to the original source.

Ferris, Timothy. *The Whole Shebang: A State-of-the-Universe(s) Report.* New York: Simon and Schuster, 1997. Modern cosmology—the Big Bang, inflation, the multiverse—plus the necessary quantum mechanics and an interesting historical perspective as well, presented concisely and elegantly by one of the best science writers today.

Feynman, Richard. *The Character of Physical Law.* Cambridge, Massachusetts: MIT Press, 1965. Indispensable. Based on a series of public lectures at Cornell University, this amazing little book offers a glimpse into how a great mind works—and what physics says *and does not say* about nature. Will never go out of date, no matter what the future brings.

————. *QED: The Strange Theory of Light and Matter*. Princeton, New Jersey: Princeton University Press, 1985. Similar in spirit to *The Character of Physical Law*, although dealing specifically with quantum electrodynamics.

Gamow, George. *Mr Tompkins in Paperback*. Cambridge, England: Cambridge University Press, 1965. Great wit is not always associated with great physicists, but Gamow was one of a kind. His fictional Mr Tompkins gets inside the atom in ways that anyone can understand. A classic.

Gardner, Martin. *Relativity Simply Explained*. Mineola, New York: Dover, 1997. Clearly written, thoughtful exposition by a master of his craft. Includes both the special and general theories.

Gleick, James. *Chaos: Making a New Science*. New York: Viking, 1987. Extraordinarily lucid introduction to chaos, both its conceptual foundations and its intellectual history.

Greene, Brian. *The Elegant Universe: Superstrings, Hidden Dimensions, and the Quest for the Ultimate Theory*. New York: W. W. Norton, 1999. The focus is on string theory, but relativity and quantum mechanics are covered strongly as well, a necessary foundation. Engagingly written by a leading contemporary researcher.

Guth, Alan H. *The Inflationary Universe: The Quest for a New Theory of Cosmic Origins*. Reading, Massachusetts: Helix / Perseus Books, 1997. Authoritative account of modern cosmology by one of the most influential figures in the field. Part scientific autobiography, the book also provides a thorough description of the particle physics relevant to the Big Bang and inflation.

Hawking, Stephen W. *A Brief History of Time: From the Big Bang to Black Holes*. New York: Bantam Books, 1988. Probably no introduction is necessary. The uncertainty principle, fundamental forces, and the arrow of time are treated in addition to the cosmological subjects suggested in the subtitle. All the more effective for its brevity.

Hey, Tony, and Patrick Walters. *The Quantum Universe*. Cambridge, England: Cambridge University Press, 1987. Accessible, nicely illustrated introduction to quantum mechanics, with emphasis on atoms and molecules. Later chapters deal with relativistic quantum fields and such topics as Feynman diagrams and the Higgs boson.

Hoffmann, Banesh. *The Strange Story of the Quantum*. New York: Harper & Brothers, 1947. Published long ago, but still worth reading.

Lindley, David. *Where Does the Weirdness Go? Why Quantum Mechanics Is Strange, But Not as Strange as You Think*. New York: BasicBooks, 1996. Useful modern exposition of quantum mechanics, especially the "measurement problem" and recent attempts to deal with it.

Morowitz, Harold J. *The Emergence of Everything: How the World Became Complex.* New York: Oxford University Press, 2002. Capsule summaries of how complexity might have emerged from simplicity, laid out in 28 case studies from the birth of the universe to the development of human thought. Philosophical and religious overtones.

Newton, Roger G. *What Makes Nature Tick?* Cambridge, Massachusetts: Harvard University Press, 1993. General principles of physical law, including classical theory as well as quantum mechanics and chaos. Some mathematics, developed as needed.

Pagels, Heinz R. *The Cosmic Code: Quantum Physics as the Language of Nature.* New York: Simon and Schuster, 1982. No superstrings, dark energy, or other developments since 1982, but nearly everything else pertaining to quantum mechanics and relativity. Well written and extremely valuable.

Peterson, Ivars. *The Mathematical Tourist: Snapshots of Modern Mathematics.* New York: W. H. Freeman, 1988. Hyperspace, chaos, fractals, and more—for the mathematically uninitiated. Many pictures, but no equations.

Polkinghorne, J. C. *The Quantum World.* Princeton, New Jersey: Princeton University Press, 1989. Serious treatment of quantum mechanics for the general reader, focusing on problems of interpretation. Occasional (brief) mathematical digressions are supplemented by a technical appendix that will be appreciated by readers with prior background in the subject.

Prigogine, Ilya. *The End of Certainty: Time, Chaos, and the New Laws of Nature.* New York: The Free Press, 1997. The case is made for reformulation of physical law on a nondeterministic foundation.

Rae, Alastair I. M. *Quantum Physics: Illusion or Reality?* Cambridge, England: Cambridge University Press, 1986. Slender but powerful volume dealing with the foundations of quantum mechanics, particularly its interpretation. The book provides a thorough discussion of the diffraction experiment.

Ruelle, David. *Chance and Chaos.* Princeton, New Jersey: Princeton University Press, 1991. Determinism and the *real* world, an insightful look into one of the great questions of physics: How can deterministic laws lead to apparently random outcomes? Has extensive notes, but no index.

Schwartz, Joseph, and Michael McGuinness. *Einstein for Beginners.* New York: Pantheon Books, 1979. A comic book dealing with special relativity! Marvelously accurate and effective, too.

Schwinger, Julian. *Einstein's Legacy: The Unity of Space and Time.* New York: Scientific American Library, 1986. Anything in the Scientific American Library is worth reading, and this volume is among the best: special and general relativity, with due deference paid to both Maxwellian electrodynamics and Galilean-Newtonian mechanics.

Taylor, Edwin F., and John Archibald Wheeler. *Spacetime Physics: Introduction to Special Relativity.* 2nd ed. New York: W. H. Freeman, 1992. Perhaps the best primer on relativity ever written, a masterpiece of expository physics. Read the 1966 first edition, if possible.

Thorne, Kip S. *Black Holes and Time Warps: Einstein's Outrageous Legacy.* New York: W. W. Norton, 1994. Thorough historical and conceptual analysis of general relativity and its consequences, written by one of the foremost experts in the field.

Weinberg, Steven. *Dreams of a Final Theory: The Scientist's Search for the Ultimate Laws of Nature.* New York: Vintage Books, 1994. Profound look into the meaning of scientific theory, focusing on quantum mechanics and the role it may play in a more comprehensive picture. Offers an especially clear explanation of gauge invariance and spontaneous symmetry breaking in general, and the electroweak interaction in particular.

————. *The First Three Minutes: A Modern View of the Origin of the Universe.* New York: Bantam Books, 1977. Enormously popular when first published, this introduction to Big Bang cosmology is still one of the best to be found.

Zee, A. *Fearful Symmetry: The Search for Beauty in Modern Physics.* New York: Macmillan, 1986. Enthusiastic, fascinating exposition of the way symmetry dictates quantum mechanical design. A book-length treatment of the ideas outlined in the second half of the present Chapter 9.

Index

405